POLYGONAL APPROXIMATION AND SCALE-SPACE ANALYSIS OF CLOSED DIGITAL CURVES

POLYGONAL APPROXIMATION AND SCALE-SPACE ANALYSIS OF CLOSED DIGITAL CURVES

Kumar S. Ray, PhD, and Bimal Kumar Ray, PhD

Apple Academic Press

TORONTO NEW JERSEY

Apple Academic Press Inc.
3333 Mistwell Crescent
Oakville, ON L6L 0A2
Canada

Apple Academic Press Inc.
9 Spinnaker Way, Waretown, NJ 08758
USA

Exclusive worldwide distribution by CRC Press, a Taylor & Francis Group

International Standard Book Number: 978-1-926895-33-8 (Hardback)

Printed in the United States of America on acid-free paper

Library of Congress Control Number: 2012951949

Library and Archives Canada Cataloguing in Publication

Ray, Kumar S.
Polygonal approximation and scale-space analysis/Kumar S. Ray and Bimal Kr. Ray.

Includes bibliographical references and index.
ISBN 978-1-926895-33-8
1. Curves--Computer simulation. 2. Geometry--Data processing.
3. Polygons. 4. Approximation theory. 5. Image processing--Mathematics.
6. Pattern recognition systems--Mathematics. 7. Computer vision-- Mathematics. I. Ray, Bimal K II. Title.

QA448.D38R39 2013 516.3'520285 C2012-906411-4

About the Authors

Kumar S. Ray, PhD

Kumar S. Ray, PhD, is a professor in the Electronics and Communication Science Unit at the Indian Statistical Institute, Kolkata, India. He has written a number of articles published in international journals and has presented at several professional meetings. His current research interests include artificial intelligence, computer vision, commonsense reasoning, soft computing, non-monotonic deductive database systems, and DNA computing.

Bimal Kumar Ray, PhD

Bimal Kumar Ray is a professor at the School of Information Technology and Engineering, Vellore Institute of Technology, Vellore, India. He received his PhD degree in computer science from the Indian Statistical Institute, Kolkata, India. He received his master's degree in applied mathematics from Calcutta University and his bachelor's degree in mathematics from St. Xavier's College, Kolkata. His research interests include computer graphics, computer vision, and image processing. He has published a number of research papers in peer-reviewed journals.

Contents

List of Abbreviation

AI	Artificial intelligence
ATM	Assumption based truth maintenance
CR	Compression ratio
CV	Computer vision
DFI	Decomposed fuzzy implication
FOM	Figure of merit
ISE	Integral square error
LMPP	Linear minimum perimeter polygon
MOM	Measure of merit
MPP	Minimum perimeter polygon
MLP	Multilayer perceptron
QT	Query table
SLS	Smoothed local symmetries

Preface

This book is divided into three parts. Part I represents polygonal approximation of closed digital curves. Part II deals with scale-space analysis with its application to corner detection. Part III demonstrates series of case studies on structural pattern classification and 2-D object recognition.

Approximation of a closed curve by piece straight line segments is known as polygonal approximation. Any curve can be approximated by a polygon with any desired degree of accuracy.

The aim of this book is to introduce to the beginner the representation of two-dimensional objects in terms of meaningful information that may be used for recognition of structural patterns, 2-D objects, scene and/or occluded scene. The book is basically an outcome of the research performed by the vision research group under Professor Kumar S. Ray at the Electronics and Communication Sciences Unit, Indian Statistical Institute, Kolkata.

This book has tremendous importance in the area of structural pattern classification, object recognition, computer vision, robot vision, medical computing, computational geometry, and bioinformatics systems. The systematic development of the concept of polygonal approximation of digital curves and its scale-space analysis are useful and attractive. Development for different algorithms of polygonal approximation and scale-space analysis and several experimental results with comparative study for measuring the performance of the algorithms are extremely useful for theoretical and application-oriented works in the above-mentioned areas. Polygonal approximation of digital curves has been cultivated by many researches, but its scale-space analysis has hardly been discussed. Hence, a systematic development of polygonal approximation and scale-space analysis of digital curve is very essential for future work. Also there is no comprehensive textbook on polygonal approximation and scale-space analysis of digital curves. Hence this book is very useful from theoretical and practical points of view. It simply fills the gap. Finally this book describes a series of applications of the methodology of polygonal approximation in the specific areas of structural pattern classification and recognition of 2-D scene and/ or 2-D occluded scene.

This book is aimed at academics, graduate, postgraduate students, and research professionals. It is a book that can be used in coursework on image processing, pattern recognition, and computer vision at graduate and post-graduate levels in computer science/electrical engineering/electronics engineering.

We have not tried to trace the full history of the subject treated; this is beyond our scope. However, we have assigned credits to the sources that are as readable as possible for one knowing what is written here. A good systematic reference is covered in the list of references of the book.

We would like to thank Mandrita Mondal for preparing the materials of the book.

— Kumar S. Ray, PhD, and Bimal Kumar Ray, PhD

1 Polygonal Approximation

CONTENTS

1.1 INTRODUCTION

In this part of the book we present a survey of some of the existing algorithms for polygonal approximation of digital curves followed by our contribution in this area.

Approximation of a closed curve by piece straight line segments is known as polygonal approximation. Any curve can be approximated by a polygon with any desired degree of accuracy. Polygonal approximation is useful in reducing the number of points required to represent a curve and to smooth data. Such representation facilitates extraction of numerical features for description and classification of curves.

Basically, there are two approaches to the problem. One is to subdivide the points into groups each of which satisfies a specific criterion function measuring the collinearity of the points. The collinearity is measured either by integral square error or by absolute error or by some other criterion function such as area deviation. This approach treats polygonal approximation as a side detection problem. Another approach to polygonal approximation is to detect the significant points and join the adjacent significant points by straight line segments. This approach treats polygonal approximation as an angle detection problem. We treat polygonal approximation as a side detection problem as well as an angle detection problem.

A number of algorithms for polygonal approximation of digital curves already exist. Ramer [118], Duda and Hart [37] propose a splitting technique which iteratively splits a curve into smaller and smaller segments until the maximum of the perpendicular distances of the points of the curve segment from the line joining the initium and terminus of the curve segment falls within specified tolerance. The curve segments are split at the point most distant from the line segment. The polygon is obtained joining the adjacent break points. The procedure needs multiple passes through data. Douglas and Peucker [34] propose a heuristic splitting technique that iteratively splits the curve into smaller segments until the distance between the points on the curve to the approximating line segment is smaller than a pre-specified value. The polygonal approximation depends on the starting point. Ballard [15] proposes a heuristic splitting algorithm based on constructing the strip tree by iteration. The idea of using strip tree is representing the curve by a hierarchical structure.

Pavlidis and Horowitz [108] use split and merge technique which fits lines to an initial segmentation of the boundary points and computes the least squares error. The procedure then iteratively splits a curve if the error is too large and merges two lines if the error is too small. Pavlidis [106] develop another procedure which is based on the concept of almost collinearity of a set of points. To check whether a set of points are collinear almost collinear the procedure computes an error of fit which is a function of two variables *C* and *T*. The *T* is the maximum of the perpendicular distances of the points being tested for collinearity from the yet-to-be obtained line segment. And *C* is a normalized variable ($0 \leq C \leq 1$) which is determined by the ratio of the number of sign changes to the total number of sign changes the perpendicular distances go through. If $T - CW_0 - T < 0$ then the line segment is acceptable else if $T - CW_0 - T_0 > 0$ then the line segment is rejected, T_0 being the acceptable error and W_0 is the weighting factor of *C*. The procedure needs multiple passes through data.

The fundamental problem in the splitting technique and in the split and merge technique process is the initial segmentation. For open curves one can start with the end points of the curve as the initial break points. Ramer [118] suggests that if the curve is closed then the top left corner and the bottom right corner can be taken as the initial break points. In the split and merge process the initial segmentation is done by decomposing the curve into suitable number of segments. Unfortunately, in this approach the approximation depends on the initial segmentation. A different mode of initial segmentation generates a different approximation. This difficulty is caused by the arbitrary initial segmentation. We believe that if the initial segmentation is done following some deterministic rules then this difficulty may be overcome.

Ansari and Delp [6] make the initial segmentation using the extreme curvature (maximum of the positive curvature and minimum of negative curvature) points as the initial set of break points. They convolve the data points with the Gaussian kernel and find the extreme curvature points of the Gaussian smoothed curve. Those points along the original boundary which correspond to the extreme curvature points are used as the initial set of break points.

The concept of a rank of a point to resolves the problem of initial segmentation [128]. The split and merge is done using the perpendicular distance of a point from the line segment joining the initium and terminus of a curve segment. The initial segmentation is computationally less expensive as compared to that in [6].

A number of sequential one-pass algorithms [120] for polygonal approximation of digital curves already exist. One such algorithm proposed by Williams [184] uses cone intersection method to find the maximal possible line segments. Circles of specified radii are drawn around each point until the intersection of the cones with their vertex at the initial point and touching the circles is an empty set. The segments are obtained by joining the initial point to the last point that passed the test. The procedure needs a single-pass through data.

Wall and Danielsson [180] develop another sequential one-pass technique which is based on the concept of area deviation per unit length of the approximating line segment. The procedure finds the maximal line segment by merging points one after another with the initial point until the area deviation per unit length of the approximating line segment exceeds a maximum allowed value. The line segment is obtained joining the initial point to the last point that passed the test.

Although Williams' procedure [184] is sequential one-pass and it needs a small and finite memory (so the procedure is highly efficient) but it has two defects. Firstly, it needs a considerably large number of arithmetic operations and secondly, it misses corners and rounds off sharp turnings (spikes). The algorithm designed by Wall and Danielsson [180] is faster than that of Williams. The Williams' procedure needs seven multiplications divisions and evaluation of one square root, whereas, Wall and Danielsson's procedure needs six multiplications and its simplified version needs only three to four multiplications. For chain coded curves Williams' procedure needs five multiplications whereas Wall and Danielsson's procedure need two multiplications and its simplified version only one. The basic version of Wall-Danielsson's procedure too rounds off sharp turnings. They introduced a peak test which retained the sharp turnings.

The sequential one-pass algorithm [120] that we propose needs no arithmetic except subtractions. The procedure is based on a proposition from numerical analysis and some concepts of regression analysis. The vertices are located by identifying some patterns exhibited by the first order finite differences of the boundary point data. Since the procedure needs no arithmetic except subtractions hence its computational load is very much lower than that of the existing sequential techniques and neither does not do it round off sharp turnings nor does it miss corners.

Though, this technique is computationally simpler than the existing ones but it holds for curves with uniformly spaced points only. We present another sequential one-pass algorithm [121] which holds for curves with uniformly spaced points as well as non-uniformly spaced points and it is computationally less expensive than the other sequential one-pass algorithms. The procedure is based on Pavlidis' concept [106] of almost collinearity of a sequence of points. The collinearity is checked by measuring the area and perimeter of the triangle formed by sequence of points triplets.

In the polygonal approximation techniques discussed so far the maximum allowable error is specified either directly or indirectly. Stone [162] considers the problem of approximating a known nonlinear function by a polygon with a specified number of line segments. The procedure minimizes the sum of squares of errors between the known function and the line segments forming the polygon and thereby determines the points of subdivision and the parameters of the line segments. Stone's approach is classical in nature. Bellman [19] uses dynamic programming [18] to solve the same problem. Bell primarily confines himself to the analytical aspects of the solution, briefly mentioning how the solution of the equation for each particular point of subdivision can be reduced to discrete search. He further suggests the extension of his method to polynomial fittings to known functions. Gluss has written a series of papers [50-52]. In [50] Gluss considers the computational aspects of Bellman's work fully, noting the similarities to some of Stone's equations and deduces an equation to determine the points of subdivision that involves an equality rather than minimization. Stone's procedure does not necessarily produce continuous approximation. The line segments can be and in general will be broken in the sense that they need not meet at the points of subdivisions. Only when the given function is quadratic, Stone obtains a continuous approximation. Bellman's procedure too does not necessarily produce continuous approximation. In [50] Gluss considers a model in which the lines are constrained to meet on the curve at the points of subdivision. In [51] Gluss considers the problem of approximating a continuous nonlinear function by a polygon where the line segments are considered to meet at the point of subdivision but not necessarily on the curve. The method of solution presented in [51] involves a functional of two variables. This makes the solution much more cumbersome that for the same problem without continuity constraint which involves a functional of a single variable [19], [50]. In [52] Gluss avoids this difficulty by introducing a new criterion function that involves derivative of known function and the slope of the line segments. Cantoni [28] solves the problem of finding a continuous polygonal approximation of a known nonlinear function by minimizing the weighted integral square error (called performance index) between the known function and the approximating line segments. The approach is classical in nature. Pavlidis [102] obtains approximation of planar curves and waveforms using integral square error as the criterion function. The breakpoints are located by finding the zero of the first order derivative of integral square error by applying Newton's vector method.

All these algorithms produce optimal polygon in the least squares sense. In all these works the user of the procedure has to specify the number of line segments and minimize the sum of squares of errors. Dunham [39] suggests an optimal algorithm which instead of specifying the number of line segments specifies the error (L_∞ norm) and determines the minimum number of line segments. Dynamic programming is used to solve the problem. The recurrence relation used to determine the minimum number of line segments is simple. A scan along implementation of the algorithm is made along the line of Williams [184] and Skanlansky and Gonzalez [158]. The approximation is continuous and the knot points are constrained to lie on the curve. Perez and Vidal [109] determine optimal polygon with a fixed number of line segments minimizing the integral square error. Their method is based on dynamic programming and the algorithmic complexity is $O(mn^2)$, where n is the number of points and m is the number of segments. Salotti [149] uses heuristic search to determine optimal polygon with a computational load of $O(n^2)$. He uses a heuristic function to estimate the cost and stop the search for optimal polygon as soon as possible. Pikaz and Dinstein [114] determine the optimal polygon keeping the maximum error fixed and looking for minimal number of sides. The city block metric is used to measure the distance between the approximation and the curve. The worst case complexity of the algorithm is $O(n^2)$. An efficient and optimal solution for the case of closed curves where no initial point is given is also proposed by the author. Yuan and Suen [197] propose an optimal algorithm for approximate chain coded curve by line segment. The algorithm turns the complicated problem of determining the straightness of digital arcs into a simple task by constructing a passing area around the pixels. It also solves the problem of detecting all straight segments from a sequence of chain codes in $O(n)$ time. The algorithm does not hold for curves with non-uniformly spaced points.

In all these algorithms the user of the procedure has to specify either the number of line segments or the maximum allowable error. The maximum allowable error or the number of line segments to used are generally determined based on trial and error process. So these procedures cannot run without operator's intervention.

We are looking for a data-driven method [122] in which neither do we specify the error nor do we specify the number of line segments. We keep both these parameters free and allow the procedure to determine them on the basis of the local topography of the curve. The procedure looks for the longest possible line segment with the minimum possible error. Integral square error is selected as the error norm to measure the closeness of the polygon to the digital curve. An objective function is constructed using the length of a line segment joining the two points of the curve and the integral square error along the line segment. The vertices are located at the points where this objective function attains local maxima. The local maxima are looked for by a discrete search. The objective function is computed by merging points one after another with an arbitrarily selected initial point until a local maxima is found. The process is restarted from the new vertex and is repeated and carried beyond the starting point till the vertex generated last coincides with one of the vertices already generated. The procedure is sequential and one-pass but neither does it miss corners nor does it round off sharp turnings. Moreover the number of arithmetic operations required by this scheme is less than that required by the Williams' scheme. Williams' scheme need seven multiplications divisions and extraction of one square root whereas this procedure needs only five multiplications divisions and extraction of one square root. Though the number of arithmetic operations required by the Wall and Danielsson's scheme is comparable to that required by this procedure but the former cannot run without operator's intervention.

All algorithms for polygonal approximation use either the integral square error or the absolute error as a measure of closeness. We present another sequential one-pass algorithm [123] where the sum of absolute errors (L_1 norm) is used as a measure of closeness. The procedure is technically and conceptually. The objective of designing this scheme is to show that, the most commonly used norms are integral square error and the absolute error but it is also possible to use L_1 norm to make polygonal approximation of digital curves.

While treating polygonal approximation as a side detection problem, the procedure looks for those points of the input curve which are vertices. Pikaz and Dinstein [113] propose method which instead of looking for vertices looks for those points of the curve that are not vertices and eliminate these points. The points that are left after elimination of non vertex points constitute the vertices of the output polygon. This approach is known as iterative point elimination. They use area deviation per unit length and perpendicular distance as the measure of closeness. Masood [83] use square of maximum error and follow the concept of iterative point elimination, calling it reverse polygonization, for polygonal approximation. In an attempt to improve this work, he [84] further uses a local optimization technique that minimizes the local integral square error. The optimization results in a new vertex which has not been obtained in any of the previous iterations.

We present a sequential two-pass algorithm. It measures the collinearity of a group of points using a metric which in contrast to the conventional perpendicular distance does not miss corners and rounds off sharp turnings (spikes). It also produces symmetric approximation from symmetric digital curves. The algorithm is compared with a recent algorithm [83] with respect to perception and a new measure of goodness of approximation.

We use reverse engineering on Bresenham's line drawing algorithm [24] to approximate digital curve by polygon. Taking cue from the Bresenham's algorithm, the sides of the approximating polygon are detected. The way the Bresenham's algorithm divides any line into four categories, in the same way the lines of the polygon are looked for. The Bresenham's algorithm selects the pixels that lie closest to the corresponding analog line, whereas the proposed method approximates a sequence of digital points by a line segment.

The clustering technique has also been adopted for polygonal approximation. Phillips and Rosenfeld [115] partitions the input data into elongated subgroups based on the well known *k* means algorithm by Anderberg [4]. It starts with an arbitrary partition of points, and the principal axis of each subgroup of points is computed. Then each point is reassigned to the subgroup of which the principal axis has the shortest perpendicular distance to that point. This process is iterated until all points no longer change their memberships. Yin [192] proposes three algorithms based on k means clustering. The first algorithm starts from an explicit partition based on the distance deviation. The approximation results obtained in different time period are consistent, and for most cases, this algorithm outputs better results than average results of Phillips and Rosenfeld's method. The second algorithm proposes a simple line fitting method instead of evaluating the principal axis to reduce the computational complexity. The third algorithm determines the number of approximating line segments automatically without any pre-specified parameter.

Since Genetic algorithms have good performance in solving optimization problems, they can also be used to find the optimal polygonal approximation [193]. The approximating polygon is mapped to a unique binary string. Thus, the string is defined as a chromosome to represent a polygon. The evolution process is iterated by using three genetic operations selection, crossover and mutation. Sun and Huang [165] present a genetic algorithm to conduct polygonal approximation. Given and error bound, an optimal solution of dominant points can be found. However, it seems to be time consuming. Yin [194] focuses on the computational efforts and proposes the tabu search technique to reduce the computation time and memory usage in polygonal approximation. Ho and Chen [62] propose an efficient evolutionary algorithm with a novel orthogonal array crossover for obtaining the optimal solution to the polygonal approximation problem. In [175], a fast algorithm for polygonal approximation based on genetic evolution with break point detection technique is proposed. It can be developed by modifying the existing genetic algorithm based method under the same desirable criteria. The break point detection on the chromosome is conducted to reduce the computations for optimization. The modification on one approach from the two main categories is demonstrated. In fact, the break point detection can be applied to the methods in both categories.

The procedures over viewed so far treat polygonal approximation as a side detection problem. The sides of a polygon are determined subject to certain constraints on the goodness of fit. Another approach to polygonal approximation is to detect the significant points and join the adjacent significant points by straight line segments. Significant points are two types, namely, curvature extreme points and points of inflexion. The concept of detecting local curvature extreme originates from Atteneave's famous observation [10] that information about a curve is concentrated at the curvature extreme points. Freeman [45] suggests that the points of inflexion carry information about a curve and so these points can be used as significant points.

A series of algorithms on the detection of significant points on digital curves already exist. Rosenfeld and Johnston [139], in an attempt to determine whether a procedure [143] designed to detect discontinuities in the average gray levels also detect discontinuities in the average slope detect significant points as the curvature extreme points. The procedure is parallel and needs an input parameter m. The value of m is taken to be 1/10 or 1/5 of the perimeter of the curve. The input parameter is introduced to determine the region of support and the *k*-cosine of the boundary points. The significant points are the local maxima of *k*-cosine.

An improved version of this procedure is given by Rosenfeld and Weszka [142]. They use smoothed *k*-cosine to determine the region of support and to detect significant points. The procedure is parallel and needs an input parameter *m* as in [139].

Freeman and Davis [48] design a corner-finding scheme which detects local maxima of curvature as significant points. The algorithm consists of scanning the chain code of a curve with a moving line segment which connects the end points of a sequence of *s* links. As the line segment moves from one chain code to the next, the angular difference between the successive segment positions are used as a smoothed measure of local curvature along the chain. The procedure is parallel and needs two input parameters s and m. Both are smoothing parameters and their assigned values determine the degree of smoothing. The greater the s, the heavier is the smoothing. The parameter m is used to allow some stray noise. For a well quantized chain s will always be a relatively small number ranging from 5 to 13. And the parameter m will take value either 1 or 2.

Anderson and Bezdek [5] devise a vertex detection algorithm which instead of approximating discrete curvature defines tangential deflection and curvature of discrete curves on the basis of the geometrical and statistical properties associated with the eigenvalue-eigenvector structure of sample covariance matrices. The vertices are the significant points in the sense that they carry information about the curve. The procedure is sequential and needs more than one parameter.

Sankar and Sharma [148] design an iterative procedure to detect significant points as points of maximum global curvature based on the local curvature of each point with respect to its immediate neighbors. The procedure is parallel. In contrast to the algorithms [139, 142, 48] and [5], it does not need any input parameter.

Each of the algorithms [139, 142, 48] and [5] need one or more input parameters. The choice of these parameters is primarily based on the level of detail of the curves. In general, it is difficult to choose a set of parameters that can successfully be used to detect the significant points of a curve which consists of features of multiple sizes. Too large a parameter will smooth out fine features and too small a parameter will generate a large number of unwanted significant points. This is the fundamental problem of scale, because the features describing the shape of curve vary enormously in size and extent and there is seldom any basis of choosing a particular value of parameter for a particular feature size [187]. Though Sankar-Sharma's procedure [148] does not need any input parameter, it does not involve determination of region of support. The procedure is iterative in nature and fails to operate successfully on curves which consist of features of multiple sizes.

To detect dominant points (curvature maxima points) Teh and Chin [170] have designed a procedure which need no input parameter and remains reliable even when features of multiple sizes are present on the curve. In contrast to the existing belief that detection of dominant points depends heavily on the accurate measures of significance (e.g. *k*-cosine, *k*-curvature, cornerity measure, and weighted curvature measure), Teh and Chin make an important observation. The detection of dominant points relies not only on the accuracy of the measure of significance, but primarily on the precise determination of the region of support. Their procedure is motivated by the Rosenfeld-Johnston angle detection algorithm [139], in which both an incorrect region of support and an incorrect curvature measure may be assigned to a point if the input parameter is not chosen correctly, and hence dominant points may be suppressed [29]. To overcome this problem they propose that the region of support and hence the corresponding scale factor or the smoothing parameter of each boundary point should be determined independently, based on its local properties. They use chord length and perpendicular distance to determine the region of support and have further shown that once the region of support of each point is determined, various measures of significance can be computed accurately for detection of dominant points. The dominant points are detected as the local maxima of *k*-cosine, *k*-curvature, and 1-curvature. All these measures of significance are found to produce almost the same results. The procedure is parallel and needs no input parameter.

We present an algorithm [124] for detection of significant points of digital curves. The procedure is motivated by the Rosenfeld-Johnston angle detection scheme [139]. The procedure need no input parameter and remains reliable even when features of multiple sizes are present. The *k*-cosine is used to determine region of support. A new measure of significance based on *k*-cosine (smoothed *k*-cosine) is introduced. The significant points are the local maxima and minima of the smoothed *k*-cosine. The procedure is parallel. The objective of this work is to show that one can use *k*-cosine itself to determine the region of support without using any input parameter.

In the works on the detection of significant points the region of support (if introduced) consists of equal number of points on either side of the point of interest. We propose to call this region of support as symmetric region of support. But the local properties of a curve may not everywhere be so as to have symmetric region of support. We believe that an asymmetric region of support consisting of unequal number of points on either side of the point of interest is more natural and more reasonable than symmetric region of support. A symmetric region of support may be looked upon as a special case of asymmetric region of support. We present a technique [125] on the detection of dominant points (curvature maxima points) which, unlike the existing algorithms, introduces the concept of region of support based on the local properties of a curve. A new measure of significance called *k-l* cosine is introduced. The dominant points are the local maxima of *k-l* cosine. The procedure is parallel. It needs no input parameter and remains reliable even when features of multiple sizes are present on the curve.

Cornic [116] suggests a dominant point detection algorithm that does not require any input parameter and does not rely on computation of curvature. He determines the left and right arm of each point using the chain code of the curve and count the number of times for which a point becomes the end point of right or left arm. The higher is this number the greater is the chance for the point to become a dominant point. Marji and Sipy [82] use the polygonal approximation technique of Ray and Ray [122] to determine the region of support. They use asymmetric region of support introduced by Ray and Ray [125]. Then they define the rank of a boundary point as the number of times it becomes the end point of support region. The strength of a boundary point is then defined based on its rank, support region and distance from the centroid of the curve. The points are sorted according to their strength, support length and distance from the centroid of the shape. From this sorted set, the algorithm tries to extract the best subset that can cover the entire shape boundary. The elements of the extracted subset that survive the following collinearity suppression stage will serve as the vertices of the approximating polygon.

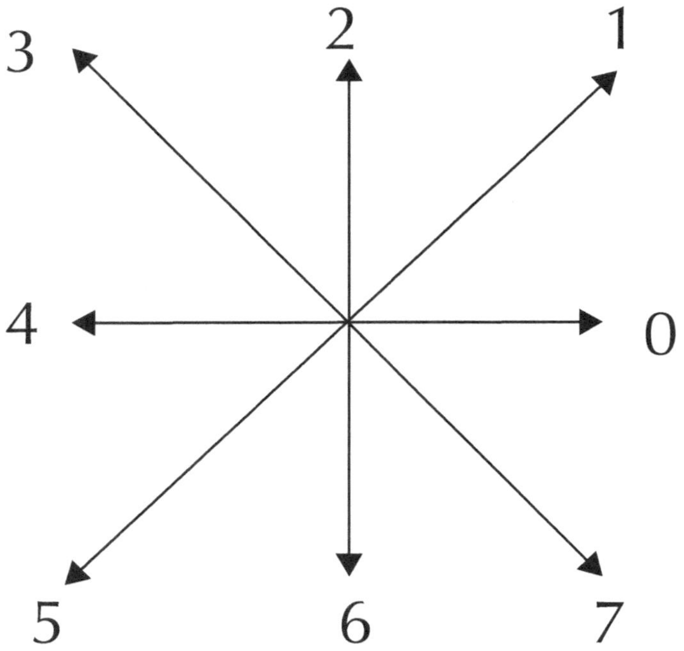

FIGURE 1 Directions in Freeman chain code.

Before we close this chapter we give the definition of a closed digital curve. A closed digital curve C_d with n points is defined by a sequence of n integer coordinate points $C_d = \{p_i = (x_i, y_i), i = 1, 2, n\}$, where p_{i+1} is a neighbor of p_i (modulo n).

The Freeman chain code of C_d consists of n vectors $c_i = p_i p_{i+1}$ each of which can be represented by an integer $f = 0, 1, 2, 7$ as shown in Figure 1, where $1/4\pi f$ is the angle between the x-axis and the vector c_i. The chain of C_d is defined by $\{c_i, \mathrm{i} = 1,n\}$ and $c_i = c_{i\pm n}$.

KEYWORDS

- **Collinearity**
- **Dynamic programming**
- **Initial segmentation**
- **One-Pass algorithm**
- **Polygonal approximation**
- **Split and merge technique**

2 A Split and Merge Technique

CONTENTS

2.1 INTRODUCTION

The fundamental problem in the existing splitting techniques and also in the split and merge techniques is the initial segmentation. Ansari and Delp [6] try to resolve this problem using curvature extreme points as the initial set of break points. An alternative approach to initial segmentation for chain coded curves. The initial segmentation is done introducing the concept of rank of a point defined in the following section. The split and merge is done using the absolute perpendicular distance of a point from the line segment joining the initium and the terminus of a curve segment as the criterion function.

2.2 PROCEDURE

Before we present the procedure we define the rank of a point. The smallest angle through which vector c_i should be rotated so that c_i and c_{i+1} have the same directions determines the rank of the point p_i and if θ_i be the smallest required angle then the rank of p_i is defined by:

$$r_i = (4/\pi)\, \theta_i, \quad i = 1,2,\ldots,n. \tag{2.1}$$

We note that the only possible values for r_i are 0, 1, 2, 3 and 4. The angle θ_i, is computed by the relation

$$\theta_i = \cos^{-1}\{(c_i \cdot c_{i+1})/ (|c_i|\, |c_{i+1}|)\} \tag{2.2}$$

As already mentioned the fundamental problem in the split and merge process is the initial segmentation. We propose to make the initial segmentation on the basis of rank of a point. The procedure initial segmentation looks for those points with rank greater than or equal to 3. If there exist at least two points with this rank then, these points are used as the initial set of break points. If there exists only one point with rank greater than or equal to 3 then this point together with those with rank 2 constitute the initial set of break points. If there is no point with rank greater than or equal to 3 and there exist at least two points with rank 2 then, these points only constitute the initial set of break points. If there exists only one point with rank 2 then this point together with those with rank 1 constitute the initial segmentation. Lastly, if there exists only two points with rank 1 then one can start with these points. We note that we give priority to those points which have higher rank. This approach reduces the false choice of vertices in initial segmentation.

Since the rank is obtained by taking the angular difference between c_i and c_{i+1} and multiplying it by $4/\pi$. Hence, the computational load is lower than that of Ansari and Delp [6] in which the total number of multiplications and additions depends on the window size of the Gaussian filter. The larger the window length is, the higher is the computational load.

Using the initial set of break points we perform a split and merge process by collinearity check. The criterion function for collinearity check is the absolute perpendicular distance of a point from the line segment joining two successive break points.

The perpendicular distance of a point p_k from the segment joining two successive break points p_i and p_j is computed by the formula:

$$d_k = |(y_j - y_i)x_k - (x_j - x_i)y_k - x_i y_j + x_j y_i| / \sqrt{\{(x_j - x_i)^2 + (y_j - y_i)^2\}}, \tag{2.3}$$

where, $k = i + 1, i + 2, \ldots j - 1$. Points are split at those points where d_k is a maximum and exceeds a specified value, which we take as one pixel. This threshold of one pixel is determined based on the fact that a slanted straight line is quantized into a set of either horizontal or vertical line segments separated by one pixel steps. In addition, we assume that the boundary noise is no more than one pixel. If this noise level is known a priori then this threshold can be adjusted accordingly [119].

Merging is done in the following manner. For each pair of adjacent line segments comprising of three successive break points p_l, p_{l+1}, and p_{l+2} (say) if the absolute perpendicular distance of all points intermediate of p_l and p_{l+2} from the line segment $p_l p_{l+2}$ does not exceed one pixel then the point p_{l+1} is merged, otherwise this point is retained.

2.3 ALGORITHM

The input are the data points (x_i, y_i), $i = 1, 2, \ldots, n$. The output is the vertices p_i of the polygon.
Step 1: Initial Segmentation

Compute

$\theta_i = \cos^{-1}\{(c_i . c_{i+1})/(|c_i|\ |c_{i+1}|)\}$, $\quad i = 1, 2, \ldots, n$
and $r_i = (4/\pi)\ \theta_i$, $\quad i = 1, 2, \ldots, n$.

- If there exists at least two values of i for which $r_i \geq 3$, then these points constitute the initial set of break points

- else if there exists only one i for which $r_i \geq 3$ then this point together with those with $r_i = 2$ constitute the initial set of break points
- else if there exists no point with rank $r_i \geq 3$ and if there exist at least two points with rank $r_i = 2$ then these points constitute the initial set of break points
- else if there exists only one point with $r_i = 2$ then this point together with those with $r_i = 1$ comprise the initial set of break points
- else if there exists no point with $r_i = 2$ but there exists at least two points with $r_i = 1$ then these points constitute the initial set of break points.

Step 2: Splitting

Compute

$$d_k = |(y_j - y_i)x_k - (x_j - x_i)y_k - x_i y_j + x_j y_i| / \sqrt{\{(x_j - x_i)^2 + (y_j - y_i)^2\}}$$

where, $k = i + 1, i + 2, \ldots, j - 1$.

- If $d_k > 1$ then the segment $p_i p_j$ is split at the point where d_k is a maximum
- else if $d_k \leq 1$ then no splitting is necessary.

After all necessary splitting are done go to step 3.

Step 3. Merging

For every three successive break points p_l, p_{l+1}, and p_{l+2} the vertex p_{l+1} is merged with the segment $p_l p_{l+2}$ if the distance of every point intermediate of p_l and p_{l+2} from the segment $p_l p_{l+2}$ does not exceed 1 else merging is not possible.

After all necessary merging is done go to step 4.

Step 4: Repeat Step 2 and Step 3 until equilibrium is reached
End.

2.4 EXPERIMENTAL RESULTS

To focus on the performance of the algorithm 2.3 we apply it on four digital curves, namely, a leaf-shaped curve (Figure 1), a figure-8 curve (Figure 2), an aircraft (Figure 3) and a screwdriver (Figure 4). The first two of these are taken from Rosenfeld and Johnston [71], the third is from Gupta and Malakapalli [58] and the fourth is from Medioni and Yasumoto [87]. The polygonal approximations are shown in Figures 1 through 4. These figures also show the approximations as obtained by the Ansari-Delp algorithm [6].

As seen from these approximations the algorithm 2.3 places the vertices at the position where they should be (as judged by perception) whereas, the Ansari-Delp algorithm sometimes fails to do so, shifting the vertices from their actual position, rounding off sharp turnings (see Figure 3 and upper right part of the Figure 1). An overview of the results of the approximations as obtained by algorithm 2.3 and those by the Ansar-Delp algorithm are displayed in Table 1.

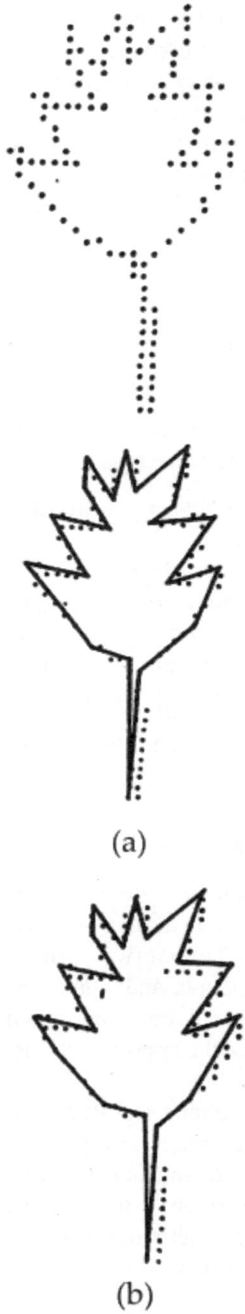

FIGURE 1 A leaf-shaped curve and its polygonal approximations. (a) Algorithm 2.3 and (b) Ansari-Delp algorithm.

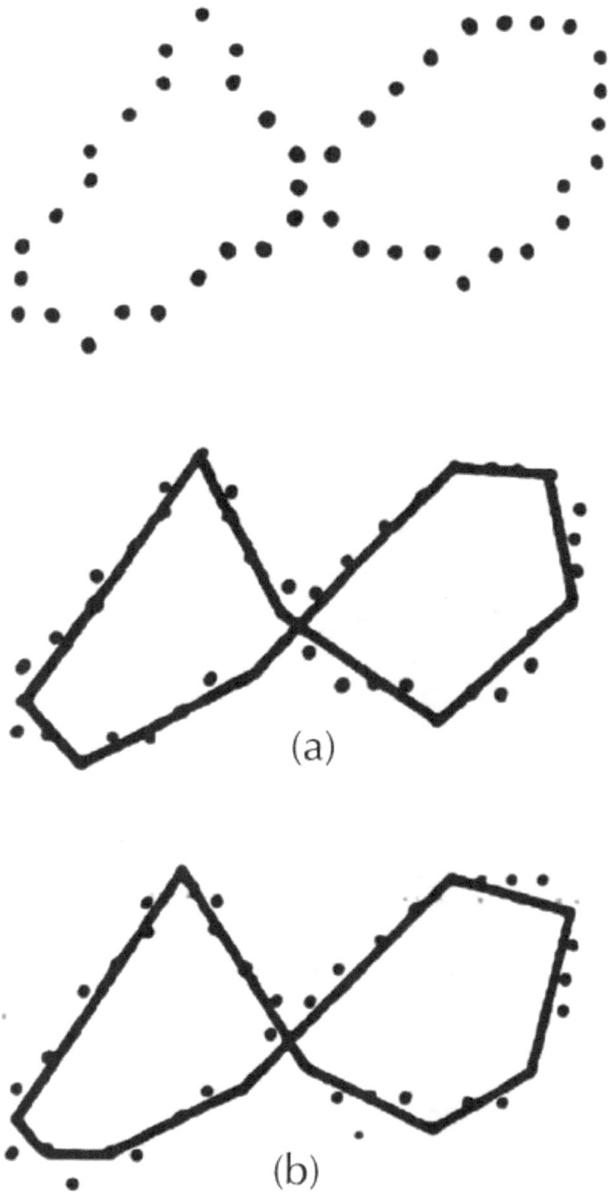

FIGURE 2 A figure-8 curve and its polygonal approximations. (a) Algorithm 2.3 and (b) Ansari-Delp algorithm.

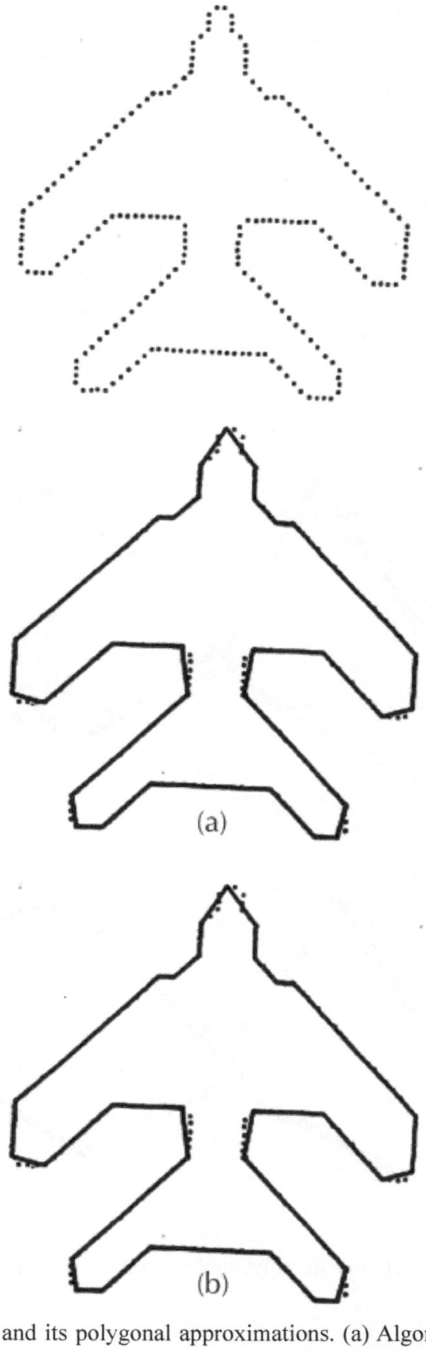

FIGURE 3 An aircraft and its polygonal approximations. (a) Algorithm 2.3 and (b) Ansari-Delp algorithm.

(a)

(b)

FIGURE 4 A screwdriver and its polygonal approximations. (a) Algorithm 2.3 and (b) Ansari-Delp algorithm.

The table shows the number of data points (n) of the digital curve, the number of vertices (n_v), the compression ratio (n/n_v), percentage of data reduction, integral square error and the maximum allowable error (which is 1.00 in all cases). As seen from this table the algorithm 2.3 consistently outperforms the Ansari-Delp algorithm with respect to integral square error, the number of vertices and compression rate of data reduction for a given maximum allowable error. For each of the digital curves the integral square error as obtained by the algorithm 2.3 is smaller than that obtained by the Ansari-Delp algorithm, though the number of vertices as obtained by algorithm 2.3 is no more than that obtained by the Ansari-Delp algorithm. Precisely speaking, the number of vertices of the figure-8 curve and of the screwdriver image as obtained by algorithm 2.3 is smaller than that obtained by the Ansari-Delp algorithm and as a consequence of it the percentage of data reduction as obtained by the algorithm 2.3 is greater than that obtained by the Ansari-Delp algorithm but the integral square error as produced by the algorithm 2.3 is less than that produced by the Ansari-Delp algorithm. As regard to the leaf shaped curve and the aircraft image the number of vertices as obtained by either of the two algorithms are the same but the integral square error as obtained by the algorithm 2.3 is lower than that obtained by the Ansari-Delp algorithm.

TABLE 1 A comparison between algorithm 2.3 and Ansari-Delp algorithm.

Digital curve	**Leaf**	**Figure-8**	**Aircraft**	**Screwdriver**
Number of points	120	45	200	267
Number of vertices (n_v)	20	9	29	41
Compression ratio (n/n_v)	6.00	5.00	6.90	6.51
Percentage of data reduction	83.33	80.00	85.50	84.00
Integral square error	21.24	5.27	8.71	55.75
Maximum error	1.00	1.00.	1.00	1.00
Results of Ansari-Delp algorithm				
Maximum allowable error	1.00	1.00.	1.00	1.00
Number of vertices (n_v)	20	10	29	43
Compression ratio (n/n_v)	6.00	4.50	6.90	6.21
Percentage of data reduction	83.33	77.80	85.50	83.90
Integral square error	28.59	5.50	14.56	62.50
Maximum error	1.00	1.00.	1.00	1.00

This shows that the algorithm 2.3 produces integral square error which is no greater than that obtained by the Ansari-Delp algorithm without producing more vertices. Moreover, algorithm 2.3 does not shift vertices; it does not round off sharp turnings and produces symmetrical approximation from a symmetrical digital curve. Whereas, Ansari-Delp algorithm does not have these merits. So we conclude that the algorithm 2.3 produces more accurate results that the Ansari-Delp algorithm without producing redundant vertices.

KEYWORDS

- **Ansari-Delp algorithm**
- **Collinearity check**
- **Initial segmentation**
- **Leaf-Shaped curve**
- **Polygonal approximations**

3 A Sequential One-pass Method

CONTENTS

3.1 INTRODUCTION

In this chapter we present a sequential one-pass technique for polygonal approximation of digital curves with uniformly spaced points. The procedure [120] is based on a result from numerical analysis and some concepts of regression analysis. Here we find the first order finite differences of the abscissa (x) and the ordinates (y) of the data points describing the digital curve. Then we try to identify some patterns (described later) in the first order finite differences Δx and Δy. By recognizing these patterns we find the sides of a polygon approximating a curve.

3.2 PATTERN RECOGNITION OF FINITE DIFFERENCES

From the boundary point data (x_i, y_i) we construct the first order finite differences Δx_i and Δy_i using

$$\Delta x_i = x_{i+1} - x_i \text{ and } \Delta y_i = y_{i+1} - y_i. \tag{3.1}$$

For digital curves with uniformly spaced points these differences take values 0, 1, –1 only. The successive differences, as we go through them exhibit one of the following patterns:

(a) Δx_i's are constant for a series of equidistant values of y.
(b) Δy_i's are constant for a series of equidistant values of x.
(c) Δx_i's take values 0 and 1 only for a series of equidistant values of y.
(d) Δx_i's take values 0 and –1 only for a series of equidistant values of y.
(e) Δy_i's take values 0 and 1 only for a series of equidistant values of x.
(f) Δy_i's take values 0 and –1 only for a series of equidistant values of x.

3.3 BASIC RESULTS AND CONCEPTS

The polygon representation of the shape of the boundary of an object can be obtained using any boundary tracking algorithm. It is not uncommon that such a representation involves a polygon having too many vertices. Since, the time required for processing a polygon is dependent on the number of vertices, this polygon is usually determined using a data smoothing technique. Our polygonal approximation too involves a smoothing technique which is based on the following proposition from numerical analysis [151] and some concepts from regression analysis.

Proposition 3.1

If the nth differences of a tabulated function are constant when the values of the independent variable are taken in arithmetic progression then the function is a polynomial of degree n.

Besides, the proposition we use some concepts of regression analysis. In the regression analysis the scatter diagram of the bivariate data exhibits either a linear or a curvilinear tendency. Either, the points of the scatter diagram exhibit a tendency to cluster around a line or they exhibit a tendency of clustering around a curve (curvilinear tendency), provided the variables are related. If the points of the scatter diagram exhibit a linear tendency then we can find a linear relationship between the variables that is the points can be approximated by a straight line. In regression analysis this line is the least square line. In our smoothing technique we exploit the concepts without using least squares line.

3.4 SMOOTHING TECHNIQUE

The smoothing technique involves reading the first order finite differences Δx_i and Δy_i, locating the break points and joining the successive break points by straight line segments. We describe how to perform the smoothing so as to locate the break points.

Following the proposition stated in the last section, the series of points where pattern (a) is observed can be smoothed out by a single straight line segments which is obtained by joining the end points of the series. The series of points revealing pattern (b) can similarly be smoothed out following the same proposition.

The series of successive points revealing pattern (c) exhibit that there is a tendency in the points to cluster around a straight line. Hence, from the concept of regression analysis stated in the last section, these points can be approximated by a straight line segment. The straight line segment is used here is not the least square line. The use of least squares line sometimes leads to unconnected boundary, since the estimated line does not necessarily have to go through any of the points of the series. The straight line segment that is used to smooth out these points is obtained by joining the end points of the series revealing pattern (c).

In order to identify pattern (a) or (b) a counter c is introduced that will count the number of finite differences of x (or y) that remain constant for a series of equidistant values of y (or x). We set the critical value of this counter to 4. Larger value of c will produce less number of vertices at the cost of higher approximation error. Thus if at least four successive differences of x (or y) remain constant for a series of equidistant values of y (or x) then the pattern (a) (or (b)) is identified. The approximation error is controlled by this counter. Setting this counter to different values different approximations can be obtained.

3.5 COMPLEXITY

The description of the smoothing technique it is clear that the procedure need no numerical computation except subtractions. For *n* boundary points 2*n* subtractions are required. The procedure mainly involves comparisons. One comparison is needed for each value of *i* and so a total of 2*n* comparisons are required. This shows that the complexity of the procedure is *O*(*n*).

3.6 EXPERIMENTAL RESULTS

We have applied the smoothing technique 3.4 on four digital curves, namely, a leaf-shaped curve (Figure 1), a figure-8 curve (Figure 2), an aircraft (Figure 3) and a screw-driver (Figure 4). The polygonal approximations of the curves are shown in Figure 1 through Figure 4. Since, this algorithm is sequential we compare it with the Williams' algorithm [184] and the Wall-Danielsson algorithm [180] both of which are sequential in nature. As already stated the smoothing technique 3.4 involves a counter *c* and the maximum error is controlled by this counter. The maximum error as obtained by this algorithm is used to run Williams' algorithm on the same digital curves. The polygonal approximations are shown in Figure 1 through Figure 4. As seen from these figures the worst approximation is obtained with the leaf shaped curve (Figure 1) where the lower part of the curve which has a sharp turning is completely wiped out by the approximation and almost none of the other turnings are approximated according to their nature. The corners are always shifted to the nearby point.

(a)

FIGURE 1 *(Continued)*

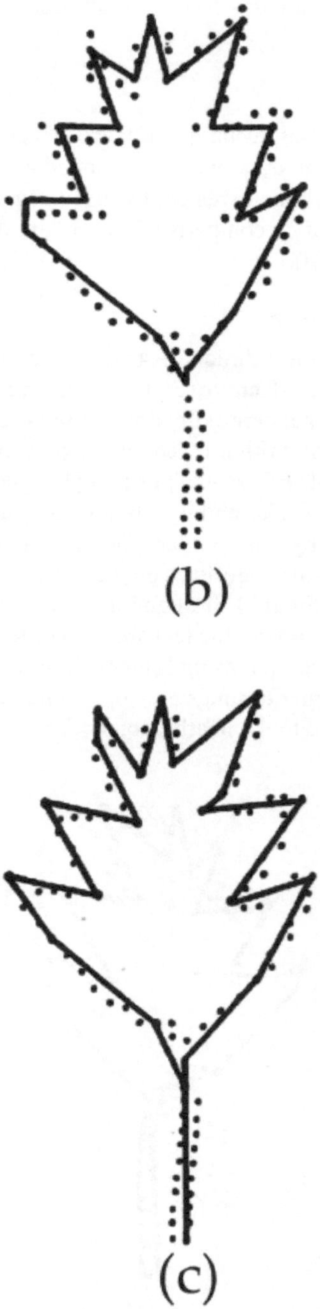

FIGURE 1 Polygonal approximations of the leaf-shaped curve. (a) Smoothing technique 3.4, (b) Williams' algorithm, and (c) Wall-Danielsson algorithm.

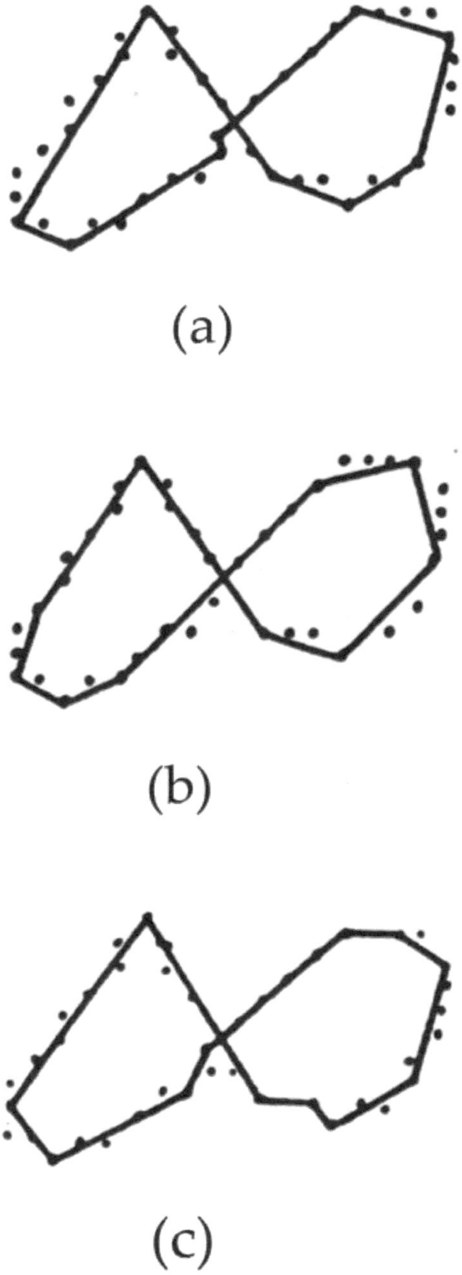

FIGURE 2 Polygonal approximations of the figure-8 curve. (a) Smoothing technique 3.4, (b) Williams` algorithm, and (c) Wall-Danielsson algorithm.

The vertices are located at a point away from where they should be. This is true for other digital curves also. But the approximations as obtained by the smoothing technique described in section 3.4 do not have any such deficiency. Neither does it round off sharp turnings nor does it miss corners. To compare our procedure with the Wall-Danielsson algorithm we use the maximum error returned by our procedure as the stopping criterion of the Wall-Danielsson algorithm. As already stated apart from the peak test.

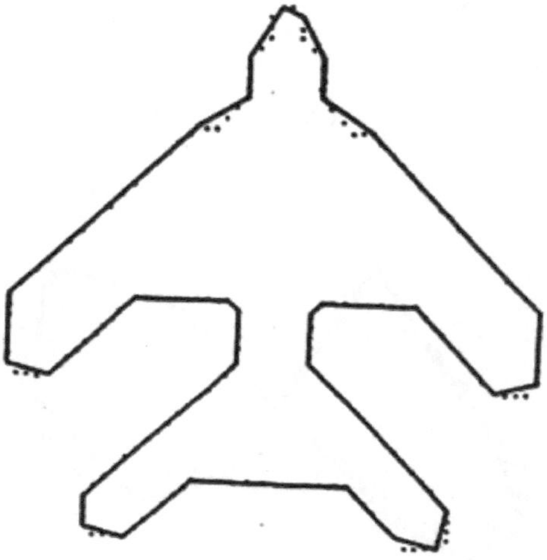

(a)

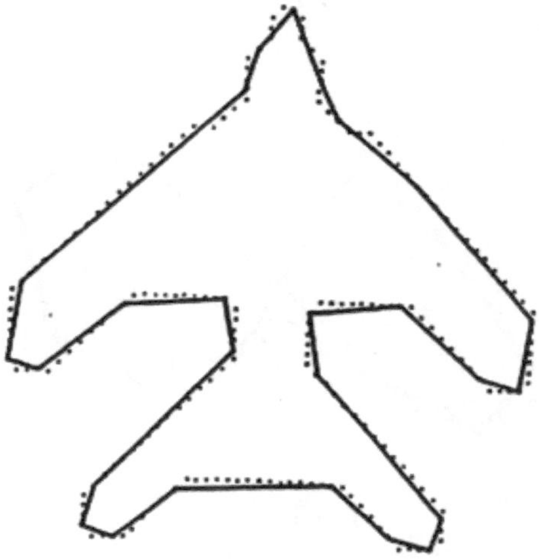

(b)

FIGURE 3 *(Continued)*

(c)

FIGURE 3 Polygonal approximations of the aircraft. (a) Smoothing technique 3.4, (b) Williams' algorithm, and (c) Wall-Danielsson algorithm.

The criterion function of the Wall-Danielsson algorithm is the area deviation per unit length. For comparison the threshold on the area deviation is initially set to a large value and is then diminished until the maximum error returned by our procedure is surpassed.

(a)

FIGURE 4 *(Continued)*

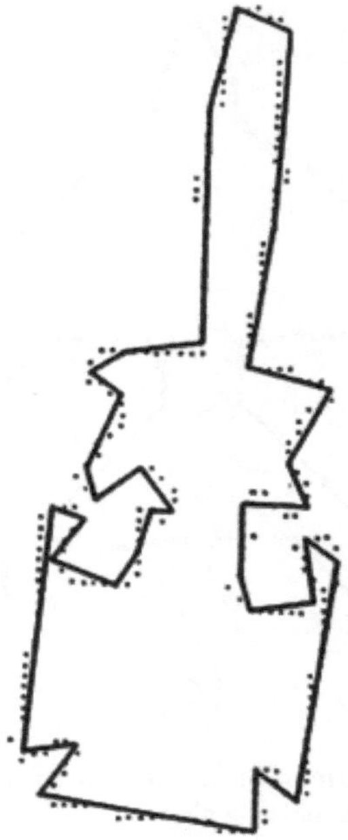

(b)

(c)

FIGURE 4 Polygonal approximations of the screwdriver. (a) Smoothing technique 3.4, (b) Williams' algorithm, and (c) Wall-Danielsson algorithm.

The polygonal approximations as obtained by the Wall-Danielsson algorithm are shown in Figure 1 through Figure 4. These approximations show that though the Wall-Danielsson algorithm retains the peaks but it fails to locate vertices at their actual position when turning is not so sharp (see Figure 3). An overview of the results of the approximations as obtained by smoothing technique 3.4 and those obtained by the Williams' and Wall-Danielsson algorithm are displayed in Table 1. The table shows number of vertices (n_v), compression ratio, percentage of data reduction, maximum error and integral square error.

TABLE 1 A comparison among smoothing technique 3.4, Williams' and Wall-Danielsson algorithm.

Digital curve	**Leaf**	**Figure-8**	**Aircraft**	**Screwdriver**
Number of points	120	45	200	267
Results of smoothing technique 3.3				
Number of vertices (n_v)	24	10	30	72
Compression ratio (n/n_v)	5.00	4.50	6.67	3.31
Percentage of data reduction	80.00	77.78	85.00	73.00
Integral square error	16.45	5.30	8.88	20.95
Maximum error	1.18	0.97	1.03	1.34
Results of Williams' algorithm				
Maximum allowable error	1.18	0.97	1.03	1.34
Number of vertices (n_v)	18	10	30	72
Compression ratio (n/n_v)	6.67	4.50	7.41	7.63
Percentage of data reduction	85.00	77.78	86.50	86.89
Integral square error	46.63	5.59	44.00	94.27
Maximum error	1.17	0.73	1.03	1.30
Results of Wall-Danielsson algorithm				
Maximum allowable error	1.18	0.97	1.03	1.34
Threshold	0.996	0.63	0.64	0.958

TABLE 1 *(Continued)*

Digital curve	Leaf	Figure-8	Aircraft	Screwdriver
Number of vertices (n_v)	21	12	38	41
Compression ratio (n/n_v)	5.71	3.75	5.26	6.51
Percentage of data reduction	82.5	73.33	81.00	84.6
Integral square error	21.075	3.22	11.22	54.54
Maximum error	0.995	0.707	0.65	1.26

KEYWORDS

- **Polygonal approximation**
- **Sequential one pass Method**
- **Smoothing technique**
- **Wall-Danielson algorithm**
- **Williams' algorithm**

4 Another Sequential One-pass Method

CONTENTS

4.1 INTRODUCTION

The procedure presented in the last chapter holds for digital curves with uniformly spaced points only. In this chapter we present another sequential one-pass algorithm [121] which holds for uniformly as well as non-uniformly spaced points. The procedure is based on Pavlidis' [106] concept of almost collinearity of a sequence of points. Initially the in-radius of triangles formed by the sequence of point's triplets is introduced as a criterion function to measure collinearity. This is an indirect approach but justified by a proposition which establishes that the higher the in-radius is, the higher is the perpendicular distance. Unfortunately, evaluation of in-radius is computationally expensive. To reduce the computational load the in radius is replaced by the area and perimeter of triangles. The vertices of the polygon are located by comparing the area and perimeter with their critical value.

4.2 APPROXIMATION TECHNIQUE

If p_i, p_j and p_k $(k = j + 1)$ be three points of a digital curve C_d, then the cross product of the vectors p_ip_j and p_ip_k is:

$$p_ip_j \times p_ip_k = \{ (x_j - x_i)(y_k - y_i) - (y_j - y_i)(x_k - x_i) \}t \tag{4.1}$$

where, t is a unit vector perpendicular the plane determined by p_ip_j and p_ip_k. The magnitude of the cross product is:

$$| p_ip_j \times p_ip_k | = |(x_j - x_i)(y_k - y_i) - (y_j - y_i)(x_k - x_i)| \tag{4.2}$$

Translating the origin of the coordinate system to the point p_i and denoting the new coordinate system by prime, the magnitude of the cross product reduces to the form

$$|op`_j \times op`_k| = |x`_j y`_k - x`_k y`_j| \tag{4.3}$$

Again the cross product of the vector $p_i p_j$ and $p_j p_k$ is:

$$p_i p_j \times p_j p_k = \{ (x_j - x_i)(y_k - y_j) - (y_j - y_i)(x_k - x_j) \} t \tag{4.4}$$

and its magnitude is

$$p_i p_j \times p_j p_k \,| = | (x_j - x_i)(y_k - y_j) - (y_j - y_i)(x_k - x_j) | \tag{4.5}$$

In prime coordinate system (4.5) reduces to:

$$|op`_j \times p`_j p`_k| = | x`_j(y`_k - y`_j) - y`_j(x`_k - x`_j)| \tag{4.6}$$

Both the magnitude (4.3) and (4.6) are twice the area (A) of the triangle with vertices at p_i, p_j, p_k that is:

$$2A = |x`_j y`_k - x`_k y`_j| \tag{4.7}$$

$$2A = | x`_j(y`_k - y`_j) - y`_j(x`_k - x`_j)| \tag{4.8}$$

The last form (4.8) is useful for chain coded curves where $y`_k - y`_j$ and $x`_k - x`_j$ take values 0, 1, and –1 only.

Proposition 4.1

If the area of a triangle with p_i, p_j and p_k as vertices be zero/almost zero (i.e. within a specified value) then the points p_i, p_j and p_k are collinear/almost collinear.

Proposition 4.2

If three points p_i, p_{i+1} and p_{i+2} be collinear/almost collinear and the three points p_i, p_{i+2} and p_{i+3} be also collinear/almost collinear then the points p_i, p_{i+1}, p_{i+2} and p_{i+3} are collinear/almost collinear. Again if p_i, p_{i+3} and p_{i+4} be collinear/almost collinear then the points p_i, p_{i+1}, p_{i+2}, p_{i+3}, and p_{i+4} are collinear/almost collinear. And this process can be carried on for any finite number of points.

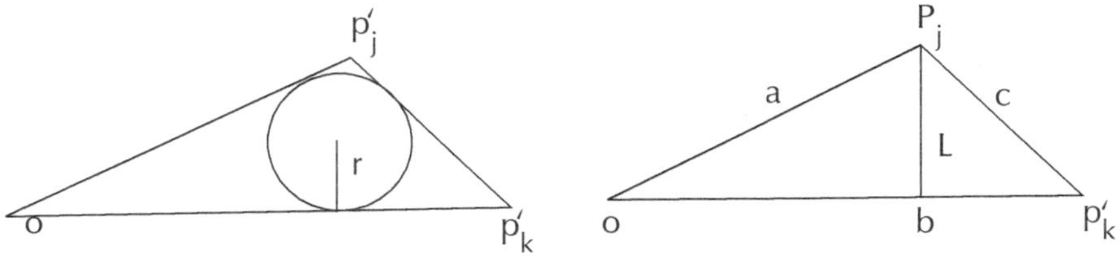

FIGURE 1 The in-radius and perpendicular distance.

Using propositions 1 and 2 and taking into account the measure (4.7) for curves with non-uniformly spaced points and (4.8) for curves with uniformly spaced points (chain-coded curve) we can check the collinearity of a sequence of points by specifying the critical value of $2A$. But the choice of this critical value is problematic. If we set it to 1, we get many redundant vertices and if we set it to 2 we miss many important vertices and so we lose the shape of the curve. So it is not effective to use $2A$ only as a measure of collinearity. In the following we introduce as an alternative measure of collinearity.

$r = A/s = (2A)/(2s) =$ (magnitude of cross product)/(perimeter of triangle $op`_j p`_k$) (4.9)

The significance of r is that it gives the in-radius of the triangle $p_i p_j p_k$ ($op`_j p`_k$). This is an indirect approach to the problem but is justified by the following proposition.

Proposition 4.3

In any triangle the in-radius (r) and absolute perpendicular distance (l) of a vertex from the opposite side are related by $r < ½l$.

Proof: Since $r = A/s$, referring to Figure 1 $r = (bl)/(a + b + c)$. In any triangle $c + a > b$, so $(c + a)/b + 1 > 2$ that is $(a + b + c)/b > 2$ that is $(bl)/(a + b + c) < ½l$ that is $r < ½l$.

This result shows that the higher the in radius is, the farther is the point $p`_j$ ($p_j$) from the line $op`_j$ ($p_i p_j$). So we may take r as a measure of prominence for the point $p`_j$ to be a vertex. We can decide upon a critical value of r and comparing r with its critical value we can locate the vertices. This procedure fails to catch the peaks of a curve. Sometimes peaks also play an important role to detect defects of an object [9]. Observation also shows that as the procedure passes a peak the in-radius maintains non-zero value which is below its critical value. So it is possible to locate the vertices accurately if we incorporate a check on two successive non-zero values of r, in addition to comparing it with its critical value.

Unfortunately, the evaluation of r is computationally expensive because evaluation of perimeter need evaluation of three square roots (the length of three vectors $op`_j$, $op`_k$ and $p`_j p`_k$. So we approximate these lengths by the maximum of the absolute value of the components of the vectors.

Thus the length of $op`_j$, $op`_k$ and $p`_j p`_k$ are approximated by:

$$|op`_j| = \max(|x`_j|, |y`_j|)$$
$$|op`_k| = \max(|x`_k|, |y`_k|)$$

$$|p`_j p`_k| = \max (|x`_k - x`_j|, |y`_k - y`_j|)$$

Using this approximation we may use r to locate the vertices by the procedure described earlier. But the evaluation of r involves division of the magnitude of the cross product by the perimeter of the triangle. We avoid this arithmetic operation by dispensing with the measure r and replacing it by two measures namely, $2A$ and $2s$. Now we can compare these two measures with their critical values 1 and 3 respectively to check the collinearity of points.

Since, comparison of r with its critical value fails to catch the peaks hence comparison of $2A$ and $2s$ with their critical value will also fail to do the same. But if we use r as a measure of collinearity, to catch peaks we incorporate a check on two successive non-zero values of r when it is below its critical value. Similarly if we use $2A$ and $2s$ as a measure of collinearity we incorporate a check on two successive values of $2s$ when it is above its critical value and $2A$ is equal to 1.

So instead of using r as a measure of collinearity we use $2A$ and $2s$ as a measure of collinearity. This replacement of r by $2A$ and $2s$ reduces the computational load significantly and at the same time they perform the same task and produce the same results as obtained using r only. Further using two measures we can make a good compromise between unwanted vertices and loss of important vertices which occur when we consider only one measure $2A$.

The collinearity check can be started from an arbitrary point and is carried beyond the starting point till the vertex generated last coincides with one of the vertices already generated.

We are now ready to present the algorithm for detecting the vertices of the polygon. We propose to denote the magnitude of the cross product (4.7) and (4.8) and the approximate value of the perimeter of the triangle formed by the points p_i, p_j, p_k ($k = j + 1$) by D_k and P_k respectively.

4.3 ALGORITHM

Comments

The inputs are the coordinates (x_i, y_i), $i = 1, 2, \ldots, n$. The outputs are the vertices. All arithmetic is performed in integer mode and is in modulo n.

Begin

Step 1: Intialte $i = 1$.

Step 2: Translate the coordinate system to the point pi so that it becomes the origin of the prime coordinate system; set $j = i + 1$; $k = i + 2$.

Step 3: Compute

$D_k = | x`_j(y`_k - y`_j) - y`_j(x`_k - x`_j)|$, for chain coded curve

$D_k = |x`_j y`_k - x`_k y`_j|$, for other curves.

Step 4: If $D_k = 0$ then the point $p`_k$ (p_k) passes the test; change j to $j + 1$; k to $k + 1$; go to step 3.

else if $D_k = 1$ than compute P_k by

$$P_k = \max(|x`_j|, |y`_j|) + \max(|x`_k|, |y`_k|) + \max(|x`_k - x`_j|, |y`_k - y`_j|).$$

Step 5: If $P_k > 3$ and $P_k \neq P_{k-1}$ then $p`_k$ (p_k) passes the test; change j to $j + 1$; k to $k + 1$; go to step 3.

else $p`_j$ (p_j) is vertex; write j; (x_j, y_j); go to step 2.

Step 6. Repeat this process till the vertex generated last coincides with one of the vertices already generated.

Step 7: Join the vertices in order to obtain the polygon.

End.

4.4 COMPUTATIONAL COMPLEXITY

As it is clear we have used two forms of the same measure D_k, namely:

$$D_k = |x`_j(y`_k - y`_j) - y`_j(x`_k - x`_j)|$$

$$\text{and } D_k = |x`_j y`_k - x`_k y`_j|$$

The second form is deducible from the first form. We use the first form for chain coded curves (curves with uniformly spaced points) where $x`_k - x`_j$ and $y`_k - y`_j$ are either 0 or 1 or –1 and so evaluation of D_k need no multiplication. For other curves we use the second form which involves two multiplications. Here if we use the first form instead of the second it will increase the total number of arithmetic operations (introducing two unnecessary subtractions). The evaluation of P_k involves no multiplication. This shows that our procedure need only two multiplications for digital curves with non-uniformly spaced points and no multiplication for chain coded curves.

4.5 DISCUSSION

The polygonal approximation proposed by Williams [184] used scan along technique which need a single pass through data. Wall and Danielsson [180] too, used scan along technique but it is faster that of Williams. For curves with uniformly spaced points Williams' technique needs seven multiplications/divisions and evaluation of a square root. Whereas, Wall and Danielsson's technique needs six multiplications and the simplified version of it requires only three to four multiplications. For curves with non-uniformly spaced points Williams' technique requires five multiplications whereas Wall and Danielsson's needs two multiplications and the simplified version needs only one multiplication. The algorithm 4.3 requires only two multiplications for curves with non-uniformly spaced points and no multiplication for uniformly spaced points.

(a)

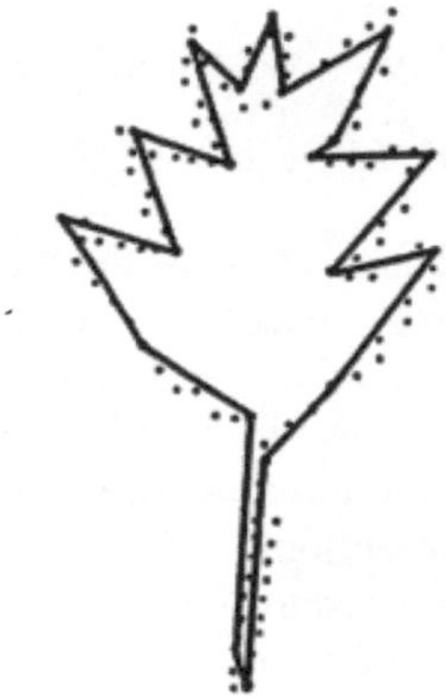

(b)

(c)

FIGURE 2 Polygonal approximations of the leaf-shaped curve. (a) Algorithm 4.3, (b) Williams' algorithm, and (c) Wall-Danielsson algorithm.

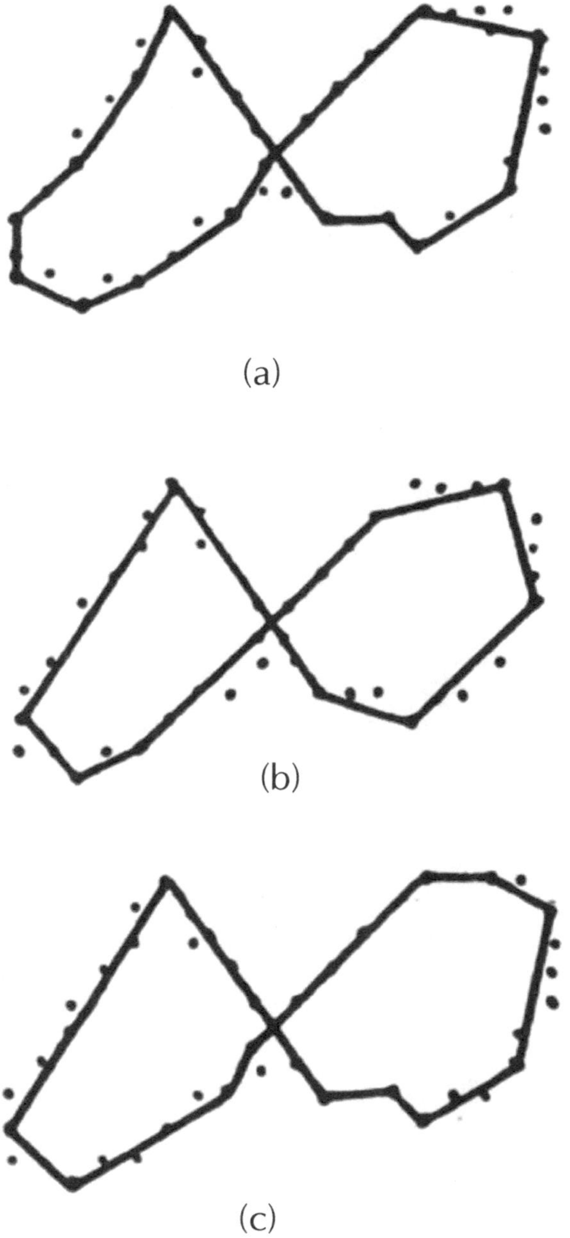

FIGURE 3 Polygonal approximations of the figure-8 curve. (a) Algorithm 4.3, (b) Williams' algorithm, and (c) Wall-Danielsson algorithm.

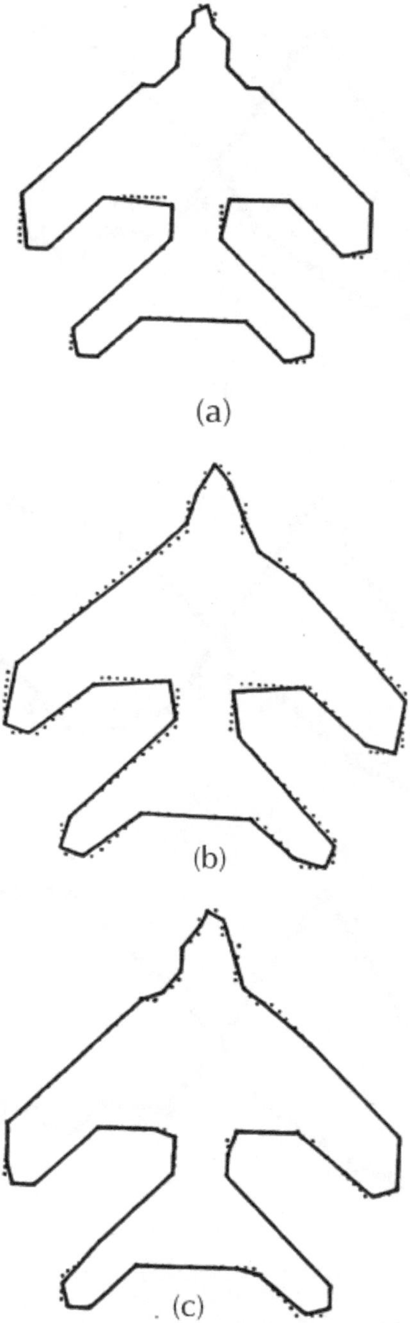

FIGURE 4 Polygonal approximations of the aircraft. (a) Algorithm 4.3, (b) Williams' algorithm, and (c) Wall-Danielsson algorithm.

4.6 EXPERIMENTAL RESULTS

The algorithm 4.3 developed is applied on the same digital curves. The digital curves and their polygonal approximations are shown in Figures 2 through 5. The data are processed in the clockwise direction. Since, this procedure is sequential and one-pass hence we compare the experimental results of this procedure with those of the Williams' algorithm [184] and the Wall-Danielsson algorithm [180]. As we have already seen this procedure controls the maximum error indirectly, so the maximum error that is obtained by applying this procedure on each of the four digital curves are used to run the Williams' and the Wall-Danielsson algorithm.

(a)

FIGURE 5 *(Continued)*

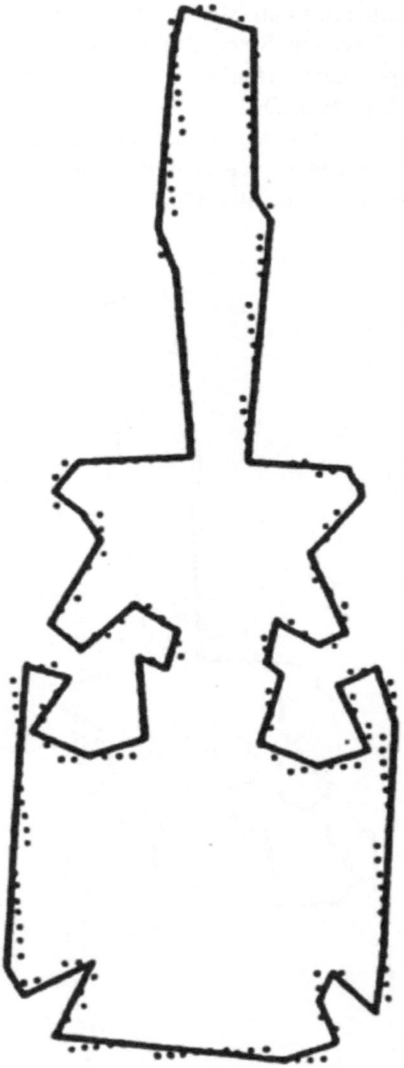

(b)

FIGURE 5 *(Continued)*

(c)

FIGURE 5 Polygonal approximations of the screwdriver. (a) Algorithm 4.3, (b) Williams' algorithm, and (c) Wall-Danielsson algorithm.

The polygonal approximations as obtained by the Williams' and Wall-Danielsson algorithm are shown in Figures 2 through 5. As seen from these figures the approximations as obtained by our procedure is better than those produced by the Williams' or the Wall-Danielsson algorithm. The Williams' algorithm misses corners and rounds off sharp turnings. It also detects false vertices. The worst approximation is obtained in the aircraft image where most of the turnings are not approximated according to their nature. The vertices are shifted to the nearby point from their actual position. Though the Wall-Danielsson algorithm retains the peaks but it misses corners (see the lower part of the leaf) and sometimes it detects false vertices (see the aircraft and screwdriver image). A comparison of the results obtained by our procedure with those of the Williams' and Wall-Danielsson are shown in Table 1.

TABLE 1 A comparison among smoothing technique 4.3, Williams' and Wall-Danielsson algorithm.

Digital curve	**Leaf**	**Figure-8**	**Aircraft**	**Screwdriver**
Number of points	**120**	**45**	**200**	**267**
Results of smoothing technique 4.2				
Number of vertices (n_v)	33	16	32	70
Compression ratio (n/n_v)	3.64	2.81	6.25	3.81
Percentage of data reduction	72.50	64.40	84.00	73.80
Integral square error	11.17	3.75	9.63	41.40
Maximum error	0.896	0.73	0.896	0.92
Results of Williams' algorithm				
Maximum allowable error	0.896	0.73	0.896	0.92
Number of vertices (n_v)	20	9	28	50
Compression ratio (n/nv)	6.00	5.00	7.14	5.34
Percentage of data reduction	83.33	80.00	86.00	81.3
Integral square error	26.61	5.99	37.50	47.85
Maximum error	0.896	0.73	0.896	0.91
Results of Wall-Danielsson algorithm				
Maximum allowable error	0.896	0.73	0.896	0.92
Threshold	0.95	0.63	0.65	0.67
Number of vertices (n_v)	21	12	38	66
Compression ratio (n/n_v)	5.71	3.75	5.26	4.05

TABLE 1 *(Continued)*

Digital curve	Leaf	Figure-8	Aircraft	Screwdriver
Percentage of data reduction	82.5	73.33	81.00	75.33
Integral square error	18.44	3.22	11.22	25.18
Maximum error	0.896	0.707	0.64	0.77

KEYWORDS

- **Collinear/almost collinear**
- **Non-Uniformly spaced points**
- **Polygonal approximations**
- **Wall-Danielsson algorithm**
- **Williams' algorithm**

5 A Data-driven Method

CONTENTS

5.1 INTRODUCTION

In this chapter we are looking for a data-driven method [122] in which neither do we specify the error nor do we specify the number of line segments. We keep both these parameters free and allow the procedure to determine the length of the segments as well as the maximum allowable error adaptively on the basis of the local topography of the curve. So the procedure does not need operator's intervention. Though the procedure is sequential and one-pass but unlike the existing sequential algorithms neither does it miss corners nor does it round off sharp turnings.

5.2 DEPARTURE FROM CONVENTIONAL APPROACH

In the polygonal approximation techniques discussed so far the maximum allowable error is specified either directly or indirectly. The optimal algorithms for polygonal approximation which use the least squares principle [19, 50-52, 102, 162] need the number of line segments to be specified. The optimal algorithm given by Dunham [39] needs error specification. In all these procedures the user has to specify either the number of line segments or the maximum allowable error. The maximum allowable error or the number of line segments to be used is determined on basis of trial and error process. So these procedures cannot run without operator's intervention.

5.3 PROCEDURE

The equation of a straight line joining the point p_i to the point p_j is:

$$(y_j - y_i)x + (x_j - x_i)x - x_i y_j + x_j y_i = 0 \tag{5.1}$$

Translating the origin of the coordinate system to the point (x_i, y_i) and denoting the new coordinate system by prime so that

$$x`_j = x_j - x_i \text{ and } y`_j = y_j - y_i \qquad (5.2)$$

The Equation (5.1) takes the form

$$y`_j x` - x_j y` = 0 \qquad (5.3)$$

The error e_k between the point p_k, $k = i + 1, i + 2, j - 1$ and the line (5.3) is the perpendicular distance of the point p_k from the line (5.3) that is,

$$e_k = \frac{y`_j x`_k - x`_j y`_k}{\sqrt{x`_j{}^2 + y`_j{}^2}} \qquad (5.4)$$

So while approximating the points p_k, $k = i + 1, i + 2, j - 1$; the integral square error along the line segment (5.3) is:

$$s_j = \sum_{k=i+1}^{j-1} \frac{(y`_j x`_k - x`_j y`_k)^2}{x`_j{}^2 + y`j^2} \qquad (5.5)$$

And the length of the line segment joining (x_i, y_i) to (x_j, y_j) is:

$$l_j = \sqrt{x`_j{}^2 + y`_j{}^2} \qquad (5.6)$$

Our objective is to make l_j as large as possible by continuously merging points one after another with the point (x_i, y_i) so that s_j is as small as possible. We note that we cannot increase l_j indefinitely nor can we decrease s_j arbitrarily. Because in the former case, the approximation error will be too large, resulting in an approximation that may fail to locate many significant vertices. And in the later case, the approximation may result in many redundant vertices. So, it is necessary to make a compromise between the length of the line segment and the approximation error. We solve this problem by combining l_j with s_j. Since our objective is to make l_j as large as possible so that s_j is as small as possible, we take the mathematical combination:

$$f_j = l_j - s_j \qquad (5.7)$$

The procedure looks for the local maximum of f_j by continuously merging points one after another with the point p_i. The value of j for which f_j attains a local maximum gives the location of the vertex.

The procedure is started from an arbitrary point p_i. The origin is shifted to the point p_i by the transformation rules (5.2), write $j = i + 1$ and compute f_j using the formula because for $j = i + 1$, $s_j = 0$.

$$f_j = \sqrt{x`_j{}^2 + y`_j{}^2} \tag{5.8}$$

Then j is changed to $j + 1$ and f_j is computed using the formula (5.7) and f_j is compared with f_{j-1}. Points are merged one after another with the point p_i giving increment to j, f_j is computed using (5.7) and compared with f_{j-1} until f_j falls below f_{j-1}. At this stage $jj = j-1$ gives the location of a vertex which is not necessarily a valid vertex. We write $i = jj$, translate the origin to the point p_i, set $j = i + 1$ and compute f_j using (5.8) j is changed to $j + 1$ and f_j is computed using (5.7) and compared with f_{j-1}. Points are merged one after another with the point p_i giving increment to j, f_j is computed using (5.7) and compared with f_{j-1} until f_j falls below f_{j-1}. At this stage another possible vertex is recorded at the point $jj = j - 1$. The coordinates of this vertex are (x_{jj}, y_{jj}). The origin is again translated to the point jj and the same computation is carried out in the same fashion as already described. The procedure is carried on along the entire digital curve beyond the starting point till the vertex generated last coincides with one of the vertices already generated.

At this point we note that Williams' [184] and Wall-Danielsson [180] too looked for the maximal possible line segments. But in their procedure it is necessary to specify the error. In Williams' technique the error is specified directly whereas in the Wall-Danielsson technique the error is specified in terms of area deviation per unit length. But, the present procedure looks for the maximal line segment without specifying the error. We keep the error unspecified and allow the procedure to determine it on the basis of the local topography of the curve.

5.4 ALGORITHM

Comments

The input are the coordinates (x_i, y_i), $i = 1, 2, n$. The output is the vertices (x_{jj}, y_{jj}). All arithmetic are performed are in modulo n.

Begin

Step 1: Initiate $i = 1$.

Step 2: Translate the origin of the coordinate system to the point (x_i, y_i) by the transformation rules.

$x`_j = x_j - x_i$ and $y`_j = y_j - y_i$

Step 3: Set $j = i + 1$

Step 4: Compute $f_j = \sqrt{x`_j{}^2 + y`_j{}^2}$

Step 5: Change j to $j + 1$.

Step 6: Compute $f_j = l_j - s_j$

Step 7: If $f_j \geq f_{j-1}$ then go to step 5 else write $jj = j - 1$ and (x_{jj}, y_{jj}) using $x_{jj} = x`_{jj} + x_i$ and $y_{jj} = y`_{jj} + y_i$; set $i = jj$; go to step 2.

Step 8: Repeat this process beyond the starting point till the vertex generated last coincides with one of the vertices already generated.

Step 9: The set of coordinates (x_{jj}, y_{jj}) that form a closed chain is the required set of vertices. These are joined in order to determine the polygon.

End.

5.5 COMPUTATIONAL COMPLEXITY

At each point of a digital curve f_j is computed once only except at the points immediately following the vertices where f_j is computed twice, once for the generated line segment and the second time for the line segment to be generated. So, if the approximation results in m line segments from a closed digital curve with n points then the number of times f_j is computed to generate the approximation is $m + n$. Since $m \leq n$ so $m + n \leq 2n$. As the computation is carried out beyond the starting point, a part of the digital curve comes under arithmetic operations twice. But, this part is much smaller than the input data size and the additional number of times for which f_j is computed on this part of the digital curve has only additive effect to the sum $m + n$. So, the computational load of the algorithm is $O(n)$.

Though the computational complexity is $O(n)$, but the program execution time will be higher if the approximation results in long line segments that if it results in short line segments. Because in the former case evaluation of s_j involves computation of a large number of terms of the form.

$$(y`_j x`_k - x`_j y`_k)^2 \tag{5.9}$$

We illustrate this with the help of an example. Let the closed digital curve consisting of 100 integers coordinate points are approximated by 10 line segments each of which approximates 11 points. Then to generate each line segment evaluation of s_j involves computation of as many as (10 × 11)/2 = 55 terms of the form (5.9). But if another closed digital curve consisting of 100 data points is approximated by 5 line segments each line segment approximating 21 points then to generate each line segment computation of s_j involves evaluation of (20 × 21)/2 = 210 terms of the form (5.9). So in the former case the total number of terns of the (5.9) to be evaluated is 55 × 10 = 550 and in the later case it is 210 × 5 = 1,050 which is almost twice of the former.

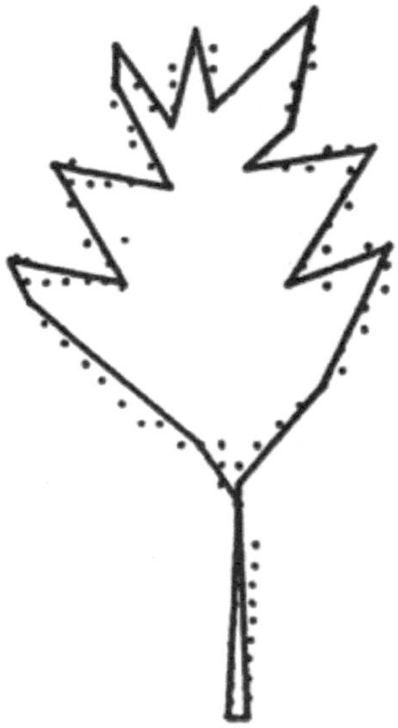

(a)

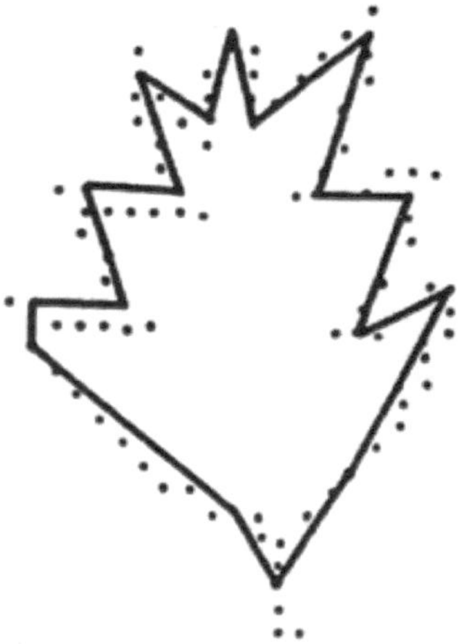

(b)

FIGURE 1 *(Continued)*

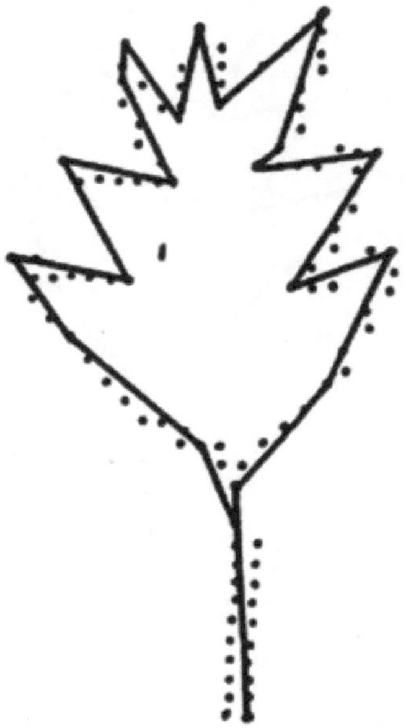

(c)

FIGURE 1 Polygonal approximations of the leaf-shaped curve. (a) Algorithm 5.4, (b) Williams' algorithm, and (c) Wall-Danielsson algorithm.

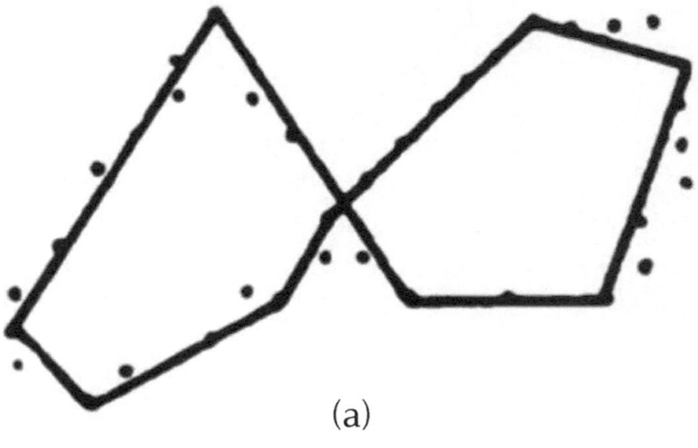

(a)

FIGURE 2 *(Continued)*

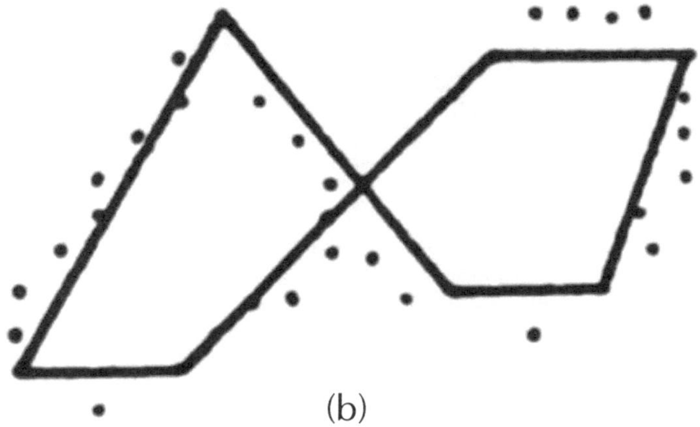

(b)

(c)

FIGURE 2 Polygonal approximations of the figure-8 curve. (a) Algorithm 5.4, (b) Williams' algorithm, and (c) Wall-Danielsson algorithm.

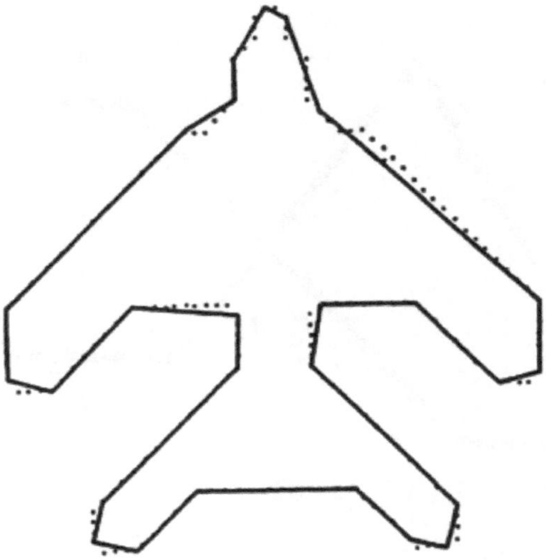

(a)

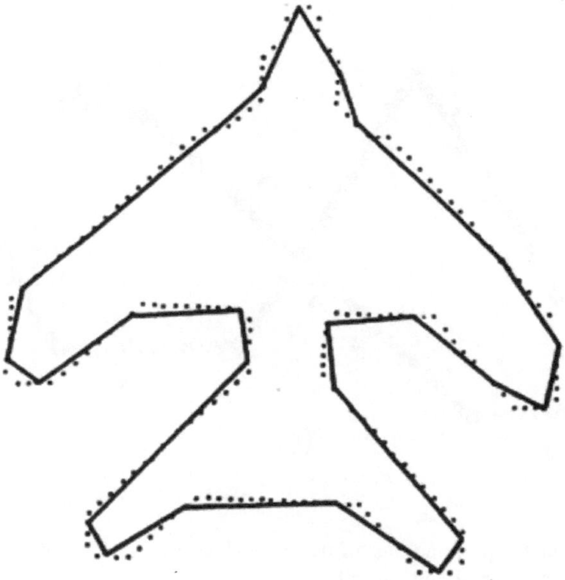

(b)

FIGURE 3 *(Continued)*

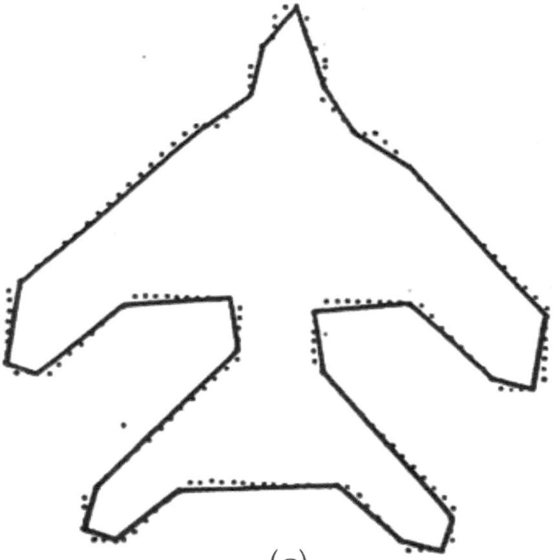

(c)

FIGURE 3 Polygonal approximations of the aircraft. (a) Algorithm 5.4, (b) William' algorithm, and (c) Wall-Danielsson algorithm.

(a)

FIGURE 4 *(Continued)*

(b)

(c)

FIGURE 4 Polygonal approximations of the screwdriver. (a) Algorithm 5.4, (b) Williams` algorithm, and (c) Wall-Danielsson algorithm.

5.6 EXPERIMENTAL RESULTS

We have applied the algorithm 5.4 on four digital curves. The data are processed in the clockwise direction starting from an arbitrary point. The polygonal approximations are shown in Figure 1 through Figure 4. Since this procedure is sequential and one-pass hence we compare it with the Williams' algorithm [184] and the Wall-Danielsson algorithm [180]. In Williams' algorithm it is necessary to specify the maximum allowable error. As already stated we do not specify the maximum allowable error in algorithm 5.4. We keep this error unspecified and allow the procedure to determine it on the basis of the local properties of the curve. The maximum error returned by the algorithm 5.4 is then used to run the Williams and Wall-Danielsson algorithm. The polygonal approximations as obtained by the Williams and Wall-Danielsson algorithm are shown in Figure 1 through Figure 4. It is clear from these figures that the algorithm 5.4 produces better results than the Williams' and Wall-Danielsson algorithm.

The worst approximation produced by the Williams' algorithm is obtained with the leaf shaped curve, where the lower part of the digital curve which has a sharp turning is completely wiped out by the approximation. Moreover, none of the other turnings are approximated according to their nature. The vertices are always placed beyond the turning point. This is true for other digital curves also. Though algorithm 5.4 is also sequential and one-pass but neither does it round off sharp turnings nor does it shifts the vertices from their actual position. The approximations obtained by the Wall-Danielsson algorithm are no better than those produced by our procedure. It produces redundant/false vertices. It also dislocates the vertices. The peak-test introduced by the Wall-Danielsson algorithm succeeds to retain the very sharp turnings but it fails to locate the vertices at their actual position when the turnings are not so sharp. An overview of the results of the approximations obtained by the three procedures is displayed in Table 1.

TABLE 1 A comparison among smoothing technique 5.4, Williams' and Wall-Danielsson algorithm.

Digital curve	**Leaf**	**Figure-8**	**Aircraft**	**Screwdriver**
Number of points	**120**	**45**	**200**	**267**
Results of smoothing technique 5.4				
Number of vertices (n_v)	22	9	26	50
Compression ratio (n/n_v)	5.45	5.00	7.69	5.34
Percentage of data reduction	81.67	80.00	87.00	81.27
Integral square error	16.95	6.01	22.80	42.85
Maximum error	1.05	1.00	1.20	0.93

TABLE 1 *(Continued)*

Digital curve	Leaf	Figure-8	Aircraft	Screwdriver
Results of Williams` algorithm				
Maximum allowable error	1.05	1.00	1.20	0.93
Number of vertices (n_v)	18	7	24	50
Compression ratio (n/n_v)	6.67	6.43	8.33	5.34
Percentage of data reduction	85.00	84.44	88.00	81.27
Integral square error	34.27	14.05	63.33	47.85
Maximum error	1.05	1.00	1.20	0.91
Results of Wall-Danielsson algorithm				
Maximum allowable error	1.05	1.00	1.20	0.93
Threshold	0.996	0.91	1.05	0.67
Number of vertices (n_v)	21	9	27	66
Compression ratio (n/n_v)	5.71	5.00	7.41	4.05
Percentage of data reduction	82.5	80.00	86.5	75.24
Integral square error	21.08	7.08	42.69	25.18
Maximum error	0.995	1.00	1.05	0.77

KEYWORDS

- **Computational complexity**
- **Data driven method**
- **Optimal algorithms**
- **Polygonal approximation techniques**
- **Williams and Wall-Danielsson algorithm**

6 Another Data-driven Method

CONTENTS

6.1 INTRODUCTION

While approximating a curve by a polygon it is necessary to have a measure of closeness. The error norms are used as a measure of closeness, Most commonly used norms are the maximum error (L_∞ norm) and the integral square error (L_2 norm). The existing algorithms for polygonal approximation of digital curve use either the maximum error or the integral square error. In the polygonal approximation schemes where approximation errors are controlled indirectly, the maximum error is controlled by the criterion function. We wish you to show that though the most commonly used norms are integral square error and the maximum error but it is also possible to use the sum of absolute errors (L_1 norm) as a measure of closeness. The procedure [123] that we present here is conceptually and technically.

6.2 PROCEDURE

The equation of a straight line joining the point p_i to the point p_j is:

$$(y_j - y_i)x + (x_j - x_i)x - x_i y_j + x_j y_i = 0. \tag{6.1}$$

Translating the origin of the coordinate system to the point (xi, yi) and denoting the new coordinate system by prime so that:

$$x`_j = x_j - x_i \text{ and } y`_j = y_j - y_i \tag{6.2}$$

The Equation (6.1) takes the form

$$y`_j x` - x_j y` = 0 \tag{6.3}$$

The error e_k between the point p_k, $k = i + 1, i + 2, j - 1$ and the line (6.3) is the perpendicular distance of the point p_k from the line (6.3) that is:

$$e_k = \frac{y`_j x`_k - x`_j y`_k}{\sqrt{x`_j{}^2 + y`_j{}^2}} \tag{6.4}$$

So while approximating the points p_k, $k = i + 1, i + 2, j - 1$; the sum of absolute errors along the line segment (6.3) is:

$$abs(s_j) = \sum_{k=i+1}^{j-1} |e_k| \tag{6.5}$$

And the length of the line segment joining (x_i, y_i) to (x_j, y_j) is:

$$l_j = \sqrt{x`_j{}^2 + y`_j{}^2} \tag{6.6}$$

Our objective is to make l_j as large as possible by continuously merging points one after another with the point (x_i, y_i) so that $\text{abs}(s_j)$ is as small as possible. We note that we cannot increase l_j indefinitely nor can we decrease s_j arbitrarily. Because in the former case, the approximation error will be too large, resulting in an approximation that may fail to locate many significant vertices. And in the later case, the approximation may result in many redundant vertices. So, it is necessary to make a compromise between the length of the line segment and the approximation error. We solve this problem by combining l_j with $\text{abs}(s_j)$. Since our objective is to make l_j as large as possible so that $\text{abs}(s_j)$ is as small as possible, we take the mathematical combination

$$f_j = l_j - \text{abs}(s_j) \tag{6.7}$$

The procedure looks for the local maximum of f_j by continuously merging points one after another with the point p_i. The value of j for which f_j attains a local maximum gives the location of the vertex. The procedure is started from an arbitrary point and is carried beyond the starting point until the vertex generated last coincides with one of the vertices already generated.

6.3 ALGORITHM

Comments

The input are the coordinates (x_i, y_i), $i = 1, 2, \ldots, n$. The outputs are the vertices (x_{jj}, y_{jj}). All arithmetic are performed are in modulo n.

Begin

Step 1: Initiate $i = 1$.

Step 2: Translate the origin of the coordinate system to the point (x_i, y_i) by the transformation rules

$$x`_j = x_j - x_i \text{ and } y`_j = y_j - y_i$$

Step 3: Set $j = i + 1$

Step 4: Compute $f_j = \sqrt{x`_j^{\,2} + y`_j^{\,2}}$

Step 5: Change j to $j + 1$.

Step 6: Compute $f_j = l_j - \text{abs}(s_j)$

Step 7: If $f_j \geq f_{j-1}$ then go to step 5 else write $jj = j - 1$ and (x_{jj}, y_{jj}) using $x_{jj} = x`_{jj} + x_i$ and $y_{jj} = y`_{jj} + y_i$; set $i = jj$; go to step 2.

Step 8: Repeat this process beyond the starting point till the vertex generated last coincides with one of the vertices already generated.

Step 9: The set of coordinates (x_{jj}, y_{jj}) that form a closed chain is the required set of vertices. These are joined in order to determine the polygon.

End.

6.4 COMPUTATIONAL COMPLEXITY

The procedure is sequential one-pass and following the same line of arguments, it can be derived that the computational complexity of the algorithm is $O(n)$ and program execution time will be higher, if the approximation results in long line segments than that if it results in short line segments.

6.5 EXPERIMENTAL RESULTS

We have applied the algorithm 6.3 on four digital curves. The data are processed in the clockwise direction. The procedure is compared with the Williams' algorithm [184] and the Wall-Danielsson algorithm [180]. The polygonal approximations as obtained by the algorithm 6.3, Williams and Wall-Danielsson algorithm are shown in Figure 1 through Figure 4. As seen from these figures the approximations generated by our procedure do not round offs

(a)

FIGURE 1 *(Continued)*

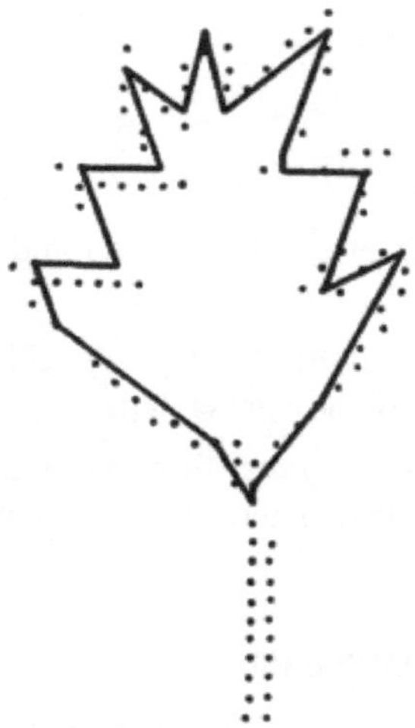

(b)

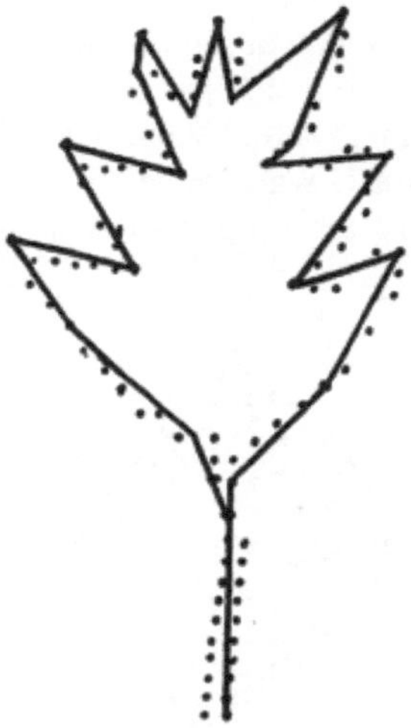

(c)

FIGURE 1 Polygonal approximations of the leaf-shaped curve. (a) Algorithm 6.3, (b) Williams' algorithm, and (c) Wall-Danielsson algorithm.

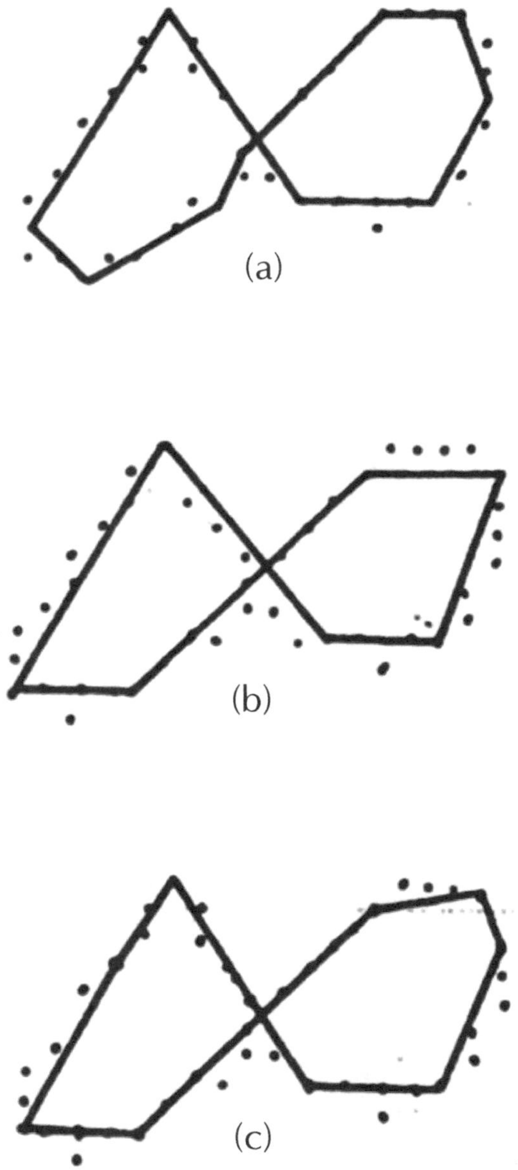

FIGURE 2 Polygonal approximations of the figure-8 curve. (a) Algorithm 6.3, (b) Williams' algorithm, and (c) Wall-Danielsson algorithm.

The sharp turnings not miss it corners. The approximations produced by the Williams' algorithm round off sharp turnings and miss corners. The approximations produced by the Wall-Danielsson algorithm are no better than our procedure. The results of the approximations as obtained by the three procedures are shown in Table 1.

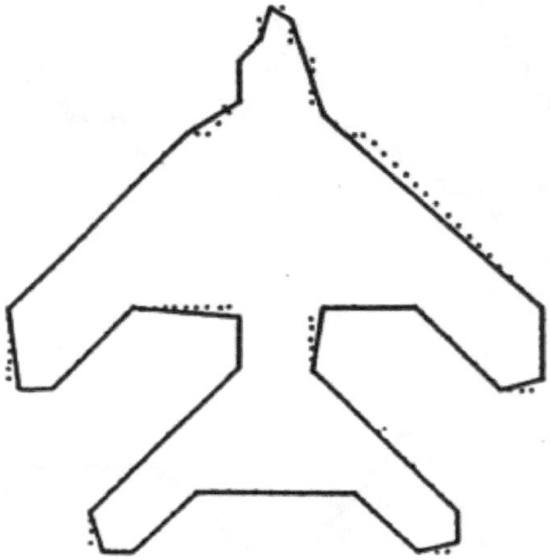

(a)

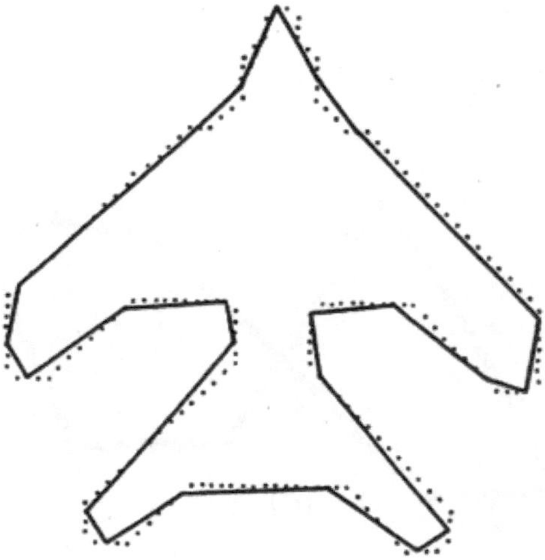

(b)

FIGURE 3 *(Continued)*

(c)

FIGURE 3 Polygonal approximations of the aircraft. (a) Algorithm 6.3, (b) Williams` algorithm, and (c) Wall-Danielsson algorithm.

(a)

FIGURE 4 *(Continued)*

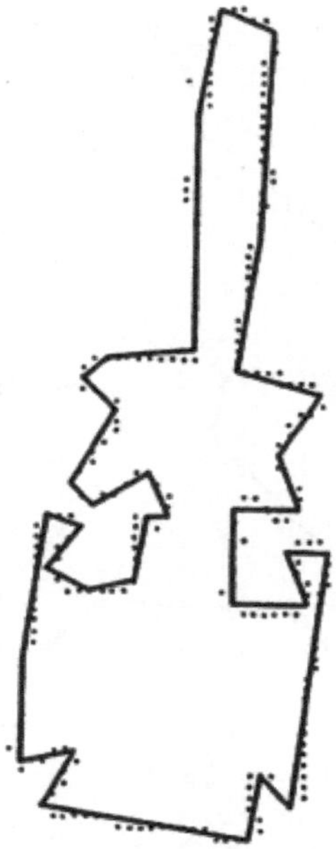

(b)

(c)

FIGURE 4 Polygonal approximations of the screwdriver. (a) Algorithm 6.3, (b) Williams' algorithm, and (c) Wall-Danielsson algorithm.

TABLE 1 A comparison among smoothing technique 6.3, Williams' and Wall-Danielsson algorithm.

Digital curve	**Leaf**	**Figure-8**	**Aircraft**	**Screwdriver**
Number of points	**120**	**45**	**200**	**267**
Results of smoothing technique 5.3				
Number of vertices (n_v)	26	10	28	56
Compression ratio (n/n_v)	4.62	4.5	7.14	4.77
Percentage of data reduction	78.33	77.80	86.00	79.03
Sum of absolute error	29.43	8.62	34.27	68.70
Maximum error	1.05	1.00	1.33	1.00
Results of Williams` algorithm				
Maximum allowable error	1.05	1.00	1.33	1.00
Number of vertices (n_v)	18	7	22	36
Compression ratio (n/n_v)	6.67	6.43	9.09	7.42
Percentage of data reduction	85.00	84.44	89.00	86.50
Sum of absolute error	45.62	17.96	105.54	101.25
Maximum error	1.05	1.00	1.31	1.00
Results of Wall-Danielsson algorithm				
Maximum allowable error	1.05	1.00	1.33	1.00
Threshold	0.996	0.91	1.05	0.78
Number of vertices (n_v)	21	9	27	61
Compression ratio (n/n_v)	5.71	5.00	7.41	4.38
Percentage of data reduction	82.5	80.00	86.5	77.13
Sum of absolute error	38.88	11.48	75.94	57.84
Maximum error	0.995	1.00	1.05	1.00

KEYWORDS

- **Another Data-Driven method**
- **Computational complexity**
- **Digital curve**
- **Polygonal approximations**
- **Williams and Wall-Danielsson algorithm**

7 A Two-pass Sequential Method

CONTENTS

7.1 INTRODUCTION

In this chapter we propose a two-pass sequential algorithm for polygonal approximation of digital planar curve. Unlike other sequential algorithms that use conventional perpendicular distance as a metric, this chapter uses distance to a point. The approximations produced are perceptually pleasing, do not round off sharp turnings, do not miss corners and are symmetrical for symmetrical input curve. This chapter also introduces a new metric to measure and compare the fidelity of an approximation with another and use this measure to compare the proposed algorithm with a recent one and show improvement over the same.

7.2 METHODOLOGY

In order to measure the collinearity of a group of points, the conventional practice is to use perpendicular distance of the intermediate points from the line segment joining the end points of the group, but use of this metric result in missing corners and rounding off sharp turnings (spikes) [122]. For instance, if it is required to check whether the three points A, B, and C, where B follows C and A follows B, are collinear (Figure 1) then using the perpendicular distance (AD) of A from the segment BC may lead to an affirmative conclusion but as evident from the arrangement of the points A, B and C.

The three points are not collinear. This is why this chapter uses distance to the segment as metric instead of perpendicular distance. This metric was first used in [39] to develop a one-pass optimal algorithm (as it is well known optimal algorithms are costly) and then in [127] to develop a split and merge technique (which is iterative in nature). But to the best of the knowledge of the author, this metric had not been used so far in sequential two-pass sub-optimal algorithm that produces perceptually pleasing approximations. The methodology is described.

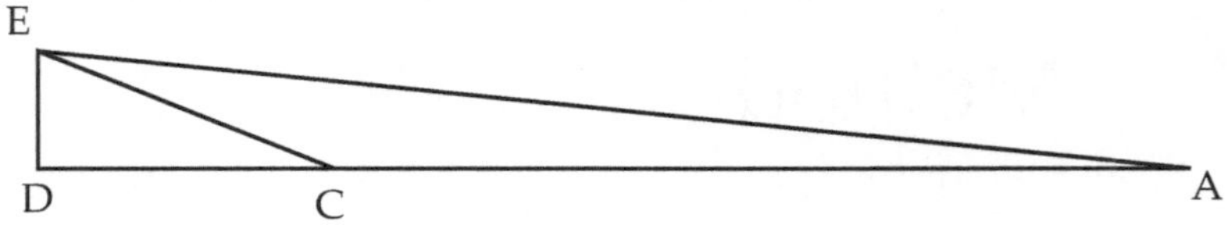

FIGURE 1 Difference between perpendicular distance of a point B from a line segment AC and the distance of B from AC. BD (shown with thick line segment) is the perpendicular distance and BC is the distance to the segment AC.

If p_i and p_j ($j > i$) are the end points of a group of points being tested for collinearity and p_k is any other point ($i < k < j$) in the group then the distance (d_k) of p_k from p_ip_j is the perpendicular distance of p_k from p_ip_j, if p_k falls between p_i and p_j, but it is the distance of p_k from p_j if p_k falls to the right of the segment p_ip_j and is the distance of p_k from p_i if p_k falls to the left of the segment. These situations are illustrated in Figure 2.

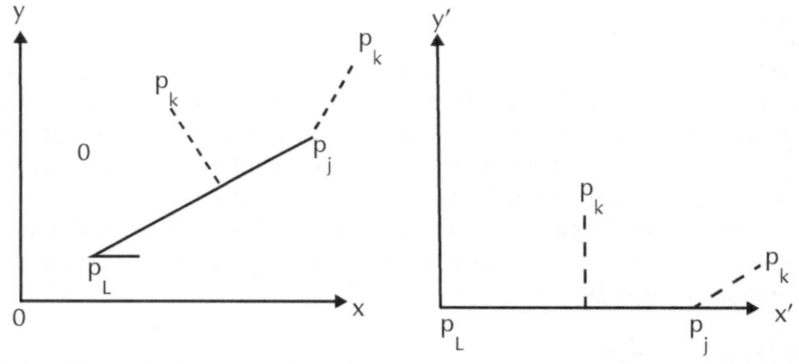

FIGURE 2 Geometrical interpretation of computation of distance to a line segment.

In order to measure the distance d_k, translate the origin to the ith point and then rotate the segment p_ip_j about the new origin so that it falls on the x-axis of the new coordinate system. Using primes to indicate the transformed coordinate system.

$$x = \{(x - x_i)(x_j - x_i) + (y - y_i)(y_j - y_i)\}/|p_ip_j|$$

$$y = \{-(x - x_i)(y_j - y_i) + (y - y_i)(x_j - x_i)\}/|p_ip_j| \qquad (7.1)$$

The distance d_k of p_k to the line segment p_ip_j is defined by:

$$d_k = \sqrt{\{(x_j - x_i)^2 + (y_j - y_i)^2\}}, \text{ if } x_k > x_j$$

$$= \sqrt{\{(x_k - x_i)^2 + (y_k - y_i)^2\}}, \text{ if } x_k < 0,$$

$$= |y_k|, \text{ otherwise.} (7.2)$$

Starting from an arbitrary point i $(1 \leq i \leq n)$ and setting $j = i + 2$, the distance d_k is computed for $k = i + 1$ using the expression (7.1) and d_k is compared with a pre-specified error value called maximum error (maxE). If d_k exceeds maxE then $k = i + 1$ is designated as a vertex of the polygon and i is assigned the value of k and j is assigned $i + 2$. If the value of d_k does not exceed maxE then j is incremented by 1 leaving i unaltered and d_k is computed for $k = i + 1$, $i + 2$. The maximum of the value of d_k is compared with maxE. If this maximum exceeds maxE then the value of k for which d_k is maximum, is the index of a vertex of the polygon. As before, i is assigned the value of k and j is assigned $i + 2$. If the maximum of d_k does not exceed maxE then j is incremented by 1 leaving i unaltered and the value of d_k is computed again for $k = i + 1$ through $j - 1$ and the maximum of d_k is compared with maxE to determine whether one of the indices $k = i + 1$ through $j - 1$ is the index of a vertex of the polygon. This process is repeated along the entire digital curve beyond the starting point until the vertex detected has the same index as that of one of the vertices already detected. Since the procedure may be started from an arbitrary point, the first vertex detected may be a false vertex (a zero/near zero curvature point), hence the procedure is carried beyond the starting point. This is the first pass of the procedure through the curve.

Though one pass of the procedure detects a set of vertices the can approximate the curve within a maximum error, nevertheless a second pass is performed moving along the curve in the opposite direction so that the polygonal approximation of a symmetric curve is also symmetrical. This is illustrated in Figure 3 and Figure 4. Both these figures show symmetrical digital curve and the results of performing one-pass and two-pass processing along the curve. The union of the two sets of vertices obtained from the two passes is the final set of vertices of the output polygon. The procedure described so far is presented in the form an algorithm.

(a)

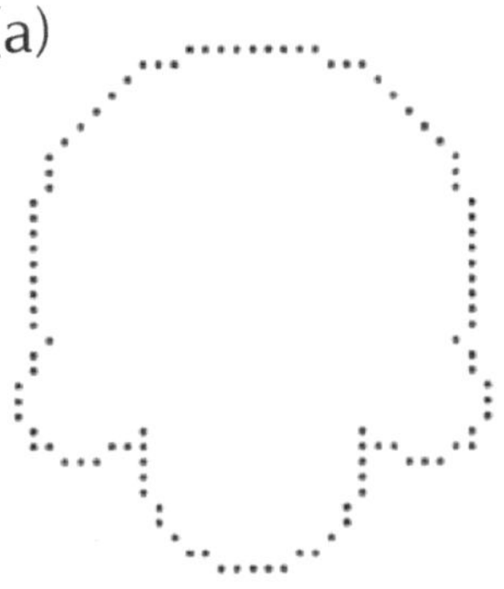

FIGURE 3 *(Continued)*

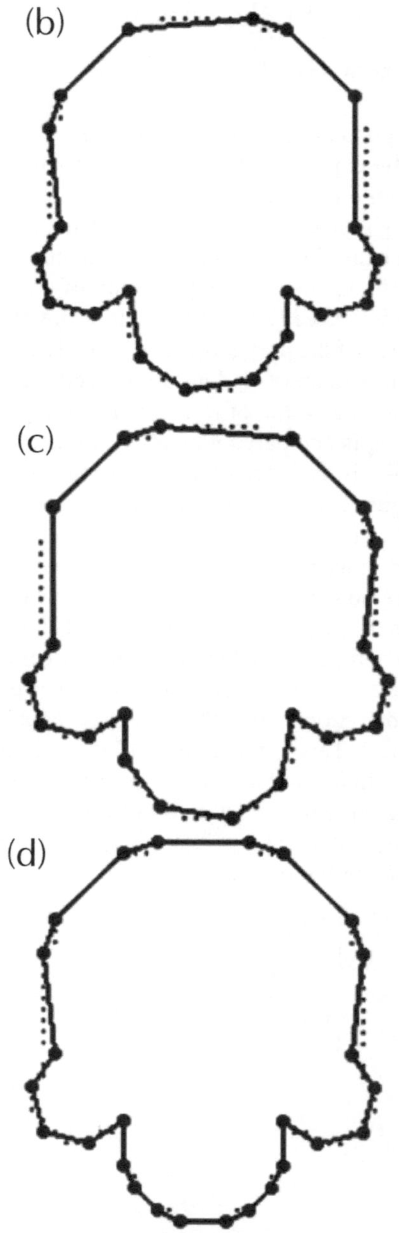

FIGURE 3 Usefulness of two-pass along a digital curve, (a) Input digital curve, a shape consisting of multiple semi circles, is symmetrical, (b) Polygonal approximation after the curve is processed in the clockwise direction, (c) Polygonal approximation after the curve is processed in the counter clockwise direction, and (d) Polygonal approximation whose vertices belong to either of the approximations (b) and (c).

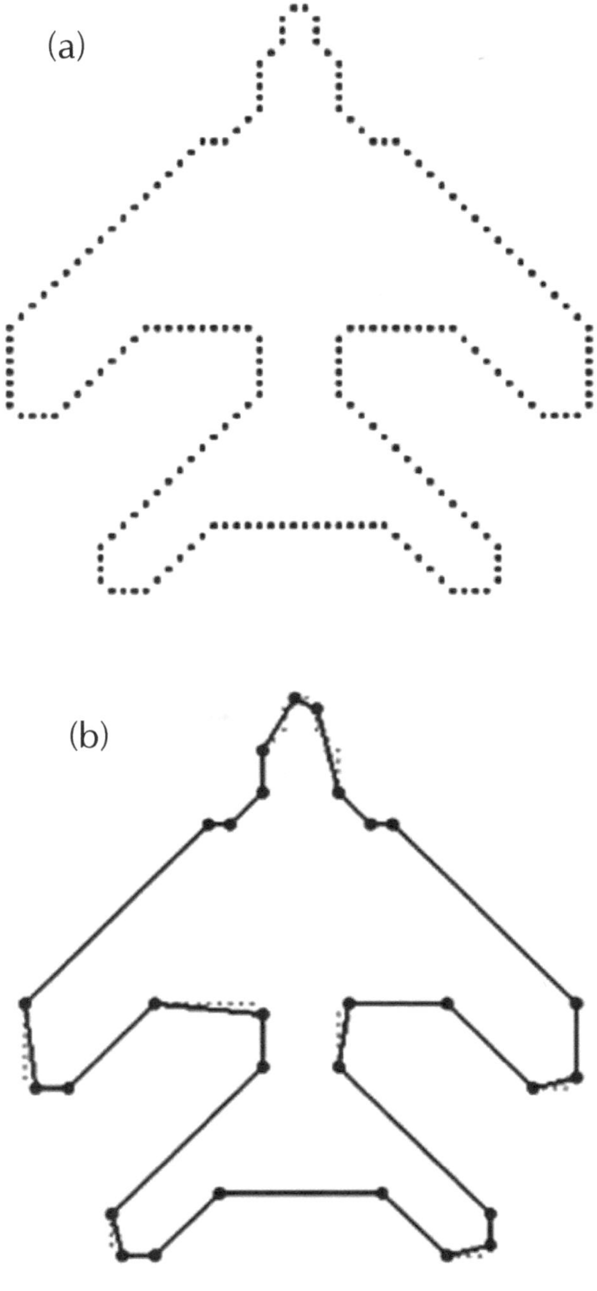

FIGURE 4 *(Continued)*

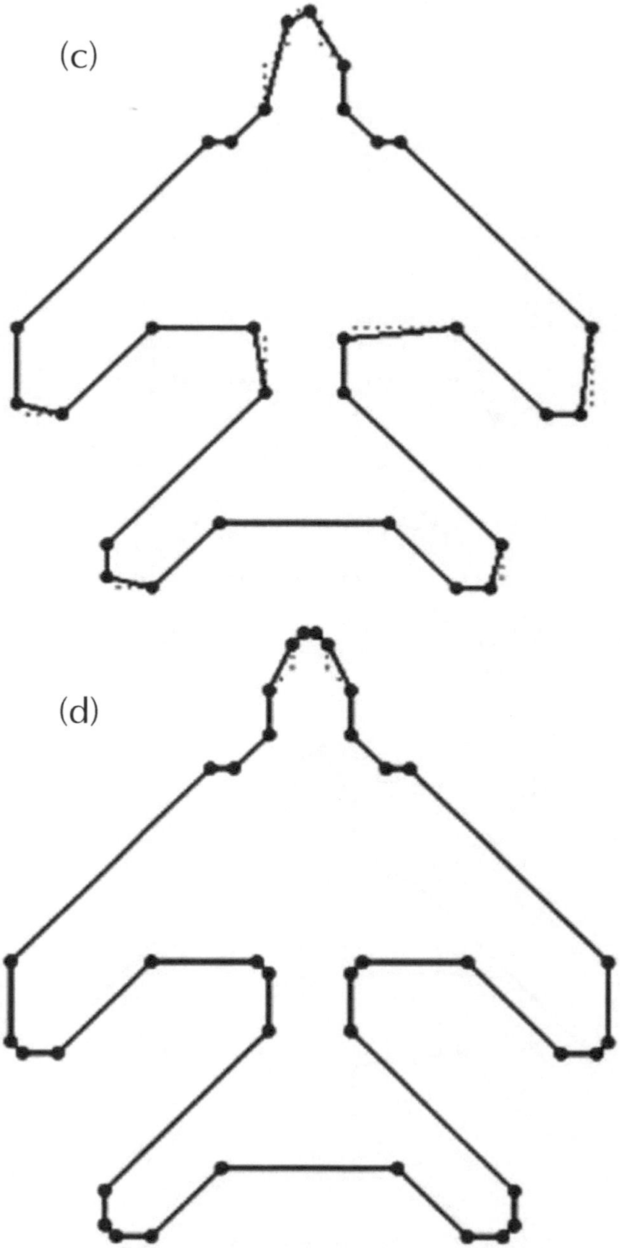

FIGURE 4 Usefulness of two-pass along a digital curve, (a) Input digital curve, an aircraft shape, is symmetrical, (b) Polygonal approximation after the curve is processed in the clockwise direction, (c) Polygonal approximation after the curve is processed in the counter clockwise direction. and (d) Polygonal approximation whose vertices belong to either of the approximations (b) and (c).

7.3 ALGORITHM POLYGONAL APPROXIMATION

The inputs are the coordinates of the points of the input digital curve and the output are index of the vertices of the approximating polygon.

Begin

Step 1: Initialize i by 1 and j by $i + 2$.

Step 2: Compute d_k using expression. (7.2)

Step 3: If dk is greater than maxE then k is the index of a vertex of the polygon, set $i = k$ and $j = i + 2$.

Step 4: If d_k is not greater than maxE then increment j by 1 and compute d_k for $k = i + 1, i + 2$.

Step 5: If maximum of d_k is greater than maxE then k is the index of a vertex of the polygon, set $i = k$ and $j = i +2$.

Step 6: If d_k is not greater than maxE then increment j by 1 and go to step 4.

Step 7: Repeat step5 and 6 moving along the curve until the index of the last vertex detected is the same as one of the vertices already detected.

Step 8: The indices produced by step 5, 6 and 7 constitute one set of vertex.

Step 9: Initialize i by n and j by $i - 2$ and move through step 2 to step 7 decrementing j.

Step 10: Perform union of the two sets of vertices.

End.

7.4 COMPUTATION COMPLEXITY

At the start of the procedure, if there are $m + 2$ points from i through j (inclusive i and j) and one of the points $i + 1$ through $j - 1$ is detected as a vertex then the number of times dk has already been computed is $m(m + 1)/2$. If the $(i + k)$th point $(k = 1,\ldots,m)$ is detected as vertex then the points $i + k + 1$ through j are to be considered again in computing d_k for detecting the next vertex. If the next approximating line segment is generated after computing dk at $m`( m` + 1)/2$ points then the total number of times it is computed at this stage is $(m - k + 1) (m - k + 2)/2 + m`(m` + 1)/2$. As m and $m`$ are both less than n and k is comparable to m, hence the computational load for generating one line segment is $O(n^2)$ and if after completion of the two-passes of the procedure the number of line segments generated is p then the computational load is $O(pn^2)$.

7.5 EXPERIMENTAL RESULTS AND COMPARISON

The algorithm proposed here is applied on four digital curves namely, a chromosome shaped curve (Figure 5), a figure-8 curve (Figure 6) a leaf-shaped curve (Figure 7) and a curve with multiple semi circles (Figure 8). These figures show the polygonal approximation overlaid on the digital curve for different values of maximum error. The vertices of the polygon are indicated by solid circles. As seen from these figures, the approximations do not miss corners and do not round off sharp that happen [15] when perpendicular distance is used as the metric for collinearity.

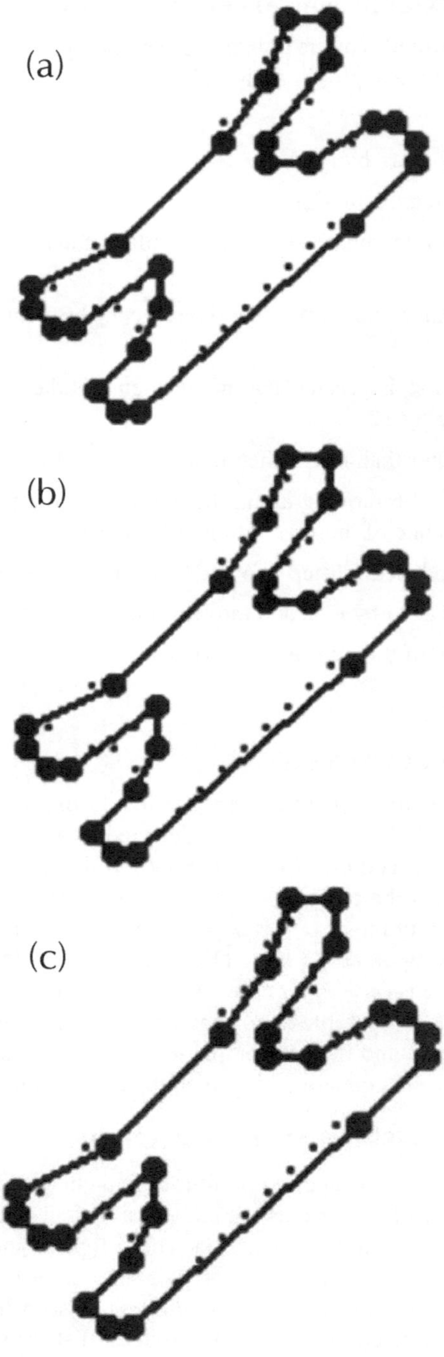

FIGURE 5 *(Continued)*

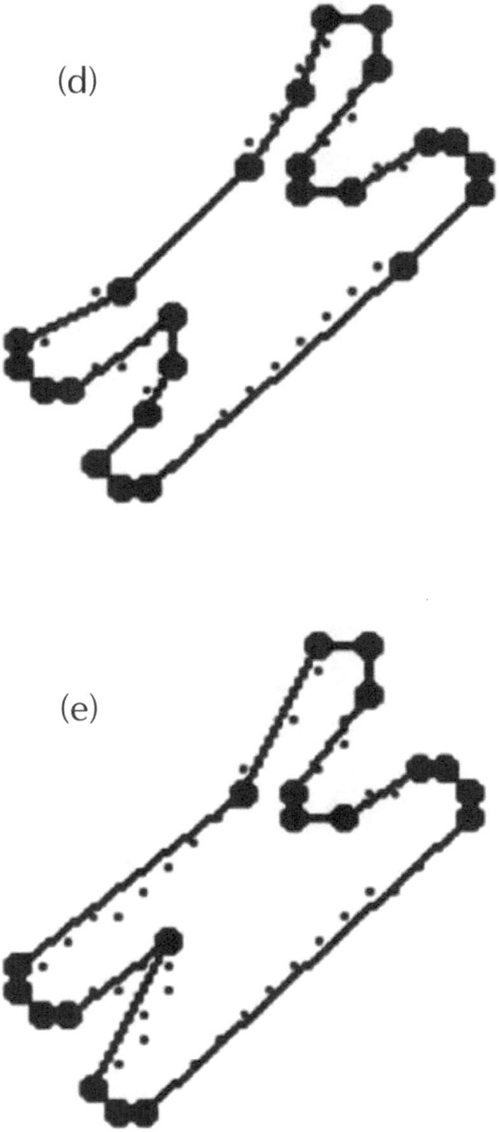

FIGURE 5 Polygonal approximation of chromosome produced by the proposed algorithm for different values of maximum error (a) maxE = 0.75 (=$\sqrt{0.56}$), (b) maxE = 0.72 (=$\sqrt{0.52}$) (c) maxE = 0.77(=$\sqrt{0.60}$), (d) maxE = 0.79 (=$\sqrt{0.63}$), and (e) maxE = 0.94 (=$\sqrt{0.88}$).

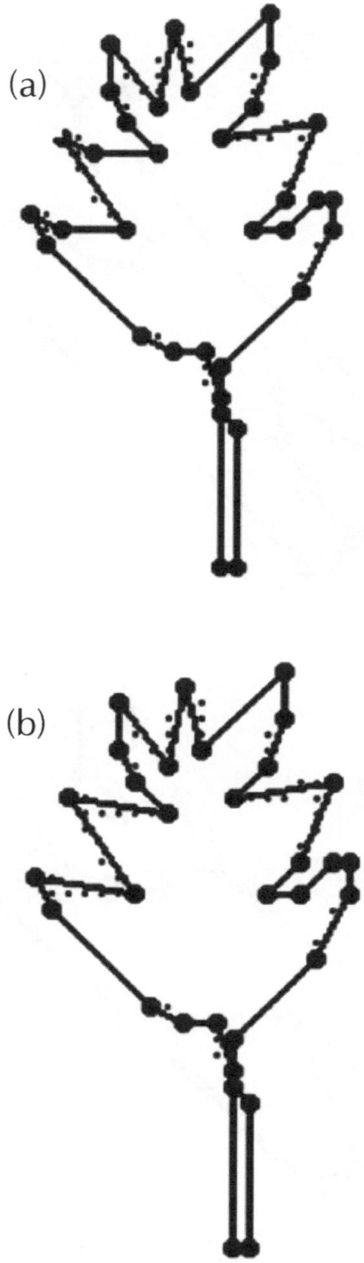

FIGURE 6 *(Continued)*

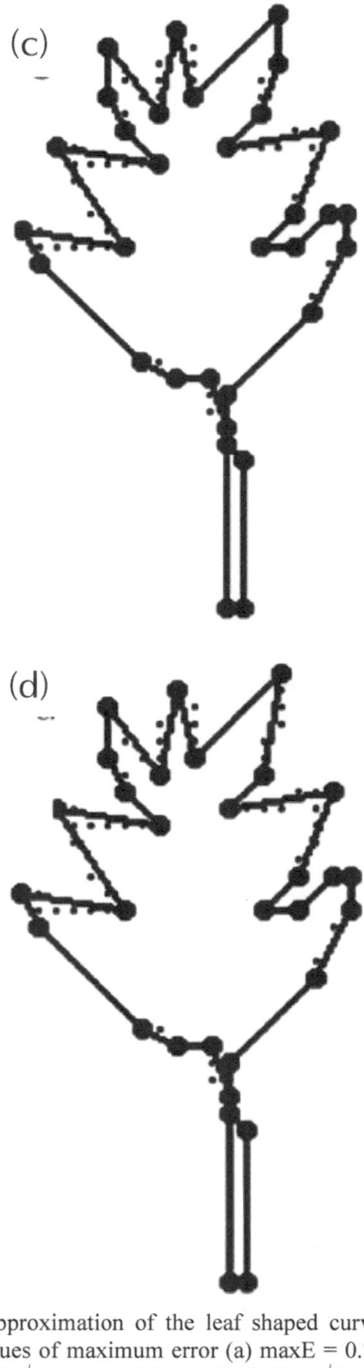

FIGURE 6 Polygonal approximation of the leaf shaped curve produced by the proposed algorithm for different values of maximum error (a) maxE = 0.77 (=√0.59), (b) maxE = 0.79 (=√0.63) (c) maxE = 0.81 (=√0.66), and (d) maxE = 0.86 (=√0.74).

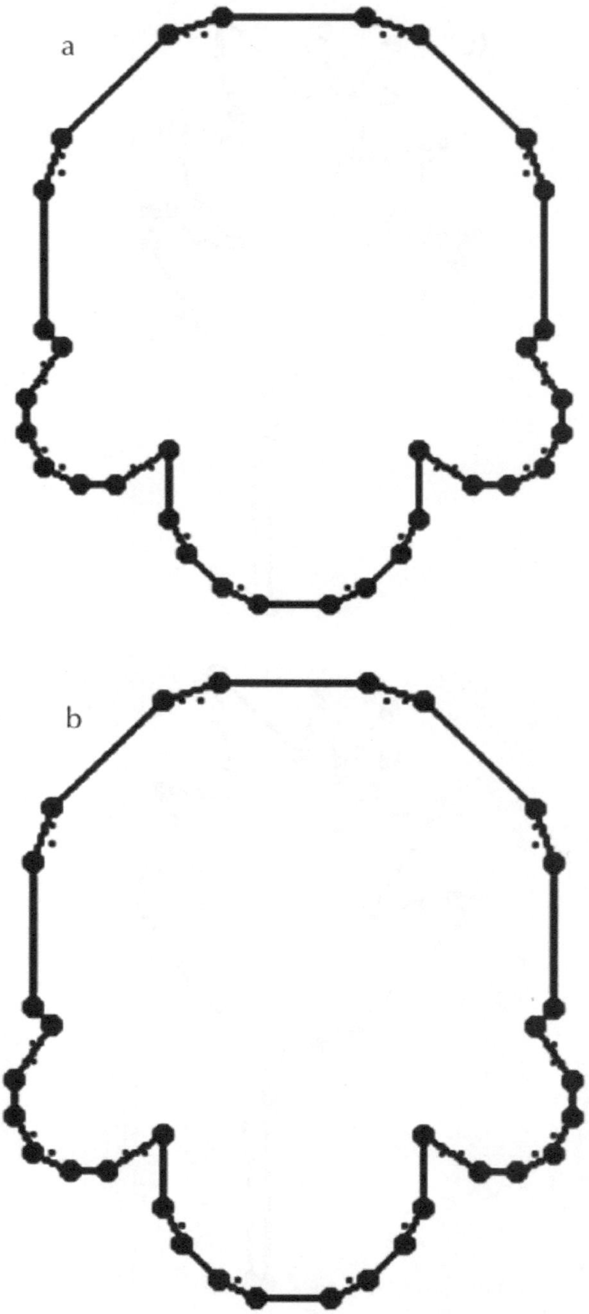

FIGURE 7 *(Continued)*

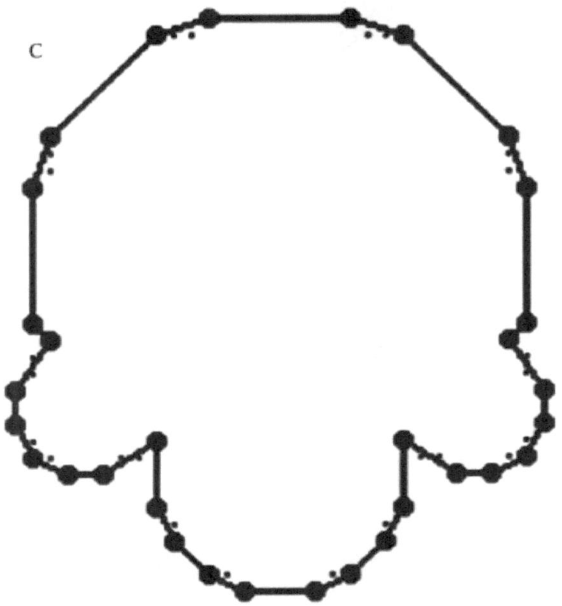

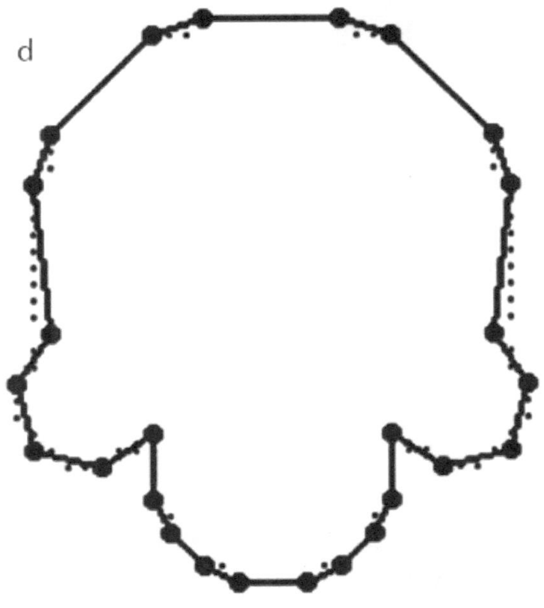

FIGURE 7 Polygonal approximation of a shape with multiple semi circles produced by the proposed algorithm for different values of maximum error (a) maxE = 0.70 (=$\sqrt{0.49}$), (b) maxE = 0.79 (=$\sqrt{0.63}$) (c) maxE = 0.85 (=$\sqrt{0.72}$), and (d) maxE = 1.00.

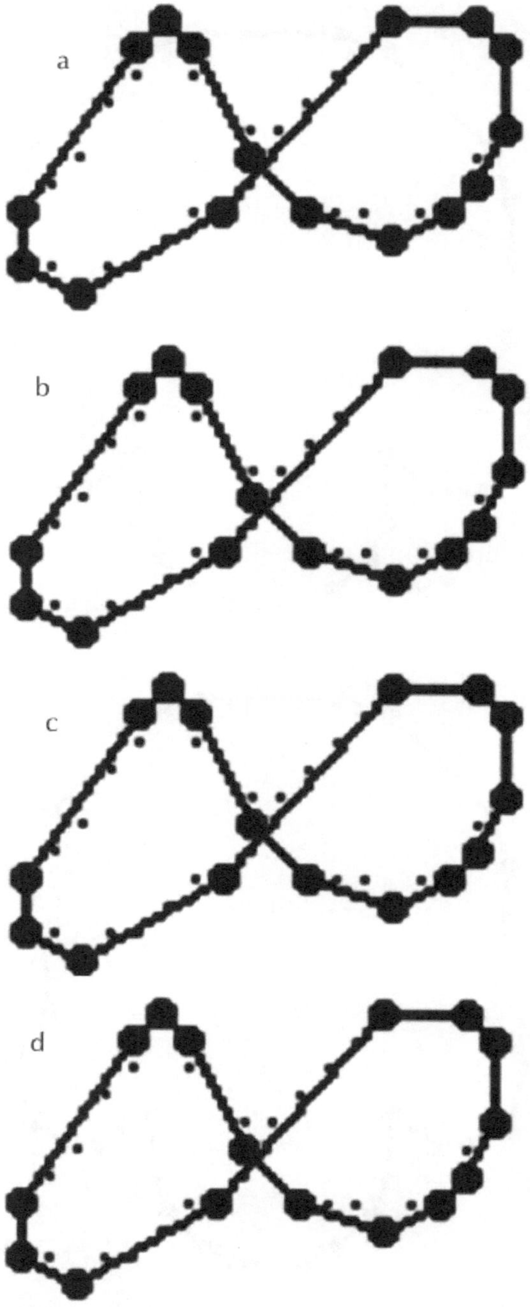

FIGURE 8 Polygonal approximation of a figure-8 curve produced by the proposed algorithm for different values of maximum error (a) maxE = 0.79 (=$\sqrt{0.63}$), (b) maxE = 0.77 (=$\sqrt{0.59}$) (c) maxE = 0.74 (=$\sqrt{0.54}$), and (d) maxE = 0.71 (=$\sqrt{0.51}$).

The approximations are perceptually pleasing and are symmetric for symmetric digital curve. The Figure 9 through Figure 12 (reproduced from [83]) show the polygonal approximation of the same curves as shown in Figure 5 through Figure 8 for the same maximum errors using the technique proposed in [84]. As seen from these figures, these approximations may miss sharp turnings (Figure 10), may be a symmetrical for symmetric curve (Figure 11) and may not be perceptually pleasing (Figure 12)

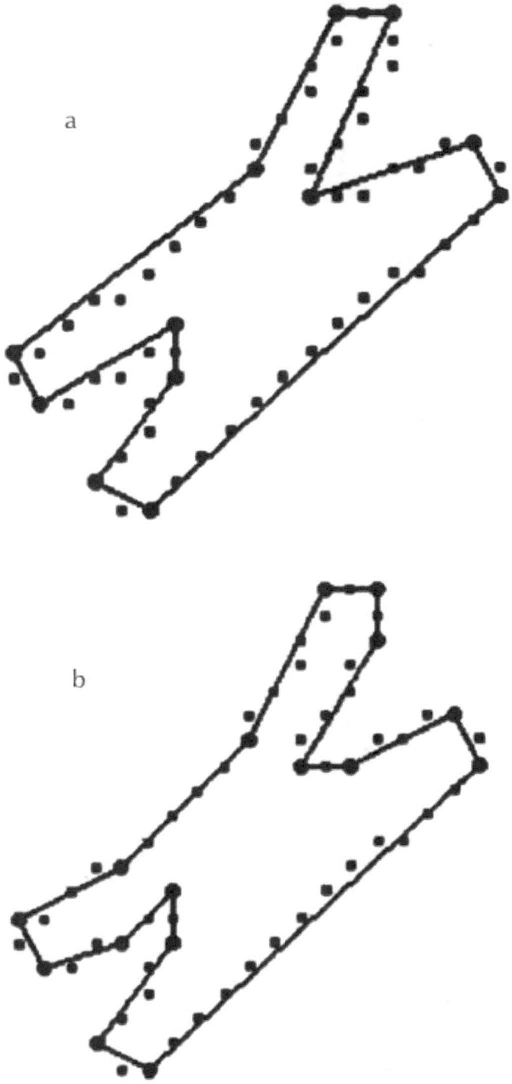

FIGURE 9 *(Continued)*

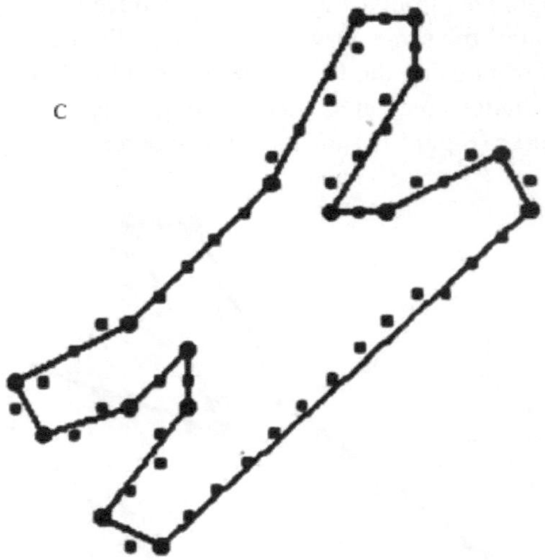

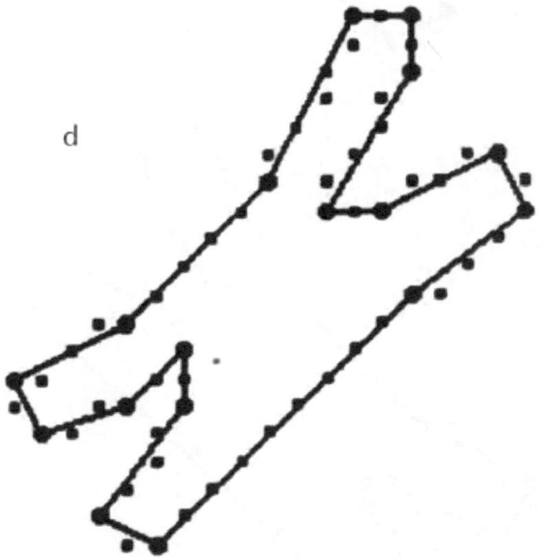

FIGURE 9 *(Continued)*

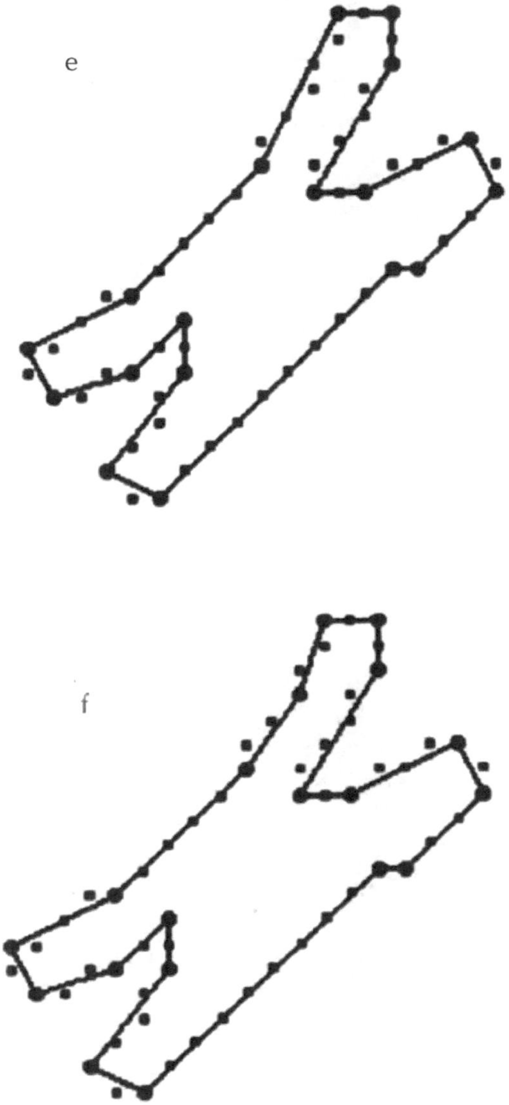

FIGURE 9 Polygonal approximation of chromosome produced by the Masood's algorithm for different number of vertices and different values of maximum error (a) m = 19, maxE = 0.75, (b) m = 18, maxE = 0.72 (c) m = 17, maxE = 0.77, (d) m = 16, maxE = 0.72, (e) m = 15, maxE = 0.79, and (f) m = 12, maxE = 0.94.

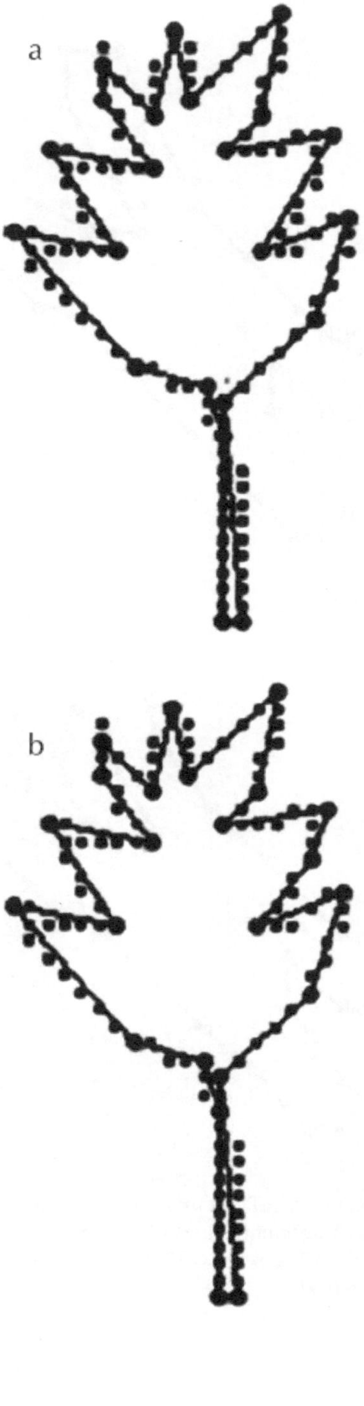

FIGURE 10 *(Continued)*

c

d

FIGURE 10 *(Continued)*

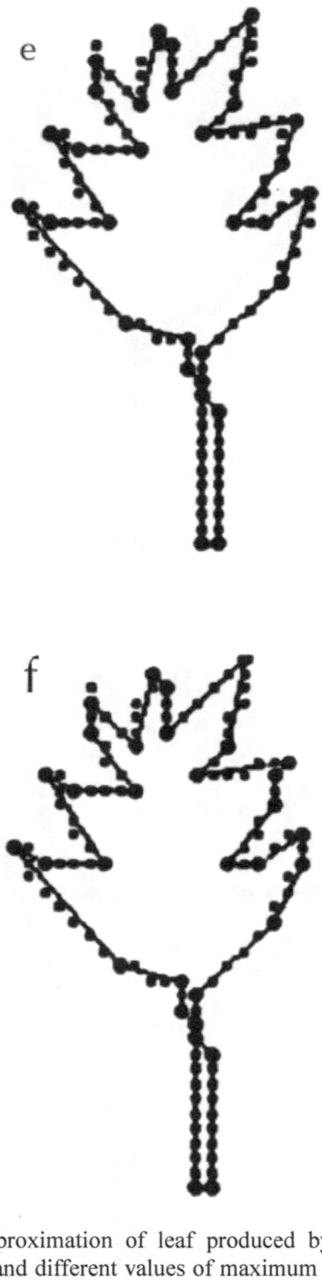

FIGURE 10 Polygonal approximation of leaf produced by the Masood`s algorithm for different number of vertices and different values of maximum error (a) m = 32, maxE = 0.77, (b) m = 30, maxE = 0.79 (c) m = 29, maxE = 0.79, (d) m = 28, maxE = 0.81, (e) m = 23, maxE = 0.86, and (f) m = 22, maxE = 0.86.

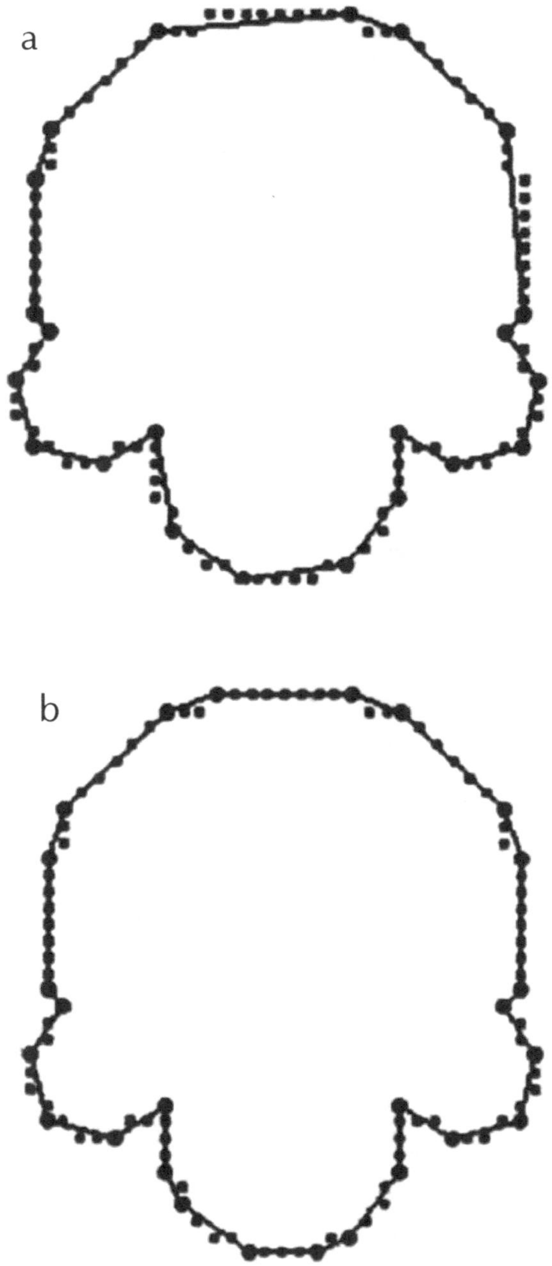

FIGURE 11 *(Continued)*

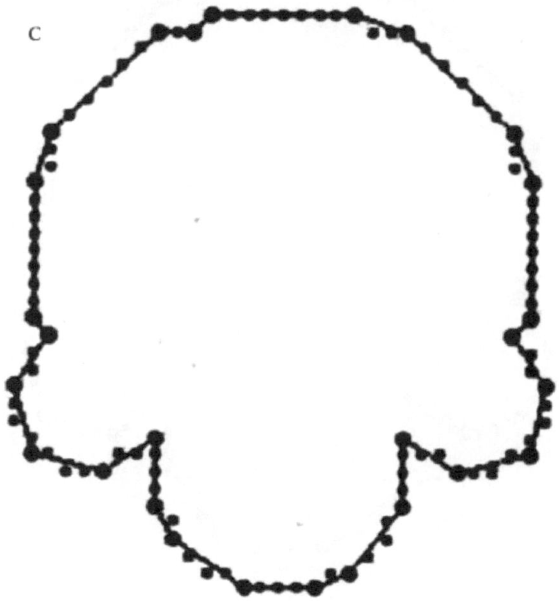

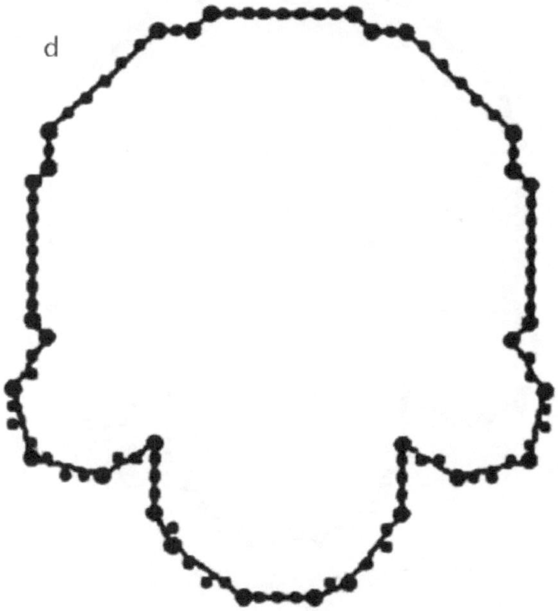

FIGURE 11 *(Continued)*

FIGURE 11 Polygonal approximation of the figure with multiple semi circles produced by the Masood`s algorithm for different number of vertices and different values of maximum error (a) m = 30, maxE = 0.70, (b) m = 29, maxE = 0.79 (c) m = 28, maxE = 0.79, (d) m = 27, maxE = 0.79, (e) m = 26, maxE = 0.79, and (f) m = 22, maxE = 0.85.

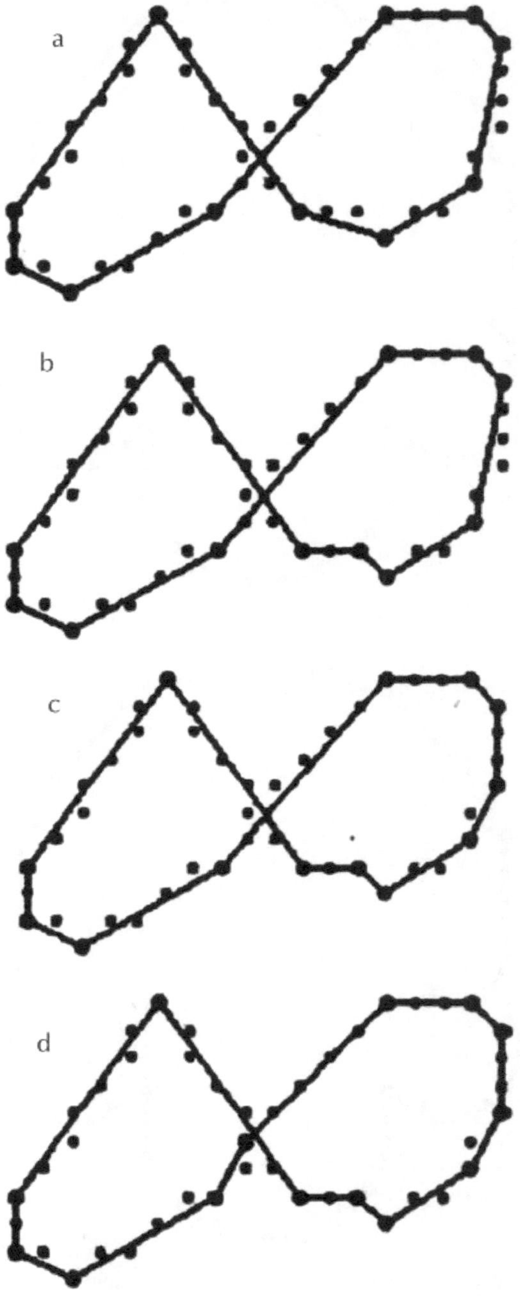

FIGURE 12 Polygonal approximation of the figure-8 curve produced by the Masood`s algorithm for different number of vertices and different values of maximum error (a) $m = 11$, maxE = 0.79, (b) $m = 12$, maxE = 0.77 (c) $m = 13$, maxE = 0.74, and (d) $m = 14$, maxE = 0.71.

Apart from visual perception and observations, the goodness of a polygonal approximation technique is also measured by the compression ratio (CR), maximum error (maxE) and integral square error (ISE). The CR is defined as the ratio of the number of points (n) in the input curve to the number of vertices (m) in the output polygon. The approximation error is measured either by the maxE or ISE. The maximum error is the maximum deviation of the approximating polygon from the input curve and is defined by maxE = max $\{ek\}$, $k = 1, n$ and ek is defined by

$$e_k = \left| \{(x_k - x_i)(y_j - y_i) - (y_k - y_i)(x_j - x_j)\} \right| / \sqrt{\{(x_j - x_i)^2 + (yj - yi)^2\}} \tag{7.3}$$

where, i and j are the indices of the coordinates of the end points of one of the sides of the approximating polygon. It may be noted that the maximum error used in the reference [83] is actually the maximum of the square of ek values. Since for any real number x with $0 < x < 1$, $x^2 < x$, so the maximum error shown in the reference [83] is less than its actual value and the values of values of maximum error shown therein are not comparable with the same by other algorithms cited for comparison in reference [83].. To alleviate this discrepancy, the maximum error of reference [83] has been square rooted (e.g. if the maximum error for an approximation is shown as 0.74 in [83] then its actual value is $\sqrt{0.74} = 0.86$) so that it can be compared with the maximum error produced by the proposed algorithm. The ISE is defined by the sum of the square of all e_k. In many research papers, CR, maxE and ISE were used to compare different methods of polygonal approximation. An approximation with higher value of CR and lower value of approximation error (maxE and/or ISE) is regarded as a better approximation. But there exists a tradeoff between CR and approximation error. As the CR increases, the approximation error also increases. To alleviate this difficulty, Sarkar [150] introduced a measure of goodness, called figure of merit (FOM) defined by the quotient of CR and approximation error (maxE or ISE) for comparing different polygonal approximation techniques. This metric has two drawbacks, namely, if the approximation error is zero then the FOM is undefined and it is not useful for comparing approximations with different number of vertices. Rosin [141] introduced another metric that can be used to compare the results of polygonal approximation with different number of vertices. This metric is based on fidelity and efficiency of approximation. According to [141], the fidelity (F) is defined by the quotient of the approximation error of an optimal algorithm (taken as the benchmark algorithm) and the approximation error produced by a suboptimal algorithm. The efficiency (E) is defined by the quotient of the number of vertices produced by the optimal benchmark algorithm and the number of vertices produced by the suboptimal algorithm. The metric to measure the goodness of approximation, called measure of merit (MOM), is defined by MOM = $\sqrt{(FE)}$, expressed in percentage. But it may be noted that the efficiency of an algorithm is measured either by its computational complexity or by its actual running time and to compare the efficiency of a method with some other, the ratio of the efficiency of the two methods should be used. The definition of MOM requires an optimal algorithm as a benchmark. If the approximating polygon consists of just break points then fidelity and efficiency are both unity and this means that the set of break points will produce prefect approximation, but as it is well known, these approximations are of no use because the CR is very low. Since both the FOM and MOM have drawbacks and since the aim of any polygonal approximation technique is to look for maximum possible line segments with minimum possible error that is to minimize the approximation error and the number of vertices of the output polygon, so the fidelity (F > 0) of a polygonal approximation is defined by the product of approximation error and the number of vertices of the output polygon. The smaller is the value of this metric, the better is the approximation. The trivial minimum of the approximation error is excluded from this definition of fidelity, because F > 0. The approximation error may be maximum error, ISE or any other error measure example area deviation of the curve from the approximating polygon. To compare the two approximations with respect to fidelity, use the ratio F*A*/F*B*, where *A* and *B* are any two different approximations. This measure is called, as coefficient of goodness. If it is greater than 1 then the approximation *B* is better than the approximation *A* and vice-versa. If *mA* and *mB* are the number of vertices produced by the approximations *A* and *B* respectively and ISE*A* and ISE*B* are the corresponding ISE then the coefficient of goodness of the approximation *B* with respect to the approximation *A* is (*mA**ISE*A*)/(*mB**ISE*B*). Moreover, as *A* and *B* are any two different approximations, this measure can be used not only to compare two approximations produced by two different algorithms but also to compare different approximations produced by the same algorithm. Two approximations are said to be different if either the number of vertices or the approximation error differs.

TABLE 1 The numerical results of the approximations produced by the proposed algorithm and the Masood's algorithm.

Shape	Method	Maximum error (maxE)	Number of vertices (*m*)	Integral square error (ISE)	Coefficient of goodness
Chromosome	Masood's	0.75	19	2.87	
	Proposed	0.75	24	1.38	1.65
	Masood's	0.72	18	2.88	
	Proposed	0.72	24	1.22	1.86
	Masood's	0.77	17	3.44	
	Proposed	0.77	24	1.38	1.77
	Masood's	0.72	16	3.84	
	Proposed	0.72	24	1.22	2.01
	Masood's	0.79	15	4.14	
	Proposed	0.79	24	1.38	1.88
	Masood's	0.94	12	7.76	
	Proposed	0.94	19	3.32	1.48

TABLE 1 *(Continued)*

Shape	Method	Maximum error (maxE)	Number of vertices (*m*)	Integral square error (ISE)	Coefficient of goodness
Leaf	Masood's	0.77	32	4.97	
	Proposed	0.77	34	3.30	1.42
	Masood's	0.79	30	5.77	
	Proposed	0.79	32	4.57	1.18
	Masood's	0.79	29	6.27	
	Proposed	0.79	32	4.57	1.24
	Masood's	0.81	28	6.91	
	Proposed	0.81	32	4.57	1.32
	Masood's	0.86	23	10.61	
	Proposed	0.86	31	4.57	1.72
	Masood's	0.86	22	11.16	
	Proposed	0.86	31	4.57	1.73
Figure with multiple semi circles	Masood's	0.70	30	2.91	
	Proposed	0.70	32	1.71	1.60
	Masood's	0.79	29	3.41	
	Proposed	0.79	32	2.21	1.40
	Masood's	0.79	28	3.91	
	Proposed	0.79	32	2.21	1.55
	Masood's	0.79	27	4.40	
	Proposed	0.79	32	2.21	1.68
	Masood's	0.79	26	4.91	
	Proposed	0.79	32	2.21	1.81
	Masood's	0.85	22	8.61	
	Proposed	0.85	32	2.21	2.68
	Masood's	1.00	19	23.90	
	Proposed	1.00	26	5.50	3.18

TABLE 1 *(Continued)*

Shape	Method	Maximum error (maxE)	Number of vertices (*m*)	Integral square error (ISE)	Coefficient of goodness
Figure–8 curve	Masood's	0.79	11	3.48	
	Proposed	0.79	16	1.67	1.43
	Masood's	0.77	12	2.98	
	Proposed	0.77	16	1.67	1.34
	Masood's	0.74	13	2.60	
	Proposed	0.74	16	1.67	1.26
	Masood's	0.71	14	2.14	
	Proposed	0.71	16	1.67	1.12

The Table 1 shows the numerical results of the approximation shown in Figure 5 through Figure 8 and those of Figure 9 through Figure 12. For each approximation, the maxE, the number of vertices (m), the ISE and coefficient of goodness of the proposed method with respect to the Masood's [83] are shown. To compute the coefficient of goodness, the ISE has been used as the approximation error as maxE is pre-specified. As evident from this table the fidelity of approximations produced by the proposed algorithm is better than those produced by the Masood's algorithm.

KEYWORDS

- **Compression ratio**
- **Digital planar curve**
- **Integral square error**
- **Masood's algorithm**
- **Split and merge technique**
- **Two-Pass sequential algorithm**

8 Polygonal Approximation Using Reverse Engineering on Bresenham's Line Drawing Technique

CONTENTS

8.1 INTRODUCTION

In this chapter we use reverse engineering on Bresenham's line drawing algorithm proposing an $O(n)$ algorithm. The algorithm, by virtue of its two-pass nature, produces symmetric approximation from symmetric digital curve. The algorithm does not require human intervention.

8.2 PROPOSED METHOD

As it is well known, Bresenham's algorithm treats lines into four categories, namely lines with positive gentle slope ($0 \leq$ slope < 1), negative gentle slope ($-1 <$ slope < 0), positive sharp slope ($1 \leq$ slope $< \infty$) and negative sharp slope ($-\infty <$ slope ≤ -1). If (x, y) is any point on an analog line then the Bresenham's algorithm works as follows. In case the slope is positive gentle then step across x and decide whether to increment y. If the slope is negative gentle then step across x and decide whether the y value should be decremented. For sharp positive slope, step across y and decide whether to increment x and for sharp negative slope, step across y and decide whether to decrement x. The decision for incrementing/decrementing x or y is based on a parameter, and its value also is updated at every step. The algorithm always starts at one specific end point and terminates at the other.

Since for positive gentle slope, x is incremented and y may be required to be incremented, so the longest line segment for which x_{i+1} is greater than x_i and y_{i+1} is either greater than or equal to than y_i is looked for. The consecutive pairs i and i + 1 constitute one line segment. This line segment has positive gentle slope because of the type of inequalities that hold between x's and y's. Since the curves are being dealt with and it is not known before hand whether x_{i+1} is greater than x_i or x_{i+1} is less than x_i so the longest possible line segment with positive gentle slope may also output by testing the inequalities $x_{i+1} < x_i$ and $y_{i+1} \leq y_i$. Similarly, the longest line segment with negative gentle slope is the output of either looping on the inequalities $x_{i+1} > x_i$ and $y_{i+1} \leq y_i$ or looping on the inequalities $x_{i+1} < x_i$ and $y_{i+1} \geq y_i$. In order to look for the longest line segment with sharp slope, it is necessary to use the technique after interchanging the role of x and y.

The methodology can be applied starting from any point of a closed digital curve and for open digital curve, it can be applied starting from one of the end points of the curve. The methodology is applied twice on the curve, traversing the curve once in the clockwise direction and then in the anticlockwise direction. Thus the procedure is a two-pass sequential in nature and because of its two-pass nature, it produces symmetric approximation from symmetric digital curve. In each pass, each point of the curve is visited only once (and the operations that are performed are integer comparison and index increment/decrement). So the computational complexity of the procedure is $O(n)$ and the procedure is fast. No arithmetic is applied on the coordinates (x, y) of the curve and this is in direct contrast to most of the other algorithms in this area.

The methodology can be applied to curve with uniformly as well as non-uniformly spaced points. In the later case, linear interpolation can be applied to convert the non-uniformly spaced points into uniformly spaced points.

8.3 ALGORITHM POLYGONAL APPROXIMATION

The input is the set of points of the curve whose coordinates are (x, y). The identifier n is the number of points the curve has, assuming that the points are equally spaced. The identifier j is the index of a point. The output is a polygon approximating the curve. All arithmetic performed is in integer mode. The procedure can be started from any point on the curve. For open digital curve the two end points are also member of the set of output vertices.

Step 1: Initiate the index i by 1 and traverse the curve clockwise.

Look for vertical line segment such that the value of y coordinates increases.

Step 2: If x_{i+1} is equal to x_i and y_{i+1} is greater than y_i then

Find out how long this line segment is.

as long as y_{i+1} is greater than y_i and x_{i+1} is equal to x_i move to the next point incrementing i.

Step 3: Store x_i and y_i as a vertex of the output polygon.

Look for vertical line segment such that the value of y coordinates decreases.

Step 4: If x_{i+1} is equal to x_i and y_{i+1} is less than y_i.

Find out how long this line segment is.

as long as y_{i+1} is less than y_i and x_{i+1} is equal to x_i move to the next point incrementing j.

Step 5: Store x_i and y_i as a vertex of the output polygon.

Look for horizontal line segment

Step 6: Same as step 2 to 5 after inter changing the role of x and y.

Look for line segment which is neither horizontal nor vertical and the slope of the line is between 0 and 1, including 0 and but excluding 1.

Step 7: If x_{i+1} is greater than x_i and y_{i+1} is greater than or equal to y_i then

Find out how long this line segment is.

as long x_{i+1} is greater than x_i and y_{i+1} is greater than or equal to y_i move to the next point incrementing i.

Step 8: Store x_i and y_i as a vertex of the output polygon.

Step 9: If x_{i+1} are less than x_i and y_{i+1} is less than or equal to y_i then

Find out how long this line segment is.

as long x_{i+1} is less than x_i and y_{i+1} is less than or equal to y_i move to the next point incrementing i.

Step 10: Store x_i and y_i as a vertex of the output polygon.

Look for line segment which is neither horizontal nor vertical and the slope of the line is between 0 and −1, including 0 and but excluding −1.

Step 11: If x_{i+1} is greater than x_i and y_{i+1} is less than or equal to y_i then

Find out how long this line segment is.

as long x_{i+1} is greater than x_i and y_{i+1} is less than or equal to y_i move to the next point incrementing i.

Step 12: Store x_i and y_i as a vertex of the output polygon.

Step 13: If x_{i+1} is less than x_i and y_{i+1} is greater than or equal to y_i then

Find out how long this line segment is.

as long x_{i+1} is less than x_i and y_{i+1} is greater than or equal to y_i move to the next point incrementing i.

Step 14: Store x_i and y_i as a vertex of the output polygon.

Look for line segments which is neither horizontal nor vertical and the slope of the line is neither between 0 and 1 nor between 0 and −1.

Step 15: Do step 7 to 14 after inter changing the role of x and y.

Step 16: Repeat the step 2 to 15 until the next vertex point is the same as the first vertex point.

Step 17: Repeat step 1 to step 16 traversing the curve counter clockwise.

8.4 EXPERIMENTAL RESULTS

The algorithm proposed here is applied on three digital curves namely, a chromosome shaped curve (Figure 1(a)), a leaf-shaped curve (Figure 2(a)) and a curve with multiple semicircular arcs (Figure 3(a)). The Figure 1(b), Figure 2(b), and Figure 3(b) show the polygonal approximation and Figure 1(c), Figure 2(c), and Figure 3(c) show the polygonal approximation overlaid on the corresponding digital curve. The vertices of the approximating polygon are indicated on the figures by solid circles.

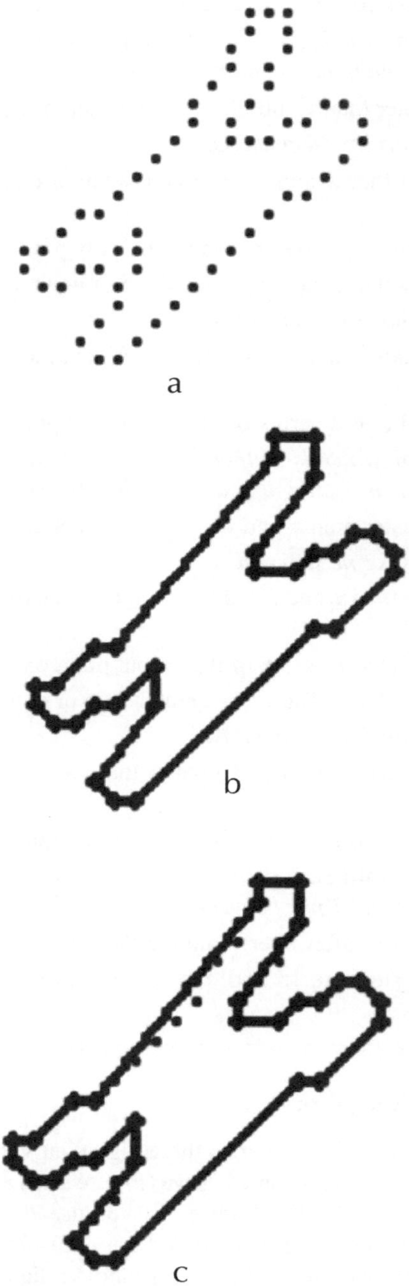

FIGURE 1 (a) A chromosome shaped curve, (b) its polygonal approximation, and (c) an overlay of the approximation on the curve.

a

b

c

FIGURE 2 (a) A leaf shaped curve, (b) its polygonal approximation, and (c) an overlay of the approximation on the curve.

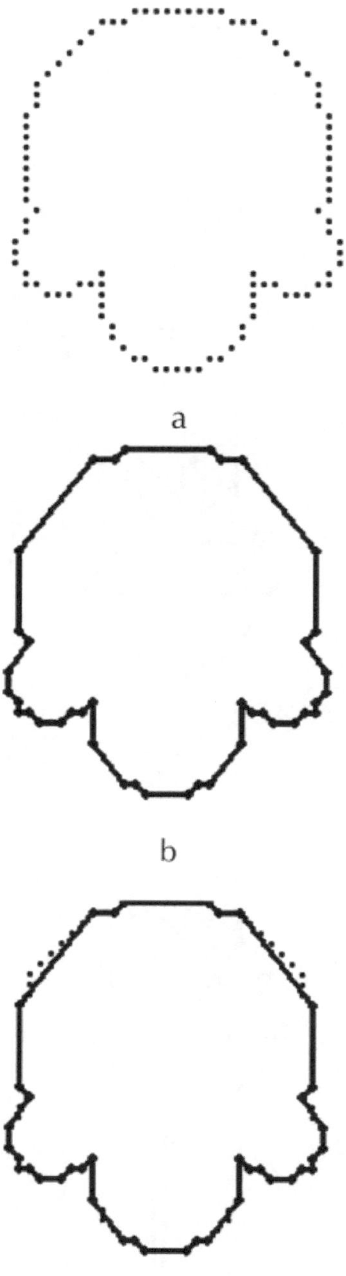

a

b

c

FIGURE 3 (a) A curve with four semi circles, (b) its polygonal approximation, and (c) an overlay of the approximation on the curve.

The Table 1 shows the number of points for each curve, the number of vertices of the output polygon, the maximum error and the integral square error (ISE) produced by the proposed algorithm for different digital curves. The Table 2 shows a comparison of the proposed method with Masood's method [83] with respect to Figure of Merit defined by the quotient of compression ratio (CR) and ISE. As evident from this table the figure of merit produced by the proposed method for the curve with semi circles is significantly less than that produced by the Masood's algorithm. It may be noted that this curve consists of semi circles and a better way of approximating this kind of curves is to use reverse engineering of mid-point circle drawing algorithm.

TABLE 1 A comparison between Masood's method and the proposed method.

Curve	Number of points	Number of vertices	Maximum error	Integral square error
Chromosome	60	29	0.78	1.95
Leaf	120	39	0.69	2.33
Curve with semi circles	102	40	1.05	6.35

TABLE 2 A comparison between Masood's algorithm and the proposed algorithm using figure of merit = CR/ISE.

Curve	Masood's method	Proposed method
Chromosome	1.29	1.06
Leaf	1.12	1.32
Curve with semi circles	1.28	0.40

There are three features of the proposed method, namely, the algorithm is sequential with complexity *O*(*n*) because each point of the curve is visited only once in each of the two passes that is made, it produces symmetric approximation from symmetric curve (please see Figure 3(b) and Figure 3(c)) and finally, it is an automatic algorithm (does not require human intervention) because neither does it require to specify the maximum allowable error nor does it require the number of sides the polygon should have.

KEYWORDS

- **Algorithm polygonal approximation**
- **Bresenham's algorithm**
- **Closed and open digital curve**
- **Proposed method**
- **Reverse engineering**

9 Polygonal Approximation as Angle Detection

CONTENTS

9.1 INTRODUCTION

In this chapter we present an algorithm [124] for polygonal approximation of digital curve which is based on discrete curvature measure. We propose that one can use *k*-cosine itself to determine the region of support without using any input parameter. A new measure of discrete curvature based on *k*-cosine called smoothed *k*-cosine is introduced. The local maxima and minima of smoothed *k*-cosine are located. We call these points as significant points. The adjacent significant points are joined to determine the polygon.

In section 9.3, we present a scheme to determine the region of support of each point of a digital curve. The procedure is parallel in the sense that the results obtained at each point do not depend on those obtained at other points. The procedure needs no input parameter. The region of support of each point is determined based on the local properties of the curve.

9.2 CURVATURE BASED POLYGONAL APPROXIMATION

The algorithm developed to treat the polygonal approximation as a side detection problem. The sides of the polygon are determined subject to certain constraints on the goodness of fit. Another approach to polygonal approximation is based on curvature estimation.

In the Euclidean plane, the curvature (κ) is defined as the rate of change of tangential angle (ψ) with respect to the arc length (*s*).

$$\kappa = \frac{d\psi}{ds} \tag{9.1}$$

In Cartesian coordinate system if the equation of a curve is expressed in the form $y = f(x)$ then the curvature at any point of the curve is defined by

$$\kappa = \frac{\dfrac{d^2y}{dx^2}}{\left(1+\left(\dfrac{dy}{dx}\right)^2\right)^{3/2}} \tag{9.2}$$

For digital curves however it is not immediately clear how to define a discrete analog of curvature. If the discrete curvature is defined by simply replacing the derivatives in (9.2) by finite differences, there is a problem that a small change in slope are impossible, since the successive slope angles on a digital curve can differ by a multiple of π/4. This difficulty is avoided by introducing a smoothed version of discrete curvature example *k*-cosine, *k*-curvature, *k* is called the region of support and it is determined by introducing an input parameter *m*. The input parameter *m* is selected on the basis of level of detail of the curve. The finer is the level of detail, the value of m should be small. Difficulty arises when a curve consists of features of multiple sizes. And this difficulty can be avoided by using different m for regions with different level of detail. Moreover, there is seldom any basis of choosing a particular value of parameter for a particular feature size. Teh and Chin [170] used chord length and perpendicular distance to determine region of support without using any input parameter. This technique is justified because the region of support for each point is determined only on the basis of the local properties of the curve.

We present an algorithm [124] for polygonal approximation of digital curve which is based on discrete curvature measure. We propose that one can use *k*-cosine itself to determine the region of support without using any input parameter. A new measure of discrete curvature based on *k*-cosine called smoothed *k*-cosine is introduced. The local maxima and minima of smoothed *k*-cosine are located. We call these points as significant points. The adjacent significant points are joined to determine the polygon.

In the following section we present a scheme to determine the region of support of each point of a digital curve. The procedure is parallel in the sense that the results obtained at each point do not depend on those obtained at other points. The procedure needs no input parameter. The region of support of each point is determined based on the local properties of the curve.

9.3 PROCEDURE DETERMINATION OF REGION OF SUPPORT

Begin

Step 1: Define the k-vectors at p_i as:

$a_{ik} = (x_{i-k} - x_i, y_{i-k} - y_i)$ (9.3)

$b_{ik} = (x_{i+k} - x_i, y_{i+k} - y_i)$ (9.4)

and the k-cosine at p_i as:

$$\cos_{ik} = \frac{a_{ik}.b_{ik}}{|a_{ik}||b_{ik}|} \tag{9.5}$$

where, $\cos_{ik}$ is the cosine of the angle between the k-vectors a_{ik} and b_{ik}, so that $-1 \leq \cos_{ik} \leq 1$.

Step 2: Start with $k = 1$. Compute $\cos_{ik}$ giving increment to k.

If $|\cos_{i,k+1}| > |\cos_{ik}|$ then k determines the region of support of p_i

else if $|\cos_{i,k+1}| = |\cos_{ik}|$ then the greatest k for which the relation holds determines the region of support of p_i.

else if $\cos_{i,k+1}$ and $\cos_{i,k}$ be of opposite sign then the least value of k for which it happens gives the region of support of p_i.

The region of support of p_i is the set of points

$D(p_i) = \{p_{i-k}, \ldots, p_{i-1}, p_i, p_{i+1}, p_{i+k}\}$.

End.

9.4 MEASURE OF SIGNIFICANCE

The last procedure determines the region of support (k_i) of the point p_i. To detect the significant points we need a measure of significance. Rosenfeld and Johnston [139] used $\cos_{i,k}$ as the measure of significance and hi as the region of support of p_i. Rosenfeld and Wezska [142] used smoothed k-cosine as a measure of significance. We propose to introduce a new measure of significance. We denote it by $\cos_i$ and define it by

$$\cos_i = \frac{1}{k_i} \sum_{j=1}^{k_i} \cos_{ij} \tag{9.6}$$

This measure of significance is kind of smoothed cosine but it is different from that given by Rosenfeld and Wezska [142]. In the following section we present a procedure for detection of significant points using the region of support k determined by the procedure given in section 9.3 and the measure of significance $\cos_i$ introduced in this section.

9.5 PROCEDURE DETECTION OF SIGNIFICANT POINTS AND POLYGONAL APPROXIMATION

Comments

As the procedure runs remove those points from consideration where $\cos_i$'s are too small ($\cos_i \leq -0.8$), because in the neighborhood of these points the curves are relatively straight and our ultimate goal is to make a polygonal approximation of the curves.

Begin

1st Pass

Retain only those points pi for which either

$$\cos_i \geq \cos_j \tag{9.7}$$

for all j satisfying

$$|i-j| \leq k_i/2, \quad k_i > 1 = k_i, k_i = 1. \tag{9.8}$$

or

(b) $\cos_i \leq \cos_j$ (9.9)

for all j satisfying

$$|i-j| \leq k_i/2, \quad k_i > 1 = k_i, \quad k_i = 1. \tag{9.10}$$

In 9.7 and 9.9 strict inequality should hold for at least one j satisfying (9.8) and (9.10) respectively. The points detected by (9.7) are the local maxima and those detected by (9.9) are the local minima of smoothed k-cosine.

2nd Pass

If minima fall within the region of support of a maxima point then the minima point is discarded and the maxima point is retained.

If two successive points p_i and p_{i+1} appear as maxima points then if both p_i and p_{i+1} have the same cosine and the same region of support then retain p_i and discard p_{i+1} else if the cosine be the same but the region of support be different then retain the point with higher region of support and discard the other.

The points obtained from 1st and the 2nd pass constitutes the set of significant points and the adjacent significant points are joined to make a polygonal approximation.

End.

Remarks 9.1

When two successive points p_i and p_{i+1} appear as maxima points and both have the same cosine the both the points are equally important for being selected as significant points and so in this situation there is a tie. We have proposed to break the tie by choosing p_i only as the significant point. On the other hand, when p_i and p_{i+1} have the same cosine but different region of support then there is no tie and the choice is deterministic.

Remarks 9.2

The 1st pass of the procedure can be carried out in parallel whereas the 2nd pass is sequential. We note that the 2nd pass is carried out only on a small number of points.

9.6 APPROXIMATION ERRORS

The shape of a digital curve is determined by its significant points. So, it is very much necessary to locate then accurately so that sufficient information about the curve is contained in the location of the significant points. In section 9.2 and section 9.4 we have described a procedure to detect significant points by point-wise error between the digital curve and the approximating polygon. We measure the error between the digital curve and the approximating polygon by the perpendicular distance of the points pi's from their approximating segment, We denote this error by e_i. Two error norms between the digital curve and the approximating polygon are:

(1) Integral square error $E_2 = \sum_{i=1}^{n} e_i^{\,2}$

(2) Maximum error $E_\infty = \max_i |e_i|$

9.7 EXPERIMENTAL RESULTS

The procedure is applied on four digital curves namely, a chromosome shaped curve (Figure 1), a figure-8 curve (Figure 2), a leaf-shaped curve (Figure 3) and a curve with four semi circles (Figure 4). The last figure taken from [170] is an example of a curve that consists of feature of multiple sizes. The chain codes of the curves are given in Table 1. The curves have been coded in the clockwise direction starting from the point marked with an arrow on each curve, using the Freeman chain code. The procedure processes data in the clockwise direction.

In an attempt to focus on the efficiency of our procedure as a significant point detector and a polygonal approximation technique we have computed (a) the data compression ratio, (b) integral square error and (c) maximum error. The results are displayed in Table 2 together with the results obtained by the Teh and Chin algorithm.

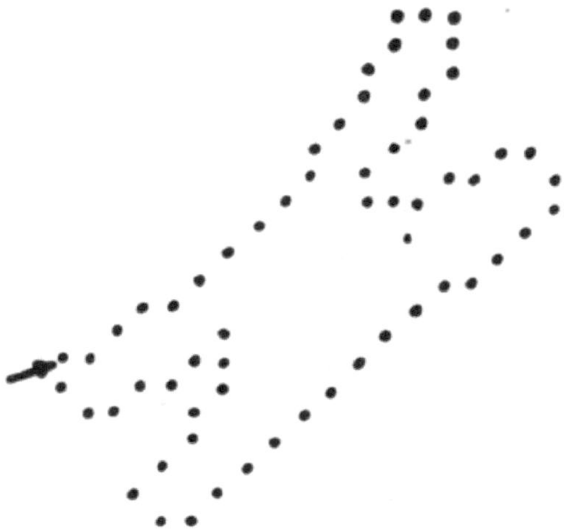

a

FIGURE 1 *(Continued)*

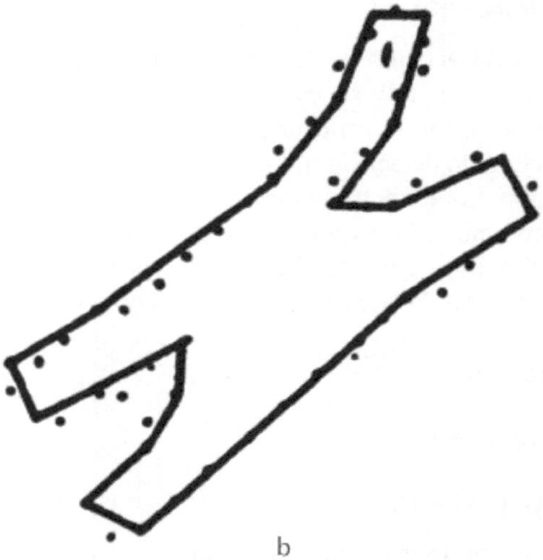

b

FIGURE 1 A chromosome shaped curve and its polygonal approximation.

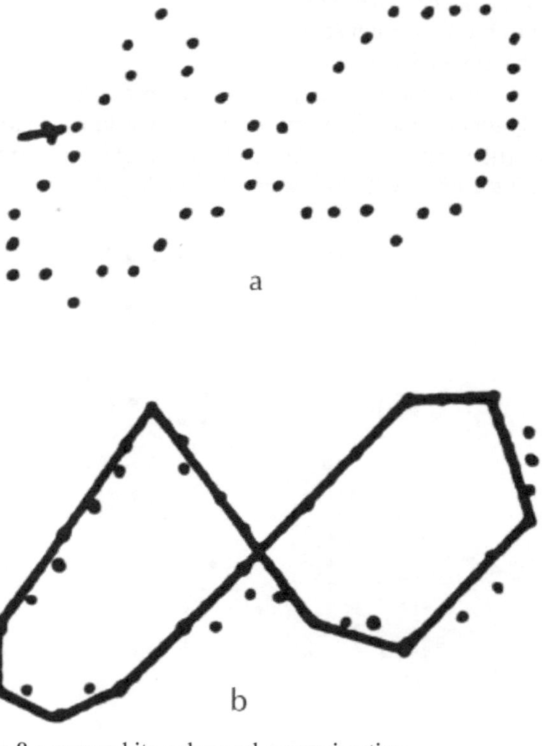

FIGURE 2 A figure-8 curve and its polygonal approximation.

a

b

FIGURE 3 A leaf-shaped curve and its polygonal approximation.

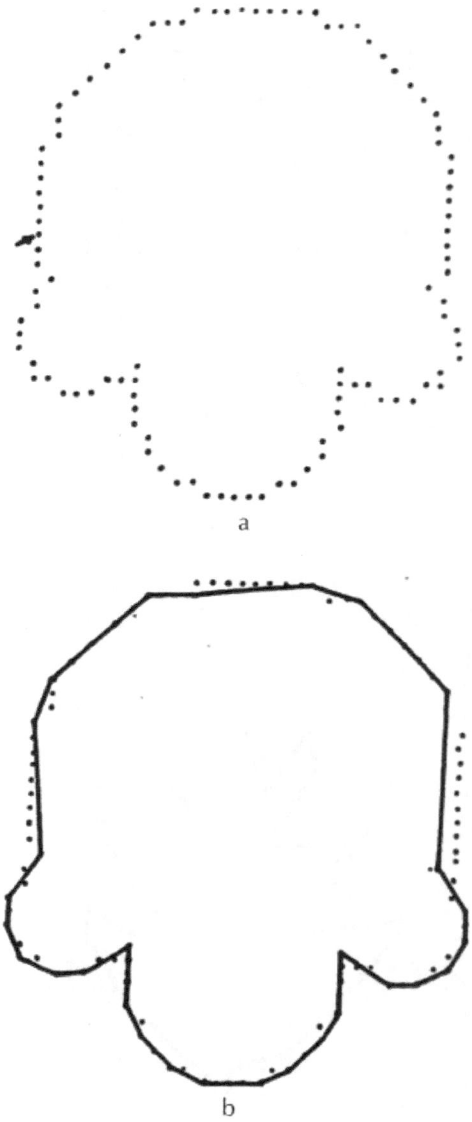

FIGURE 4 A curve with four semi-circles and its polygonal approximation.

9.8 DISCUSSION

We have made no attempt to compare our procedure with those in [5, 48, 139, 142, 148]. A very good comparison of these algorithms with that in [170] can be found in [170].

We only focus on the following features of our procedure:

(a) Our procedure like that of Sankar-Sharma [148] and Teh and Chin [170] does not require any input parameter.
(b) Sankar—Sharma's procedure does not require any input parameter but it does not determine the region of support, whereas our procedure like that of Teh and Chin does determine the region of support.
(c) Our procedure detects more significant points that the Teh and Chin algorithm.
(d) All processing are done locally, so it is suitable for parallel processing.

TABLE 1 Chain code of digital curves.

Chromosome shaped curve
01101 11112 11212 00665 65560 01010 76555 45555 55555 43112 12255 45432
Figure-8 curve
11217 67767 70071 01212 22344 45555 56545 54534 22112
Leaf shaped curve
33333 32307 00003 32323 07000 03323 22267 77222 12766 61111 16665 66550 00100 56656 55001 10665 65655 55566 67666 66666 64222 22222 22232 24434
Curve in Figure 4
22222 21221 11111 00100 00000 07007 77777 66766 66666 65767 66564 54434 36666 56554 54444 34332 32222 54544 34232 21213 22

TABLE 2 A comparison between algorithm 9.5 and Teh and Chin algorithm.

Digital curve	**Chromosome**	**Figure-8**	**Leaf**	**Figure 4**
Number of points	**60**	**45**	**120**	**102**
Results of algorithm 9.4				
Number of significant points (n_v)	18	14	29	30
Compression ratio (n/n_v)	3.33	3.21	4.14	3.40
Integral square error	5.25	2.51	15.34	17.19
Maximum error	0.686	0.728	0.996	1.00
Results of Teh & Chin algorithm				
Number of significant points (n_v)	15	13	29	22

TABLE 2 *(Continued)*

Digital curve	**Chromosome**	**Figure-8**	**Leaf**	**Figure 4**
Number of points	**60**	**45**	**120**	**102**
Compression ratio (n/n_v)	4.00	3.50	4.10	4.60
Integral square error	7.20	5.93	14.96	20.61
Maximum error	0.74	1.00	0.99	1.00

KEYWORDS

- **Digital curve**
- **Integral square error**
- **Maximum error**
- **Polygonal approximation**
- **Significant points**

10 Polygonal Approximation as Angle Detection Using Asymmetric Region of Support

CONTENTS

10.1 INTRODUTION

In the last chapter we have shown that it is possible to use *k*-cosine to determine the region of support without using any input parameter. The region of support as determined by this procedure is symmetric in the sense that it consists of equal number of points on either side of the point of interest. The region of support as determined by Rosenfeld-Johnston [139], Rosenfeld-Wezska [142] and Teh and Chin [170] is also symmetric. But there is no reason why the region of support should be symmetric. We believe that an asymmetric region of support is more reasonable and more natural to occur than a symmetric region of support. A symmetric region of support may be looked upon as a special case of asymmetric region of support.

10.2 ASYMMETRIC REGION OF SUPPORT

We propose to determine asymmetric region of support without using any input parameter. The region of support is determined on the basis of the local properties of the curve. We call the regions on either side of the point p_i as the arms of the point. The curve is described in the clockwise direction. The arm extending from the point p_i to the forward direction is regarded as the right arm of p_i and the arm extending from p_i to the backward direction is regarded as the left arm of the point p_i. The size of the arms is the number of points comprising the arms. We propose to denote the size of the right arm by k and that of the left arm by l. The region of support of p_i comprise of the points in its arms and the point p_i itself.

This concept of asymmetric region of support is utilized to detect dominant points on digital curves. The dominant points are the points with high curvature. To determine the curvature using the concept of asymmetric region of support, we introduce the concept of *k-l* cosine defined as the cosine of angle between the *k*-vector and *l*-vector at the point p_i. The dominant points are those points at which the *k-l* cosine is a local maximum with respect to its immediate neighbors. The polygonal approximation is made joining the adjacent dominant points [125].

10.3 DETERMINATION OF ARM

We define the vectors R_i and R_{ij} at p_i by:

$$R_i = (x_{i+1} - x_i, y_{i+1} - y_i)$$

$$R_{ij} = (x_j - x_i, y_j - y_i) \tag{10.1}$$

where, $j = i + 2, i + 3, i + 4$, If θ_j denotes the angle between R_i and R_{ij} (Figure 1) then we propose to compute θ_j by the relation:

$$\theta_j = \cos^{-1}\left(\frac{R_i . R_{ij}}{|R_i||R_{ij}|}\right) \tag{10.2}$$

θ_j is computed giving increment to j until for some j

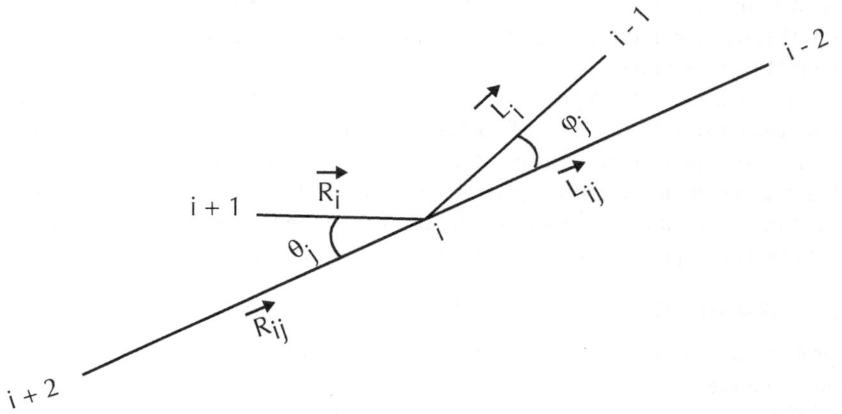

FIGURE 1 Left and right vector and angles θ and φ.

$$\theta_{j-1} < \theta_j \leq \theta_{j+1} \tag{10.3}$$

When the last relation holds for three consecutive values of j ($j-1, j, j+1$) then the set of points $\{p_{i+1}, p_{i+2}, p_{j-1}\}$ is said to constitute the right arm of the point p_i and the length of the arm is $k = j - 1$.

Again we define two vectors L_i and L_{ij} at p_i by:

$$L_i = (x_{i-1} - x_i, y_{i-1} - y_i)$$

$$L_{ij} = (x_j - x_i, y_j - y_i) \tag{10.4}$$

where, $j = i-2, i-3, i-4$, If φ_j denotes the angle between L_i and L_{ij} (Figure 1) then we propose to compute φ_j by the relation:

$$\varphi_j = \cos^{-1}\left(\frac{L_i.L_{ij}}{|L_i||L_{ij}|}\right) \tag{10.5}$$

φ_j is computed giving decrement to j until for some j

$$\varphi_{j+1} < \varphi_j \leq \varphi_{j-1} \tag{10.6}$$

When the last relation holds for three consecutive values of j ($j+1, j, j-1$) then the set of points $\{p_{i-1}, p_{i-2}, p_{j+1}\}$ is said to constitute the left arm of the point p_i and the length of the arm is $l = i - j - 1$.

The region of support of p_i is then given by the set of points

$$\{p_{i-l}, \ldots, p_{i-1}, p_i, p_{i+1}, p_{i+2}, p_{i+k}\}$$

This technique for determination of arms is illustrated with the help of the figure-8 curve shown in Figure 2. For the ith point as shown in the figure, the vector R_i has components (0, 1) and the vector $R_{ij} = R_{i, i+2}$ has components (0, 2) and so using (10.2) we get $\theta_{i+2} = 0°$. Similarly $\theta_{i+3} = 18.4°$, $\theta_{i+4} = 26.6°$. And since $\theta_{i+2} < \theta_{i+3} < \theta_{i+4}$ hence the points p_{i+1} and p_{i+2} comprise the right arm of the points p_i and the length of the arm is $k = j - i - 1 = i + 3 - i - 1 = 2$.

Again for the ith point L_i has components (1, 0) and $L_{ij} = L_{i, i-2}$ has the components (2, –1) and hence using (10.5) we get $\varphi_{i-2} = 26.6°$. Similarly $\varphi_{i-3} = 0°$, $\varphi_{i-4} = 0°$, $\varphi_{i-5} = 11.3°$, $\varphi_{i-6} = 18.6°$ and hence $\varphi_{i-4} < \varphi_{i-5} < \varphi_{i-6}$. So the points $p_{i-1}, p_{i-2}, p_{i-3}$, and p_{i-4} comprise the left arm of the point p_i and $l = i - j - 1 = i - i + 5 - 1 = 4$ is the length of the left arm.

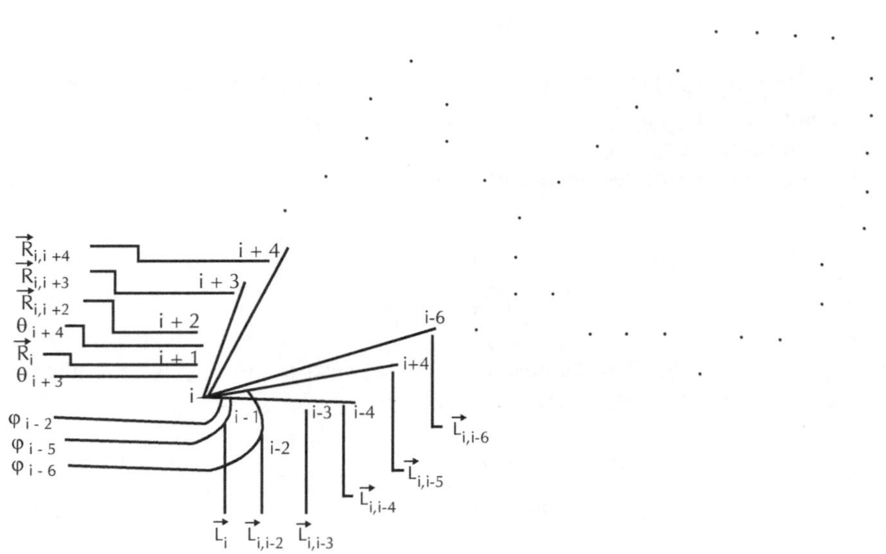

FIGURE 2 Illustrating computation of θ`s and φ`s.

The region of support of p_i is the set of points:

$$\{p_{i-4}, p_{i-3}, p_{i-2}, p_{i-1}, p_i, p_{i+1}, p_{i+2}\}.$$

The region of support of other points is computed in a similar manner.

10.4 PROCEDURE DETERMINATION OF REGION OF SUPPORT

Begin

Step 1: Define

$R_i = (x_{i+1} - x_i, y_{i+1} - y_i)$

$R_{ij} = (x_j - x_i, y_j - y_i)$

where $j = i + 2, i + 3, i + 4, \ldots$

Compute

$$\theta_j = \cos^{-1}\left(\frac{R_i.R_{ij}}{|R_i||R_{ij}|}\right)$$

giving increment to j until for some j

$\theta_{j-1} < \theta_j \leq \theta_{j+1}$

$k = j - i - 1$ is the length of the right arm and the set of points $\{p_{i+1}, p_{i+2}, p_{i+k}\}$ comprise the right arm of the point p_i.

Step 2: Define

$L_i = (x_{i-1} - x_i, y_{i-1} - y_i)$

$L_{ij} = (x_j - x_i, y_j - y_i)$

where $j = i - 2, i - 3, i - 4,$

Compute

$$\varphi_j = \cos^{-1}\left(\frac{L_i.L_{ij}}{|L_i||L_{ij}|}\right)$$

giving decrement to j until for some j

$\varphi_{j+1} < \varphi_j \leq \varphi_{j-1}$

$l = i - j - 1$ is the length of the left arm and the set of points $\{p_{i-1}, p_{i-2}, \ldots, p_{i-l}\}$ comprise the left arm of the point p_i.

Step 3: The region of support of p_i is the set of points

$\{p_{i-l}, p_{i-1}, p_i, p_{i+1}, p_{i+2}, p_{i+k}\}$.

End.

10.5 DETECTION OF DOMINANT POINTS

Once the right arm and the left arm of each point are determined by the last procedure we define the right vector of pi. The right vector is denoted by a_{ik} and is defined by

$$a_{ik} = (x_{i+k} - x_i, y_{i+k} - y_i) \tag{10.7}$$

and the left vector at p_i denoted by b_{il} is defined by

$$b_{ik} = (x_{i-l} - x_i, y_{i-l} - y_i) \tag{10.8}$$

We define *k-l* cosine at p_i by

$$\cos_{ikl} = \frac{a_{ik}.b_{il}}{|a_{ik}||b_{il}|} \tag{10.9}$$

A two-stage procedure is applied to detect dominant points. At the first stage some input threshold is applied to the *k-l* cosine to eliminate those points from consideration whose *k-l* cosine is too small (≤ -0.8). At the second stage, a process of non-maxima suppression is applied to the remaining points to eliminate points whose *k-l* cosine are not local maxima with respect to its immediate neighbors. The points remaining after these two stages are the dominant points. To improve the compression ratio if two successive points of a curve appear as dominant points, the point with smaller region of support is suppressed. The polygonal approximation is made by joining the adjacent dominant points.

10.6 PROCEDURE DETECTION OF DOMINANT POINTS AND POLYGONAL APPROXIMATION

Begin

Step 1: Define k-vector at p_i as:

$a_{ik} = (x_{i+k} - x_i, y_{i+k} - y_i)$

and an l-vector at p_i as

$b_{il} = (x_{i-l} - x_i, y_{i-l} - y_i)$

Step 2: Define k-l cosine at p_i as:

$$\cos_{ikl} = \frac{a_{ik} . b_{il}}{|a_{ik}||b_{il}|}$$

Step 3: Suppress those points whose k-l cosine ≤ -0.8.

Step 4: If $\cos_{i, kl}$ lies between $\cos_{i-1, kl}$ and $\cos_{i+1, kl}$ then suppress p_i else p_i is a dominant point.

Step 5: If two successive points of a curve appear as dominant points then the point with smaller region of support is suppressed.

Step 6: The points remaining after step 5 are joined successively to make a polygonal approximation of the curve.

End.

10.7 APPROXIMATION ERRORS

The accuracy of the location of dominant points and closeness of the polygon to a digital curve can be determined by the point-wise error between digital curve and the approximating polygon. We measure this error by the perpendicular distance of the points p_i`s from their approximating segment and denote it by e_i.

Two error norms are defined by:

(1) Integral square error $E_2 = \sum_{i=1}^{n} e_i^2$

(2) Maximum error $E_\infty = \max_i |e_i|$

10.8 EXPERIMENTAL RESULTS

We have applied the procedure developed on four digital curves. The digital curves and their corresponding approximations are shown in Figure 3 through Figure 6. In an attempt to focus on the efficiency of the procedure for detection of dominant points and polygonal approximation, we have computed the data compression ratio, the maximum error and integral square error. The results are displayed in the Table 1 which also contains the results of the Teh and Chin algorithm.

FIGURE 3 The chromosome shaped curve and its polygonal approximation.

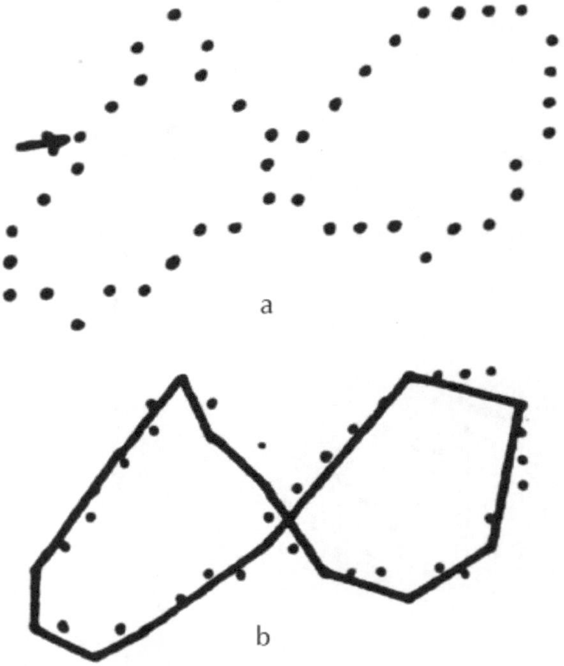

FIGURE 4 The figure-8 curve and its polygonal approximation.

a

FIGURE 5 *(Continued)*

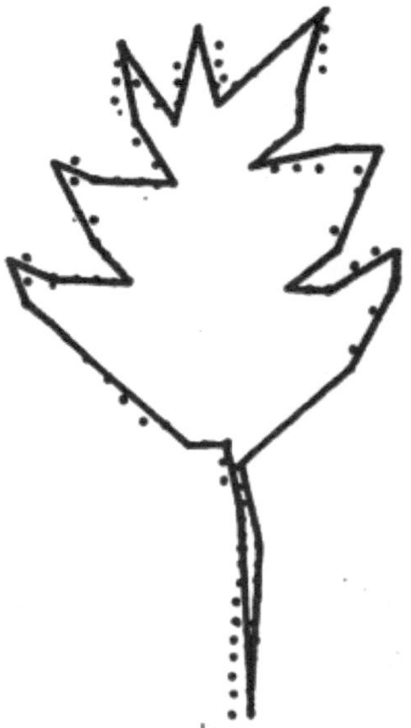

b

FIGURE 5 The leaf-shaped curve and its polygonal approximation.

a

FIGURE 6 *(Continued)*

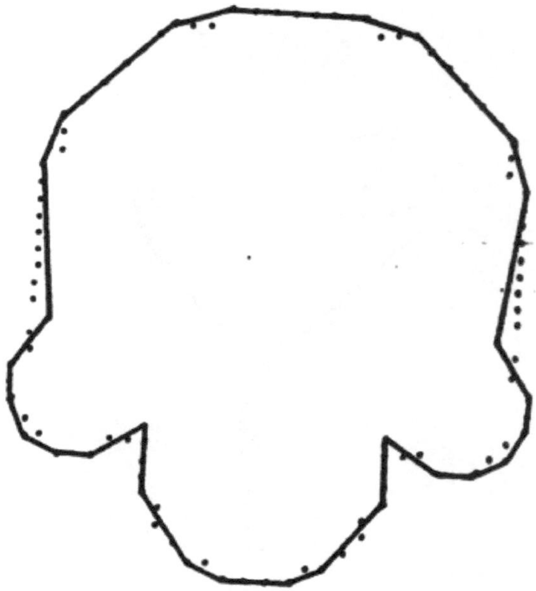

FIGURE 6 A curves with four semi circles and its polygonal approximation.

TABLE 1 A comparison between algorithm 10.6 and Teh and Chin algorithm.

Digital curve	**Chromosome**	**Figure-8**	**Leaf**	**algorithm 7.4**
Number of points	**60**	**45**	**120**	**102**
Results of algorithm 7.4				
Number of significant points (n_v)	18	14	32	28
Compression ratio (n/n_v)	3.33	3.21	3.75	3.64
Integral square error	5.42	4.81	13.40	9.27
Maximum error	0.64	0.73	0.996	0.88
Results of Teh and Chin algorithm				
Number of significant points (n_v)	15	13	29	22
Compression ratio (n/n_v)	4.00	3.50	4.10	4.60
Integral square error	7.20	5.93	14.96	20.61
Maximum error	0.74	1.00	0.99	1.00

10.9 DISCUSSION

We have made no attempt to compare this procedure with the existing ones except with the Teh and Chin algorithm.

We find the following features of the procedure:

(1) It uses asymmetric region of support.
(2) It requires no input parameter.
(3) It determines the region of support.
(4) It detects more dominant points than the Teh and Chin algorithm.
(5) All processing are done locally so it is suitable for parallel processing.

KEYWORDS

- **Approximation errors**
- **Asymmetric region of support**
- **Digital curves**
- **Dominant points**
- **Teh and Chin algorithm**

11 Scale Space Analysis with Application to Corner Detection

CONTENTS

11.1 INTRODUCTION

As already stated the fundamental problem in data smoothing is the choice of input parameter. Too large a parameter will smooth out many important features (corners, vertices, zero-crossings) and too small a parameter will produce many redundant features. This is the fundamental problem of scale, because features appearing on a curve vary enormously in size and extent and there is seldom any basis of choosing a particular parameter for a feature of particular size. This problem may be resolved by automatic parameter tuning. The parameter size can be tuned on the basis of the local properties of the curve using suitable criterion function. Teh and Chin [170] resolve this problem using chord length and perpendicular distance as the criterion function to determine the region of support of each point based on the local properties of the curve.

Another approach to the problem of scale is the scale space analysis. The concept of scale space was introduced by Iijima [59] and popularized by Witkin [187] and Koenderink and Van Doorn [75]. Scale space analysis of a signal is made by convolving it with the Gaussian kernel treating the parameter σ of the kernel as a continuous scale parameter. The zero-crossings of curvature/extreme curvature points of the smoothed signal are located by varying the parameter continuously. The arc length is shown along the x-axis (horizontal) and the scale parameter along the y-axis (vertical). The image on the xy half plane showing the location of curvature extrema or the zero-crossings of curvature at varying scale is called scale space map. Mocktarian and Mackworth [91] extend this work to two-dimensional shapes locating zero-crossings of curvature over scales.

Asada and Brady [9] use the concept of scale space filtering to extract primitives such as corners, smooth joins, crank, bump/dent from the bounding contours of planar shapes. They make scale space analysis of the behavior of these primitives by varying the parameter of the Gaussian kernel in one octave, corresponding to multiplying by $\sqrt{2}$. A tree representation showing the movement of the position of the local positive maxima and negative minima in the first and second derivatives of the Gaussian smoothed curve is constructed by varying the parameter of the kernel. The primitives are detected and located in a process of parsing the tree. Using the location of the primitives at each scale as a set of knot points they have made polygonal approximation, circular spline approximation and cubic spline approximation and B-spline approximation of the planar shapes.

Saint-Marc et al. [147] suggest adaptive smoothing leading to construction of scale space map without using Gaussian kernel. The smoothing is done using a decaying exponential window which is a function of a smoothing parameter k and a measure of signal discontinuity. The underlying concept of the procedure is to keep the window size of the smoothing kernel constant and to apply the kernel iteratively on the signal. They construct two types of scale space map. In one kind they keep the parameter k fixed and use the number of iterations as the scale parameter. Hence, the parameter k determines the magnitude of the edges (corners) to be preserved during the smoothing process. They call this scale space map as the Gaussian scale space map. The other scale space map that they construct has been referred to as adaptive scale space map. Here the number of iterations to be performed is held fixed and the parameter k is varied to construct the scale space map. Without making an attempt to give a multi scale interpretation of the map they detect corners on planar shapes for different values of k. They also show application of the procedure to edge detection from gray level image and to range image segmentation.

Meer et al. [89] suggest a method to detect dominant points by first determining the optimal scale of a Gaussian-like convolution multiple scale representation of the boundary. Then a measure of optimality which is directly proportional to the total curvature of the boundary is defined. The optimal scale is determined such that the difference in the measure of optimality between two successive scales is the smallest. The corners are detected at the optimal scale. The procedure does not take into account various levels of detail of the curve.

Rattarangsi and Chin [119] use the concept of scale space filtering to design a corner detector which takes into account various levels of detail of a curve. They make an analysis of the scale space behavior of different corner models such as Γ models, END models and STAIR models. The extreme curvature points are detected and located by convolving the curve with the Gaussian kernel with varying window size. The scale space map shows the movement of the extreme curvature points over scales. The scale space map is converted into a tree representation with the help of two assumptions namely, identification, and localization. A number of stability criteria are derived on the basis of the scale space behavior of the corner models. The tree is interpreted using these stability criteria and corners are detected and located. The procedure works well on curves which consists of features of multiple sizes and is robust to noise.

Mokhtarian and Suomela introduce [93] curvature scale space and thereby detects corners. The corners are detected at high scale and localized at the finest scale through tracking in the neighborhood of the point where corner is detected. Since the search for the location of the corner is restricted in the neighborhood of the point where the corner is detected so, convolution of the entire curve with the Gaussian kernel is not necessary and hence the performance is high. The curvature scale space map is invariant to rotation, translation and scaling. The method is also found to be robust to noise. The curvature scale space technique has been used in a wide range of applications such as curve matching, feature extraction and object recognition. It has also been selected as a shape descriptor for image indexing and retrieval in the MPEG-7 standard [92].

Zhong and Lioa [201] introduced direct curvature scale space wherein they convolve the curvature of a curve with the Gaussian kernel. They make analysis of the behavior of different corner models as in Rattarangsi and Chin [119] and based on this analysis detect and locate corners. Since curvature is inherently sensitive to noise they suggest a hybrid approach where the curve is initially smoothed with Gaussian kernel with a small value of the Gaussian parameter and then the curvature value computed from this slightly smoothed curve is convolved with Gaussian kernel.

Following the concept of curvature scale space, Zhang et al. [202] convolve a curve with the Gaussian kernel with varying parameter and use local extremes of the product of curvature values from fine scale to coarse scale as a measure of cornerity when the value of the product exceeds a threshold. Since the finest scale is involved in computation of curvature (product), tracking is not necessary and since many scales are involved, false positive/negative detections are unlikely even with a single threshold. The corner detector is found to be robust, simple and effective. Wang et al. [186] use local extrema of multi scale curvature product in the framework of B-spline curvature scale space. Corners are constructed as the local maxima by thresholding the curvature product results across several scales.

Zhong and Ma [200] find that a planar curve shrinks and collapse as it is smoothed with the increasing value of the parameter of the Gaussian kernel. They study the shrinkage and collapse of various corner models as in Rattaragnsi and Chin, to investigate how the local structures of curves shrink and collapse and what the smoothed curve may converge to. They define the sawtooth model to simulate the effect of local fluctuation caused by noise. The convergence property of the model helps to understand how curve noise is suppressed by the Gaussian filter. It is also shown that a closed curve collapse to its center of mass as the scale parameter goes to infinity.

In Chapter 12 we present scale space analysis of digital curves using one of our polygonal approximation schemes and show its application to corner detection. The procedure [126] presented in this chapter holds for curves with uniformly spaced points only. The scale space analysis is made without convolution with a smoothing kernel. The corner detection is done without estimating curvature.

In Chapter 13 we present scale space analysis using convolution with the Gaussian kernel and corner detection is done *via* curvature estimation [128]. In Chapter 14 we present scale space analysis using convolution with a discrete scale space kernel and show its application to corner detection [129].

In contrast to the existing methods of smoothing an entire curve at various levels of detail in Chapter 15 we suggest an adaptive method of corner detection. This procedure does not require construction of the complete scale space map and it is also not necessary to convert the map into a tree representation. The procedure has been applied on a number of digital curves and the experimental results have been compared with existing work. Finally, in Chapter 16 we draw conclusion.

KEYWORDS

- **Adaptive method**
- **Corner detection**
- **Corner models**
- **Optimal scale**
- **Sawtooth model**
- **Scale space analysis**

12 Scale Space Analysis and Corner Detection on Chain Coded Curves

CONTENTS

12.1 INTRODUCTION

As already stated in the opening chapter the fundamental problem in data smoothing is the choice of input parameter. Too large a parameter will smooth out many important features (corners, vertices, zero-crossings) and too small a parameter will produce many redundant features. This is the fundamental problem of scale because features appearing on any curve vary enormously in size and extent and there is seldom any basis of choosing a particular parameter for a particular feature size. This problem may be resolved by automatic parameter tuning. The parameter size can be tuned on the basis of the local properties of the curve using suitable criterion function. Teh and Chin [170] resolve this problem using the chord length and perpendicular distance as the criterion function to determine the region of support. In Chapter 9 and 10 we too have developed two such schemes in which automatic parameter tuning on the basis of the local properties of a curve is suggested. Another approach to solution of the problem of scale is scale space analysis.

In this chapter we propose to scale space analysis of digital curves using the polygonal approximation scheme presented in Chapter 3 and show its application to corner detection [125]. Our scale parameter is discrete in nature. As the scale parameter is varied the location of the vertex points of the polygon is plotted against the ordinal number of the points of the digital curve. The map showing the location of the vertex points over scales is called a scale space map. This scale space map is used to detect and locate corners on digital curves.

12.2 SCALE SPACE MAP

The polygonal approximation discussed to involves a counter *c* which is responsible for imparting different degrees of smoothing to a curve. As the counter is varied different sets of vertex points are obtained. For small values of the counter the approximation results in a large number of vertices and for large values of the counter the approximation results in a small number of vertices. For small values of *c* the vertices are located at fine levels of detail and for large values of *c* the vertices are located at coarse level of detail. As *c* is varied the curve is analyzed at different levels of detail. So *c* is called a scale parameter [13]. This scale parameter is discrete in nature.

As the value of *c* is increased from zero in a step size of one, different sets of vertices result and different polygonal approximations are obtained. For $c = 0$ all points of a digital curve are vertex points. For $c = 1$ for each point where the curve changes its direction is a vertex point. The number of vertex points for $c = 1$ is less than that for $c = 0$. As *c* is increased further the number of vertex points decreases resulting in higher smoothing though two or more successive values of *c* may produce the same number of vertices.

The striking feature is that as *c* is varied from small values to large values no new vertex point is introduced but as *c* decreases from large value to small value new vertex point may be introduced, but existing ones never vanish. As a digital curve is smoothed out with different values of c, different polygonal approximations result, each of which reflects a specific level of detail of the curve. This information is integrated in the form of scale space map.

The *x*-axis (horizontal) indicates the ordinal number *i* of the point p_i of a digital curve and the *y*-axis (vertical) indicates the values of the counter *c*. As the counter is varied from small values to large values the position of the vertex points are plotted on the *xy* half plane. The map showing the position of the vertex points at different values of the counter is called scale space map. The map consists of a series of vertical lines at each vertex point. Some of these lines grow indefinitely and some other terminates as the counter increases. The vertical lines that grow indefinitely are indicative of those vertices which are detected at all levels of detail, fine as well as course. Whereas, those lines that terminate are indicative of those vertices that are detected at fine level of detail but disappear as the degree of smoothing increases. The location of the vertices at any scale is obtained by the orthogonal projection of the vertical lines on the *x*-axis.

We have constructed a scale space map of the digital screwdriver. The digital curve and its polygonal approximations for different values of *c* are shown in Figure 1 and the scale space map is shown in Figure 2. The counter has been varied from zero to 32 ($=2^5$). As the counter increases the number of vertices decrease. We find that for $c \geq 15$ the number of vertices remains unaltered indicating the existence of a stable scale. The scale space map is obtained by combining the different sets of vertices obtained through the variation of *c*. The curve is described in the clockwise direction starting from the point marked with an arrow on the digital curve of Figure 1.

FIGURE 1 *(Continued)*

FIGURE 1 The screwdriver and its polygonal approximation for varying values of c, (a) c = 8, (b) c ≥ 15.

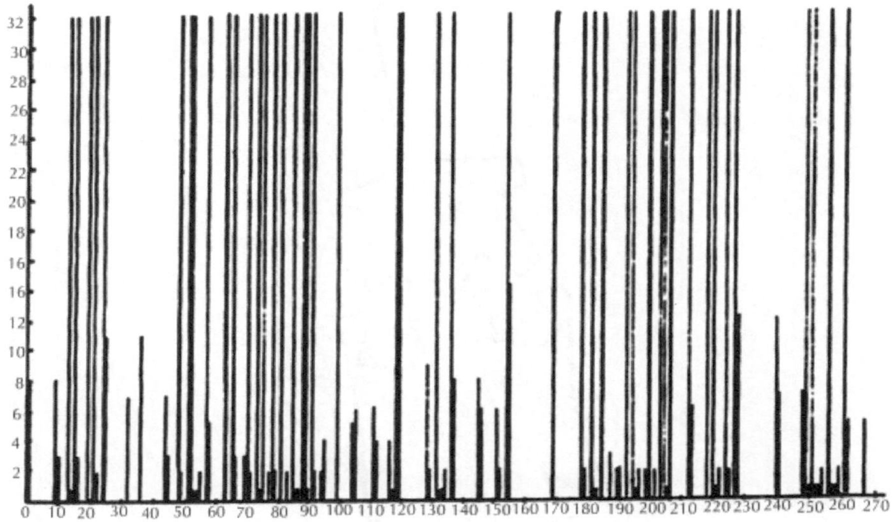

FIGURE 2 Scale space map of the screwdriver.

12.3 SCALE SPACE BEHAVIOR OF CORNER MODELS AND CORNER DETECTION

In this chapter we are studying the scale space behavior of different corner models, namely, Γ model, END model and STAIR model as presented in [118]. We consider these models on digital domain where each model is represented by a sequence of integer coordinate points.

Figure 3 shows Γ models with different included angle and their respective scale space map. As we are considering eight-connected curves, the included angle can be a multiple of π/4. As seen from these figures, if the included angle be 3π/4 then the scale space map of the model consists of a single line pattern which vanishes ultimately as the counter increases.

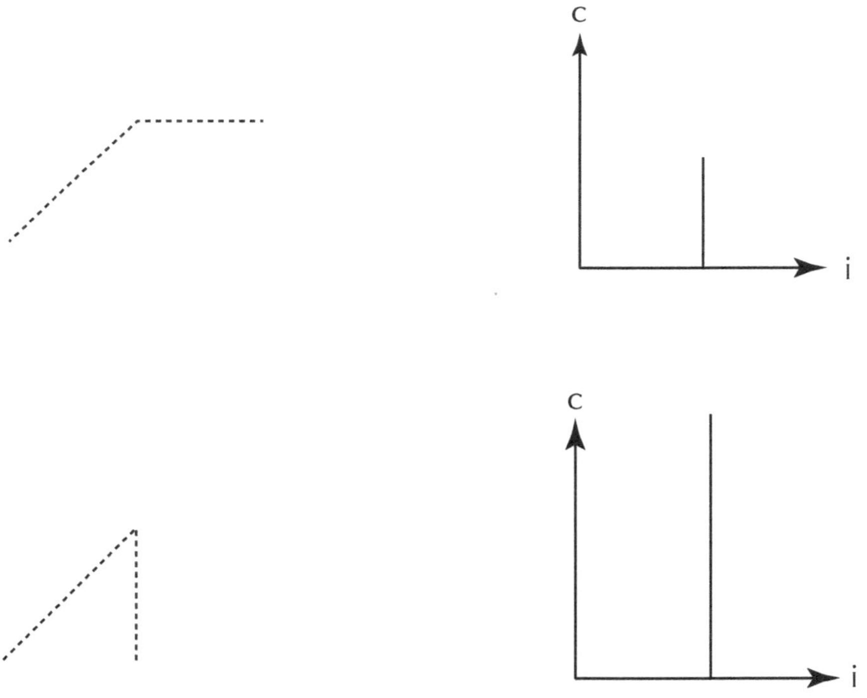

FIGURE 3 Γ models and their scale space map. The left figure is a model and the right figure is its scale space map.

If the included angle be either π/4 or π/2 then the scale space map consists of a persistent line pattern which never disappears no matter how large the counter is.

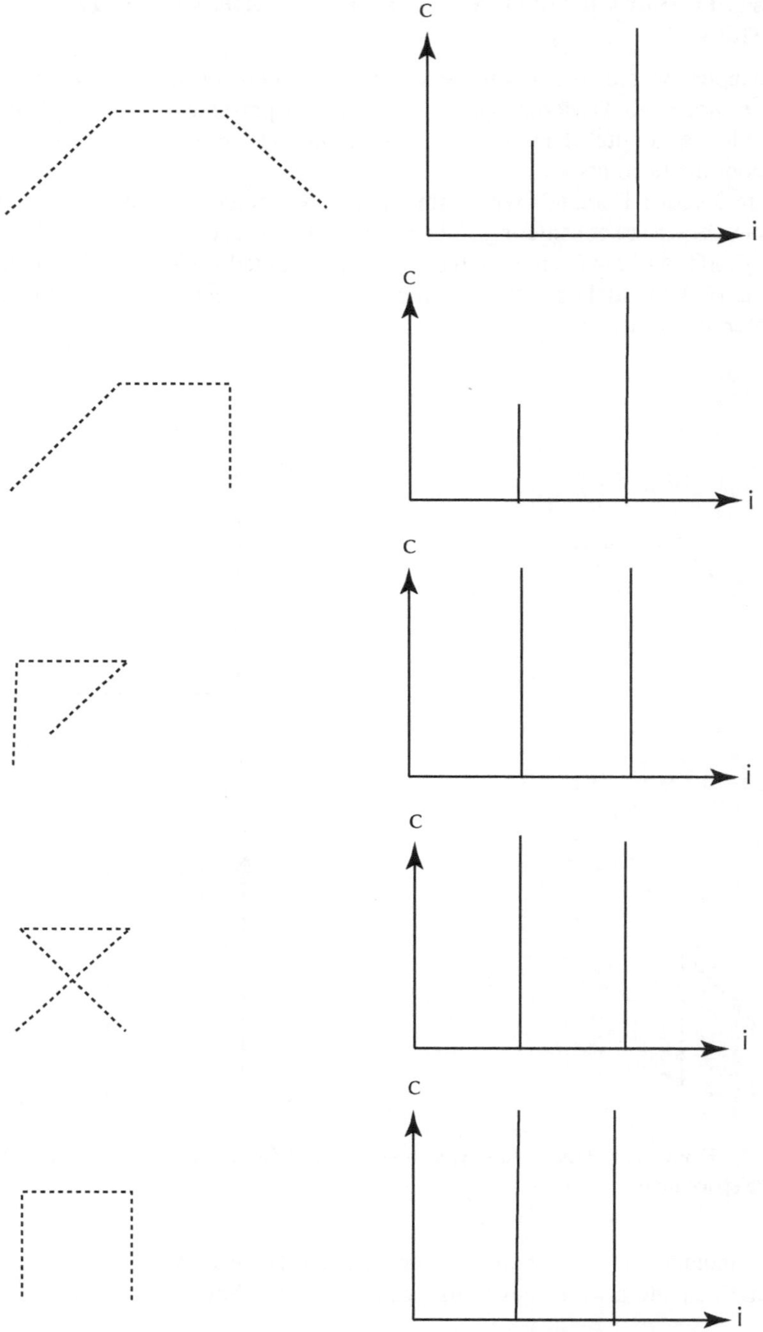

FIGURE 4 END models and their scale space map. The left figure is the model and the right figure is its scale space map.

Figure 4 shows END models with different included angles and their respective scale space map. Here too, the included angle can be a multiple of $\pi/4$. As seen from these figures, if the included angle be $3\pi/4$ then the scale space map of the model consists of two line patterns one of which survives and the other disappears as the counter increases. If one angle be $3\pi/4$ and the other be $\pi/4$ or $\pi/2$ then also the scale space map exhibits the same behavior except that in the former case (when both the included angles are $3\pi/4$) the line pattern which is further from the starting point persists whereas in the later case the line pattern that corresponds to included angle not equal to $3\pi/4$ only persists. If none of the included angles is $3\pi/4$ then the scale space map consists of two persistent line patterns none of which disappears no matter how large the counter is, showing that the END model is a combination of two Γ models with included angle other than $3\pi/4$.

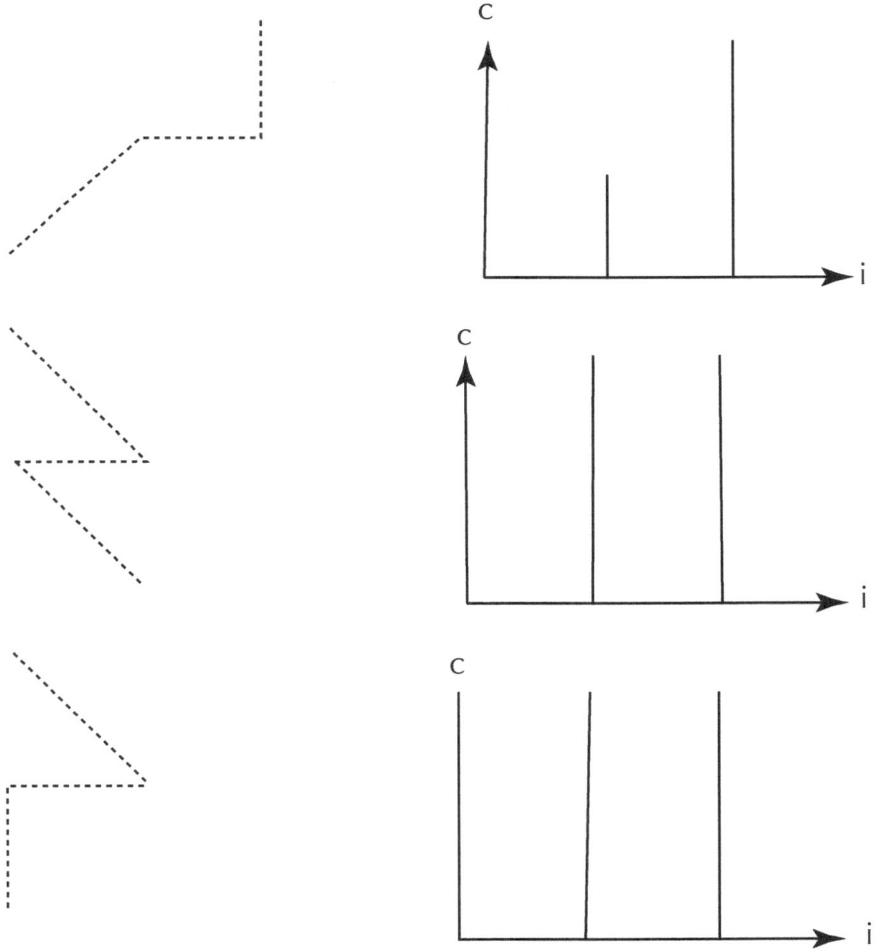

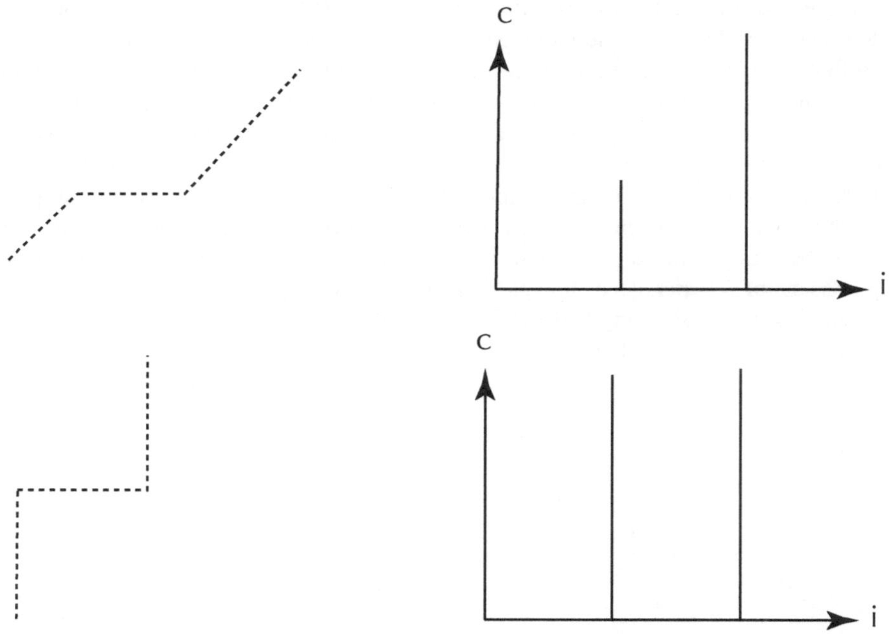

FIGURE 5 STAIR models and their scale space map. The left figure is a modle and the right figure is its scale space map.

Figure 5 shows STAIR models with different included angles and their respective scale space map. Here too, the included angle can be a multiple of $\pi/4$. As seen from these figures, if the included angle be $3\pi/4$ then the scale space map of the model consists of two line patterns both of which vanish as the counter increases. If the leading arm, trailing arm and the base consist of equal number of points then both the line patterns disappear together, otherwise one vanishes earlier than the other depending on the length of the arms and the base. If one angle be $3\pi/4$ and the other be $\pi/4$ or $\pi/2$ then the scale space map consists of two line patterns and the line pattern corresponding to the angle $3\pi/4$ disappear and the other persists. If none of the included angles be $3\pi/4$ then the scale space map consists of two persistent line patterns none of which disappears no matter how large the counter is, showing that the STAIR model is a combination of two Γ models with included angle other than $3\pi/4$.

The central problem is how far the counter should be varied. One can observe that the maximum value of the counter cannot exceed the number of points of the curve. But it is not necessary to make the counter vary up to this limit. To find the maximum value of the counter we look for a series of adjoining STAIR models with included angles $3\pi/4$. The maximum value of the counter can be taken to be equal to the total number of points in this longest series. This observation is in direct consequence of the scale space behavior of different corner models.

We derive the following stability criteria for corner detection from the scale space map.

(1) The lines that persist up to the maximum value of the counter corresponds the corner points.
(2) The lines that do not survive all scales but in the immediate neighborhood of the persistent lines also correspond to corner points provided these non-surviving lines are separated from the persistent lines by more than unit length in the scale space map.
(3) The pair of lines that do not survive all scales correspond to corner points provided that they are separated by more than unit length in the scale space map.

These stability criteria are the direct consequence of the scale space behavior of different corner models.

In order to detect corners a scale space map of the digital curve is constructed and the corners are detected and located by analyzing the scale space map with the help of stability criteria. The corners that are detected by this process are not necessarily true corners; some of them may be redundant due to quantization noise and boundary noise. In order to remove these redundant corners we introduce a two-stage cleaning process. At the first stage if two points p_i and p_{i+1} appear as corner points then the point p_i is retained and p_{i+1} is discarded. At the second stage each corner point we find out the perpendicular distance of the corner point from the line joining the two adjacent corner points. If this distance exceeds unity then the corner point is retained otherwise it is discarded. This threshold unity is selected based on the fact that a slanted straight line is quantized into a set of either horizontal or vertical line segments separated by one pixel steps. In addition, we assume that the boundary noise is no more than one pixel and if the noise level is known a priori then this threshold can be adjusted accordingly [118].

(a)

FIGURE 6 *(Continued)*

FIGURE 6 The chromosome shaped curve. The corners are indicated by bold solid circles. (a) Present method, (b) Rattarangsi-Chin method.

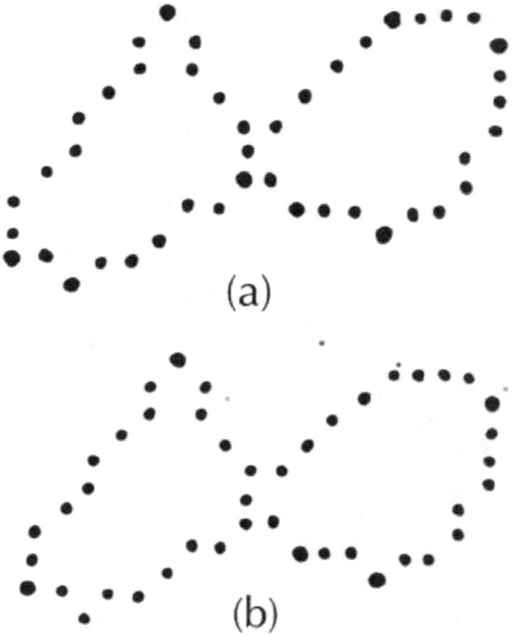

FIGURE 7 The figure-8 curve, the corners are indicated by bold solid circles. (a) Present method, (b) Rattarangsi-Chin method.

(a)

(b)

FIGURE 8 The leaf-shaped curve, the corners are indicated by bold solid circles. (a) Present method, (b) Rattarangsi-Chin method.

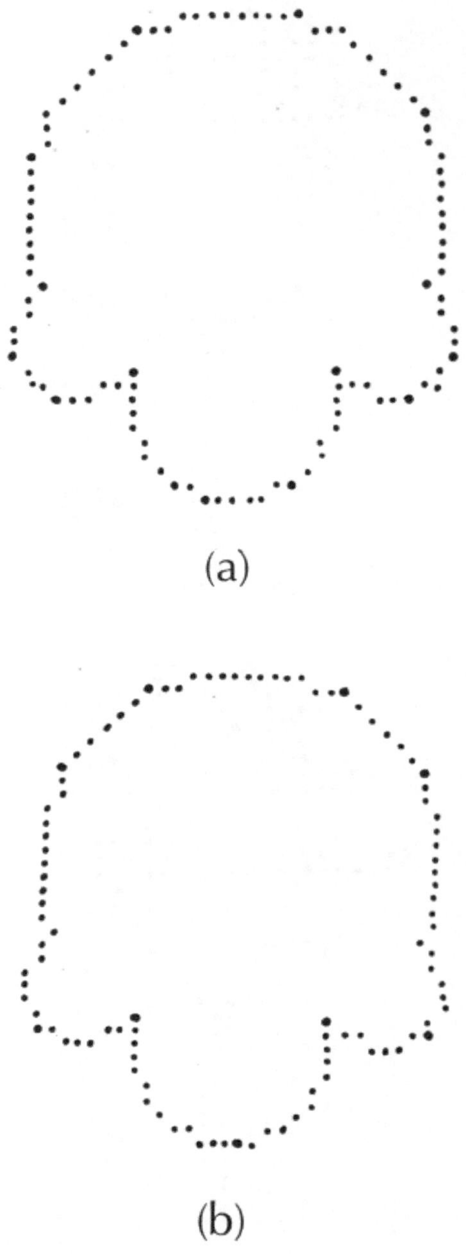

FIGURE 9 The curve with four semi circles. The corners are indicated by bold solid circles. (a) Present method, (b) Rattarangsi-Chin method.

12.4 EXPERIMENTAL RESULTS

The corner detector developed is applied on four digital curves as shown in Figure 6 through Figure 9. The corners are indicated by bold circles. These figures also show the results obtained by the Rattarangsi-Chin algorithm [119]. We find that our corner detector detects more corner points that the Rattarangsi-Chin algorithm.

KEYWORDS

- **Chain coded curves**
- **Corner detection**
- **Digital curve**
- **Polygonal approximations**
- **Rattarangsi-Chin method**
- **Scale space analysis**

13 Scale Space Analysis and Corner Detection Using Iterative Gaussian Smoothing with Constant Window Size

CONTENTS

13.1 INTRODUCTION

The scale space analysis and corner detection scheme presented in the last chapter holds for curves with uniformly spaced points only. The scale space analysis does not involve convolution of the curve with Gaussian kernel. Corner detection is done without estimating curvature. In this chapter we propose another scale space analysis technique followed by corner detection. The procedure involves convolution of a digital curve with a smoothing kernel. Corner detection is done *via* curvature estimation. The procedure holds for curves with uniformly as well as non-uniformly spaced points. Rattarangsi and Chin [119] make scale space analysis using Gaussian kernel with varying window size. So the space requirement for the Gaussian filter coefficients is of the order of the square of data size. In this chapter we present an alternative approach to multiscale corner detection using iterative Gaussian smoothing with constant window size. As the window size is held constant, the space requirement for the Gaussian filter coefficients is finite and independent of data size. In the following chapter we present the iterative Gaussian smoothing process, show its convergence and the maximum allowable number of iterations that can be performed on a closed digital curve without wrap around effects. A scale space map showing the location of the maxima of absolute curvature is proposed. The map is shown to enjoy scale space property. An analysis of the scale space behavior of corner models is presented. Based on this analysis a tree organization is designed and corners are detected and located in a process of interpreting the tree. The space requirements and computational load is discussed and compared with [119]. Experimental results are presented to show the performance of the corner detector.

13.2 GAUSSIAN SMOOTHING AND CURVATURE MEASUREMENT

For a continuous and smooth curve, the curvature at a point is defined as the rate change of tangential angle (ψ) with respect to the arc length (s). So the curvature at an arc length of s is given by:

$$\kappa = d\psi/ds \tag{13.1}$$

In the Cartesian coordinate system, if the equation of the curve is described by y = f(x), the curvature at a point (x, y) is given by:

$$\kappa(x) = \frac{\dfrac{d^2y}{dx^2}}{\left(1+\left(\dfrac{dy}{dx}\right)^2\right)^{3/2}} \tag{13.2}$$

If the equation of the curve is expressed parametrically with the arc length s as the parameter so that the curve is described by x = $x(s)$ and y = $y(s)$ then the curvature at a point s is given by:

$$\kappa(s) = \frac{\dfrac{d^2y}{ds^2}\dfrac{dx}{ds} - \dfrac{d^2x}{ds^2}\dfrac{dy}{ds}}{\left(\left(\dfrac{dx}{ds}\right)^2 + \left(\dfrac{dy}{ds}\right)^2\right)^{3/2}} \tag{13.3}$$

but

$$\frac{dx}{ds} = \cos\psi \text{ and } \frac{dy}{ds} = \sin\psi$$

so

$$\left(\frac{dx}{ds}\right)^2 + \left(\frac{dy}{ds}\right)^2 = 1$$

and

$$\kappa(s) = \frac{d^2y}{ds^2}\frac{dx}{ds} - \frac{d^2x}{ds^2}\frac{dy}{ds} \tag{13.4}$$

The curve ($x(s)$, $y(s)$) that arise in computer vision and other allied problems may be continuous but not smooth (a digital curve is neither continuous nor smooth). So smoothing is, in general necessary before we detect the significant local events of a curve. Smoothing is usually done with the Gaussian kernel as it has some attractive properties [13, 196]. To smooth a curve ($x(s)$, $y(s)$) it is convolved with the Gaussian kernel:

$$g(s,\sigma) = \frac{1}{\sqrt{2\pi\sigma^2}} e^{-s^2/2\sigma^2}, -\infty < s < \infty \tag{13.5}$$

We assume that the curve in question is closed with arc length S so that $0 \le s \le S$ and the parameter σ is sufficiently small so that three times σ does not exceed $S/2$ so as to avoid aliasing effects. The convolution of ($x(s)$, $y(s)$) with the Gaussian kernel (13.5) is defined by:

$$X(s, \sigma) = x(s) * g(s, \sigma)$$

$$= \int_{u=s-S/2}^{u=s+S/2} x(u) g(s-u,\sigma) du \tag{13.6}$$

$$Y(s, \sigma) = y(s) * g(s, \sigma)$$

$$= \int_{u=s-S/2}^{u=s+S/2} y(u) g(s-u,\sigma) du \tag{13.7}$$

As $g(s, \sigma)$ is maximally often differentiable and the convolution and differentiation are commutative so the derivatives of $X(s, \sigma)$ and $Y(s, \sigma)$ with respect to s do exist and in particular, their first and second derivatives are given by:

$$\frac{dX}{ds} = x(s) * \frac{dg(s,\sigma)}{ds},$$

$$\frac{dY}{ds} = y(s) * \frac{dg(s,\sigma)}{ds}, \tag{13.8}$$

$$\frac{d^2X}{ds^2} = x(s) * \frac{d^2g(s,\sigma)}{ds}$$

$$\frac{d^2Y}{ds^2} = y(s) * \frac{d^2g(s,\sigma)}{ds} \tag{13.9}$$

And the curvature of the Gaussian smoothed curve is given by:

$$k(s,\sigma) = \frac{d^2Y}{ds^2}\frac{dX}{ds} - \frac{d^2X}{ds^2}\frac{dY}{ds} \tag{13.10}$$

Rattarangsi and Chin [119] find the first and second order derivatives of $\kappa(s, \sigma)$ with respect to s to locate the extreme curvature points for varying σ. Since the Gaussian kernel has negligible contribution beyond the 3σ limits hence the maximum value up to which the parameter σ is to be varied is determined by $3\sigma_{max} = S/2$ which leads to a $\sigma_{max} = S/6$ so as to avoid aliasing effects [58].

13.3 ITERATIVE GAUSSIAN SMOOTHING

To convolve a digital curve with the Gaussian kernel, Rattarangsi and Chin [119] use the digital Gaussian filter coefficients of [26], $c_{-1} = 0.2236$, $c_0 = 0.5477$ and $c_1 = 0.2236$ for window size $w = 3$. These filter coefficients has been mentioned in [88] and [27] as the best approximation of the Gaussian distribution for $w = 3$. The digital Gaussian filter coefficients for window size higher than $w = 3$ are obtained by repeated convolutions of these coefficients with themselves. The maximum window size should not exceed the length of the curve so as to avoid the wrap around effects. Each window size corresponds to a specific value of the parameter σ of the Gaussian distribution. The greater the window size is, the higher is the value of the parameter.

We make an alternative approach to the problem. Instead of smoothing the curve with varying window size, we repeatedly convolve the curve keeping the window size constant at $w = 3$. Intuitively speaking, this approach finds repeatedly the weighted average of the coordinates, the weights being the digital Gaussian filter coefficients for window size $w = 3$. This approach has an advantage over that presented in [119]. In the later approach as the varying window size is used, the space requirement by the

Gaussian filter coefficients is of the order of the square of the data size. But in the present approach the space requirements by the filter coefficients is finite, small and does not depend on the data size. Moreover, it is shown later that the computational load of the smoothing process in the Rattarangsi and Chin [119] algorithm is $O(n^2)$ whereas in the present approach it is $O(n)$.

The iterative convolution is performed with the digital Gaussian filter $c_{-1} = 0.2236$, $co = 0.5477$, and $c_1 = 0.2236$ using the iterative process:

$$X_i(t) = \sum X_{i+m}(t-1)c_m$$

$$Y_i(t) = \sum Y_{i+m}(t-1)c_m \tag{13.11}$$

$(X_i(t), Y_i(t))$ denote the Gaussian smoothed coordinates of the ith point (x_i, y_i) at the tth iteration, t = 1, 2, 3, and $X_i(0) = x_i$, $Y_i(0) = y_i$, $i = 1, 2, 3, n$.

Although Saint-Marc et al. [147] use repeated weighted averaging process but they do not use the Gaussian filter coefficients. The kernel they use depends on input data. The kernel involves gradient of the signal. As gradient is orientation dependent so the smoothing process too, will depend on the orientation of the signal. A particular point of a signal will be subjected to different degree of smoothing for different orientation of the signal. But it is essential that the extent of smoothing should not depend on the orientation of the signal. On the other hand the Gaussian kernel is shift invariant (does not depend on data). The degree of smoothing subjected to a curve/signal depends only on the Gaussian filter coefficients.

13.4 CONVERGENCE

The repeated convolution of a digital curve with the digital Gaussian filter coefficients can be written in the matrix form as:

$$X(t) = AX(t-1) \text{ and } Y(t) = AY(t-1) \tag{13.12}$$

where, A is the convolution matrix of order $(n+2)\times(n+2)$. The matrix is tridiagonal, symmetric and diagonally dominant and its elements a_{ij} are given by

$a_{ij} = c_0$ for $i = j$

$= c_1$ for $j = i + 1$

$= c_{-1}$ for $j = i - 1$

and

$\mathbf{X}(t) = [X_n(t), X_1(t), X_2(t), X_n(t), X_1(t)]$

$\mathbf{Y}(t) = [Y_n(t), Y_1(t), Y_2(t), Y_n(t), Y_1(t)]$

are the column matrices of order n + 2. The iterative process will converge if for some norm, **A** satisfies $||\mathbf{A}|| \leq 1$. Since the infinity norm of a matrix is simply the maximum sum of the moduli of the elements of the matrix, hence the infinity norm of the convolution matrix **A** is $||\mathbf{A}||_\infty = c_{-1} + c_0 + c_1 < 1$. Again, the largest of the moduli of the eigen values of a square matrix cannot exceed its infinity norm that is $\rho(\mathbf{A}) \leq ||\mathbf{A}||_\infty$. So the *2- norm* of the matrix **A** is also bounded by unity. So the convolution process converges as $||\mathbf{A}||_\infty < 1$ [147].

13.5 MAXIMUM NUMBER OF ITERATION

For a closed digital curve with n points the number of iterations is determined by the data size n in order to avoid the aliasing effects. If the curve has an odd number of points (n odd) then the maximum number of iterations that can be performed is $(n - 1)/2$ and if n be even then the maximum number of iterations that can be performed is $n/2 - 1$. To show that we decompose the Gaussian smoothed coordinates $(X_i(t), Y_i(t))$ in terms of(x_i, y_i). For this analysis we propose to write $c_{-1} = c_{-1}(1)$, $c_0 = c^1_{0'}(1)$ and $c^1_1 = c_1.(1)$.
So

$$X_i(t) = \sum_{m=-1}^{1} X_{i+m}(t-1)c_m(1)$$

$$Y_i(t) = \sum_{m=-1}^{1} Y_{i+m}(t-1)c_m(1)$$

But

$$X_i(t-1) = \sum_{m=-1}^{1} X_{i+m}(t-2)c_m(1)$$

so

$$X_i(t) = \sum_{m=-2}^{2} X_{i+m}(t-2)c_m(2)$$

where
$c_{-2}(2) = \{c_{-1}(1)\}^2$, $c_{-1}(2) = 2c_0(1)c_{-1}(1)$, $c_0(2) = \{c_{-1}(1)\}^2 + \{c_0(1)\}^2 + \{c_1(1)\}^2$, $c_1(2) = 2c_0(1)c_1(1)$, $c_2(2) = \{c_1(1)\}^2$.

Decomposing $X_i(t-2)$ in terms of $X_i(t-3)$ we get

$$X_i(t) = \sum_{m=-3}^{3} X_{i+m}(t-3)c_m(3)$$

$c_m(3)$ is a homogeneous function of c_{-1}, c_0 and c_1 of degree 3, for each value of m = –3, –2, –1, 0, 1, 2, 3. So using mathematical induction, at the lth decomposition we get,

$$X_i(t) = \sum_{m=-l}^{l} X_{i+m}(t-l)c_m(l)$$

where $c_m(l)$ is a homogeneous function of c_{-1}, c_0 and c_1 of degree l. So $X_i(t)$ in terms of x_i is

$$X_i(t) = \sum_{m=-t}^{t} x_{i+m} c_m(t) \qquad t = 1, 2$$

where $c_m(t)$ is a homogeneous function of c_{-1}, c_0, c_1 of degree t. Similar expression holds for $Y_i(t)$

This result shows that at any iteration t the iterative convolution process actually takes into account the effects of the points i – t, i – t + 1, i – 1, i, i + 1, i + t – 1, i + t. So in order that a particular point of a curve is not evaluated twice during the convolution process that is to avoid the aliasing effects, none of the member of the set {i –t, i – t + 1 i – 1 }should be a member of the set {i + 1, i + 2, i + t – 1,i + t} and vice-versa. This condition is satisfied if and only if

$$t_{max} = n/2 - 1,\ n \text{ even}$$

$$= (n-1)/2,\ n \text{ odd.} \qquad (13.13)$$

So the maximum number of iterations that can be performed on a closed digital curve with n points is

$$t_{max} = n/2 - 1,\ n \text{ even}$$

$$= (n-1)/2,\ \text{n odd.} \qquad (13.14)$$

Though Saint-Marc et al. [147] have shown the convergence (which is too slow) of their iterative process but they could not suggest the number of iterations that should be performed on a closed digital curve. For open digital curves the number of iterations to be performed should be so chosen that the end effects do not come into play. This can be done by choosing t_{max} properly so that it does not exceed the number of points on the right/left of any point being convolved. One can as well take the open curve to be sufficiently large, so that one can perform a large number of iterations.

13.6 CURVATURE ESTIMATION AND SCALE SPACE MAP

To compute the curvature at a point of digital curve Rattarangsi and Chin [119] define the first and second order finite differences of (x_i, y_i) as:

$$\Delta x_i = \frac{x_{i+1} - x_{i-1}}{\sqrt{(x_{i+1} - x_{i-1})^2 + (y_{i+1} - y_{i-1})^2}} \qquad (13.15)$$

$$\Delta y_i = \frac{y_{i+1} - y_{i-1}}{\sqrt{(x_{i+1} - x_{i-1})^2 + (y_{i+1} - y_{i-1})^2}} \tag{13.16}$$

$$\Delta^2 x_i = \frac{\dfrac{x_{i+1} - x_i}{\sqrt{(x_{i+1} - x_i)^2 + (y_{i+1} - y_i)^2}} - \dfrac{x_i - x_{i-1}}{\sqrt{(x_i - x_{i-1})^2 + (y_i - y_{i-1})^2}}}{\dfrac{1}{2}\sqrt{(x_{i+1} - x_{i-1})^2 + (y_{i+1} - y_{i-1})^2}} \tag{13.17}$$

$$\Delta^2 y_i = \frac{\dfrac{y_{i+1} - y_i}{\sqrt{(x_{i+1} - x_i)^2 + (y_{i+1} - y_i)^2}} - \dfrac{y_i - y_{i-1}}{\sqrt{(x_i - x_{i-1})^2 + (y_i - y_{i-1})^2}}}{\dfrac{1}{2}\sqrt{(x_{i+1} - x_{i-1})^2 + (y_{i+1} - y_{i-1})^2}} \tag{13.18}$$

And the curvature measure is:

$$\kappa = \Delta x \qquad \Delta^2 y - \Delta y \Delta^2 x \tag{13.19}$$

We use these expressions for Δx, Δy, $\Delta^2 x$, $\Delta^2 y$, and κ with (x, y) being replaced by the corresponding Gaussian smoothed coordinates (X, Y). So the curvature measure of the Gaussian smoothed curve at the tth iteration and at the ith point is given by:

$$\kappa_i(t) = \Delta X_i(t)\Delta^2 Y_i(t) - \Delta Y_i(t)\Delta^2 X_i(t) \tag{13.20}$$

At each iteration t = 1, 2, 3 the local maxima of the absolute curvature $|\kappa_i(t)|$ are detected. For a fixed t, the value i for which $|\kappa_i(t)|$ exceeds $|\kappa_{i-1}(t)|$ and $|\kappa_{i+1}(t)|$ gives the location of a local maximum of the absolute curvature. As the iteration proceeds different sets of absolute curvature maxima for different values of t are obtained.

This information is integrated in the form of a scale space map of the digital curve. Along the *x*-axis (horizontal) the ordinal number *i* of the points is shown, along the y-axis (vertical) the number of iterations t is shown. As the iteration proceeds, location of the maxima of absolute curvature are plotted on the *xy* half plane. The dot diagram showing the location of the absolute curvature maxima at different iterations is a scale space map which we propose to call Iterative Gaussian scale space map. The map consists of a series of dot patterns some of which grow reaching the maximum iteration scale and some other terminate as the iteration scale increases. A pair of dot patterns may grow and merge to become a single dot pattern which may reach the maximum iteration scale. The dot patterns that survive the maximum number of iterations are indicative of those local maxima which are detected at all levels of detail, fine as well as coarse. The dot patterns that appear but terminate after a number of iterations are indicative of those local maxima that are detected at fine levels of detail but disappear as the degree of smoothing increases. The dot patterns that appear at all scales are the more compelling determiners of the global shape of a curve than those that appear at fine scales, but disappear as the iteration scale increases. As the iterations proceed the dot patterns either remain stationary or interact with each other. Two neighboring dot patterns may either attract or repel.

A number of digital curves and their scale space map are shown in Figure 1 through Figure 4. The starting point is indicated with an arrow on each digital curve and the curve as described in the clockwise direction. The scale space map of each curve consists of an aggregate of dot patterns some of which persist- surviving the maximum iteration scale and some other disappear as the iteration proceeds. Two dot patterns may interact either attracting or repelling each other. Some of the dot patterns merge to become a single dot pattern which may persist and reach the maximum number of iteration.

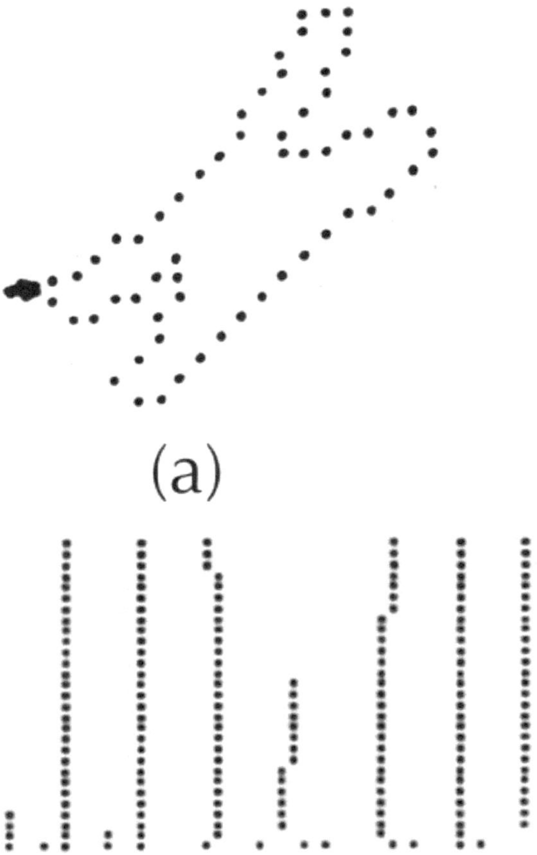

(a)

(b)

FIGURE 1 *(Continued)*

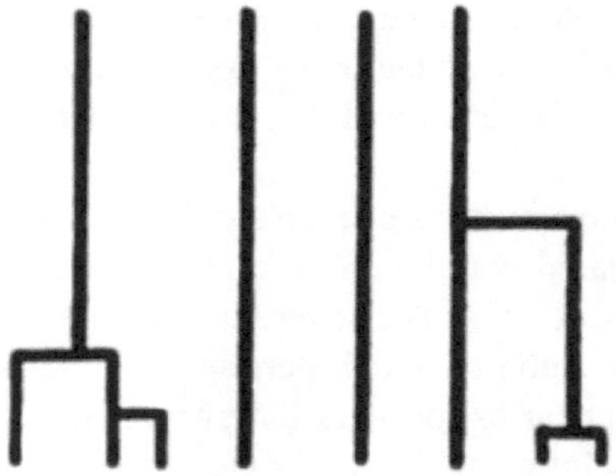

(c)

FIGURE 1 (a) The chromosome shaped curve, (b) its scale space map and (c) tree organization.

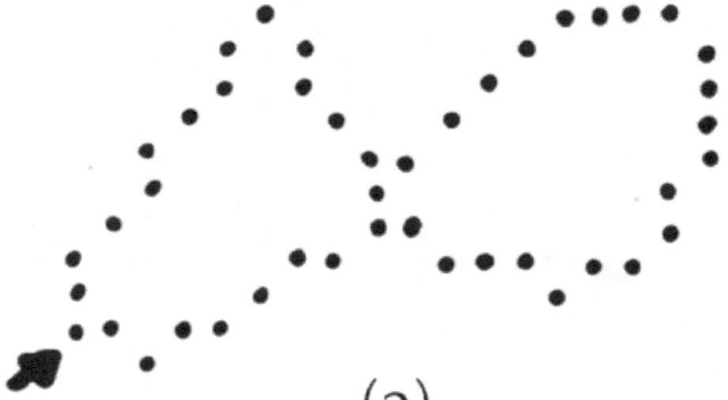

(a)

(b)

FIGURE 2 *(Continued)*

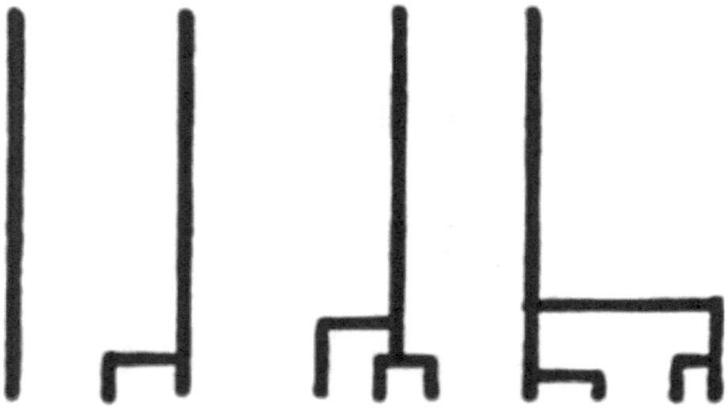

(c)

FIGURE 2 (a) The figure-8 curve, (b) its scale space map, and (c) tree organization.

(a)

FIGURE 3 *(Continued)*

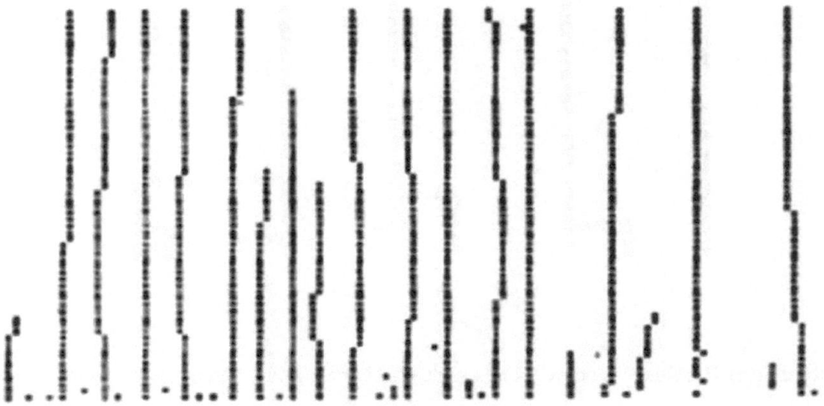

(b)

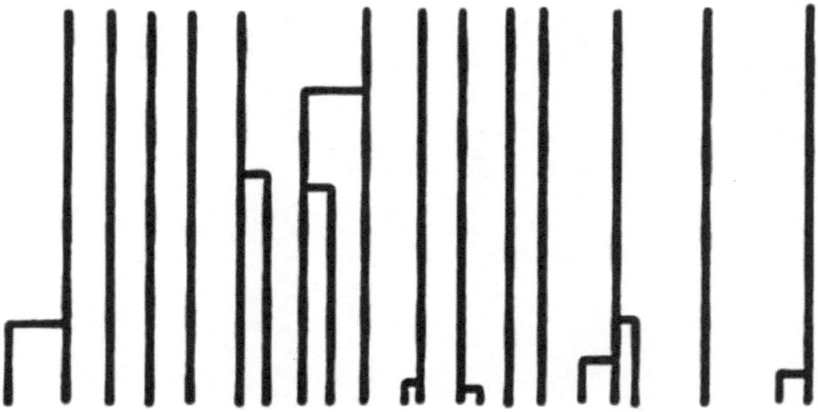

(c)

FIGURE 3 (a) The leaf-shaped curve, (b) its scale space map, and (c) tree organization.

(a)

(b)

FIGURE 4 *(Continued)*

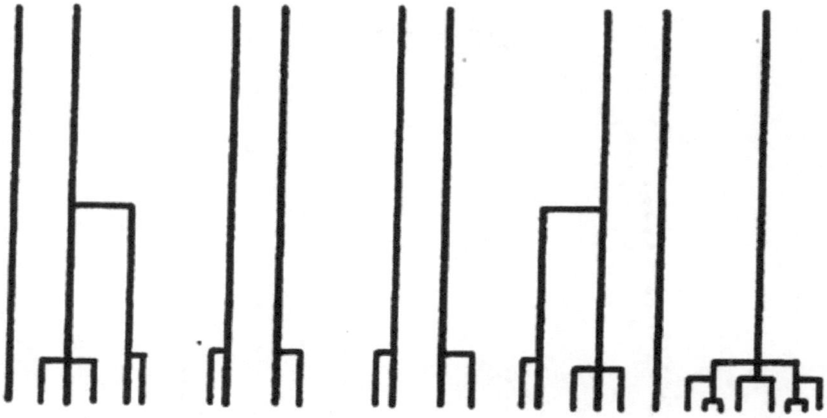

(c)

FIGURE 4 (a) A curve with four semi circles, (b) its scale space map, and (c) tree organization.

13.7 SCALE SPACE PROPERTY

The digital Gaussian filter coefficients c_{-1}, c_0, and c_1 satisfy the scale space property which demands that the number of local maxima of absolute curvature should not increase as the iteration scale increases. This conjecture follows from a proposition presented in [76] which states that a three-kernel with positive elements c_{-1}, c_0 and c_1 is a scale space kernel if and only if $c_0 > 2\sqrt{(c_{-1}c_1)}$. We find that the coefficients c_{-1}, c_0 and c_1 are all positive and the relation $c_0 > 2\sqrt{(c_{-1}c_1)}$ is satisfied. As the digital curve is iterative convolved with the same kernel at each iteration the scale space property is preserved.

13.8 SCALE SPACE BEHAVIOR OF CORNER MODELS

In this section we propose to make an analysis of the scale space behavior of isolated corner models, namely Γ model, END model and STAIR model as presented in the last chapter. The models being open curves, the curvature measurements on them are affected by the ends of the curve. In order to avoid the end effects the arm length of the models (given by the number of points on the leading and the training side of the model) is taken to be sufficiently large (each side of the model should consist of at least 100 points) so that 100 iterations can be performed on each model. The search for extrema is restricted in the neighborhood of the angular points of the model. In order to avoid the spurious numerical absolute curvature maxima, a very small input threshold 0.0001 is used. An absolute curvature maxima is regarded as a true maximum, if the absolute curvature exceeds 0.0001. The scale space map of a model is constructed neglecting the spurious numerical maxima of absolute curvature and restricting the search for true extreme near the angular point(s) of the models. The map shows the location of the true absolute curvature maxima over the iteration scales.

Figure 5 shows Γ models with different included angles and their scale space map. As seen from these figures, the scale space map of a Γ model with included angle $\pi/2$ consists of a single persistent and stationary dot pattern located at the angular point of the model. The scale space map of a Γ model with included angle of either $\pi/4$ or $3\pi/4$, too consists of a single persistent dot pattern located at the angular point of the model. The dot pattern is not stationary. It exhibits movement as the iteration scale increases.

Figure 6 shows END models with different included angles and their scale space map. The scale space map of an END model with both the included angles equal to *π/2* (Figure.6 (a)

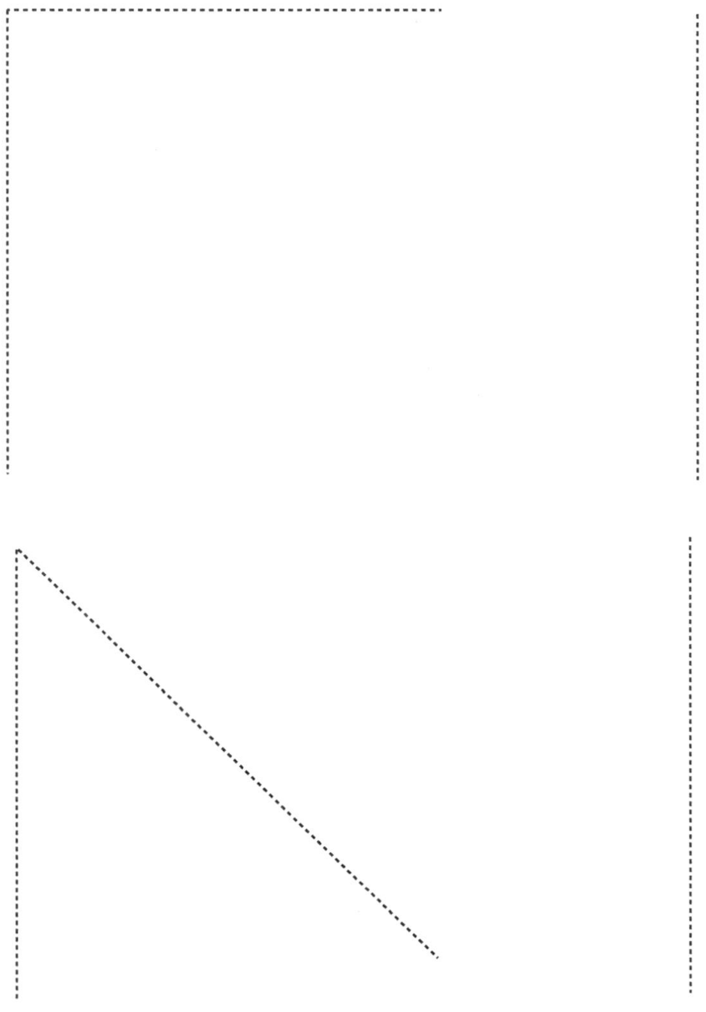

FIGURE 5 *(Continued)*

FIGURE 5 The Γ models and their scale space map. The left figure is the model the right figure is its scale space map.

Initially, consists of two dot patterns each located at the angular points of the model. But as the iteration proceeds the two dot patterns merge to become a single persistent dot pattern. The persistent is stationary if the number of points forming the width of the END is odd otherwise it exhibits an oscillatory movement (Figure 6(b) Figure 6(c)) shows another END model with both the included angles equal to $\pi/2$ but the width of this model is half the width of the model in

(a)

FIGURE 6 *(Continued)*

(b)

(c)

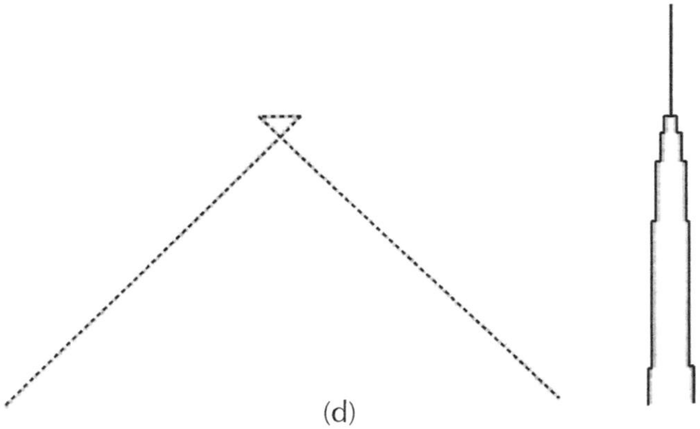

(d)

FIGURE 6 *(Continued)*

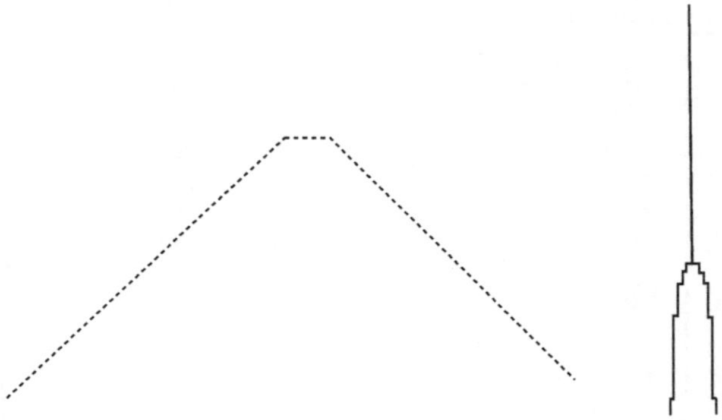

(e)

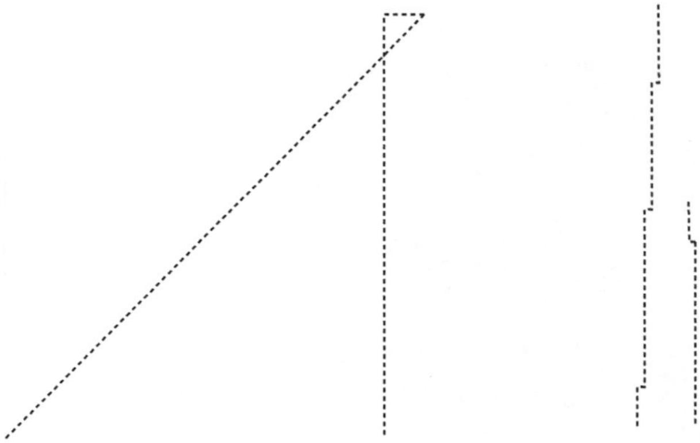

(f)

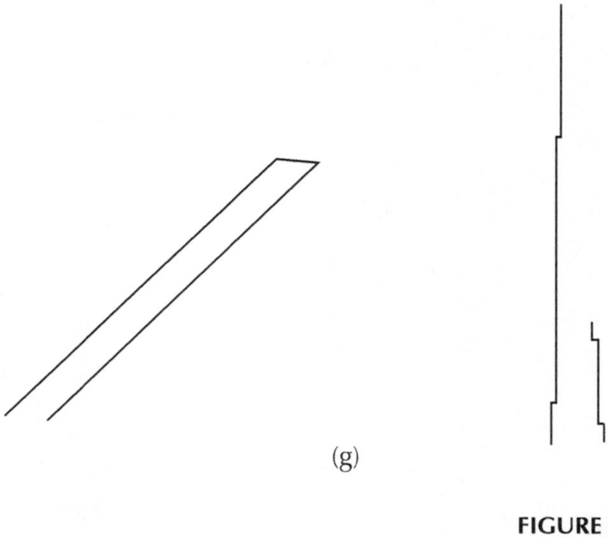

(g)

FIGURE 6 *(Continued)*

(h)

Figure 6 END models and their scale space map. The left figure is the model and the right figure is its scale space map.

Figure 6(a) the scale space map of this model exhibits the same behavior as that of the former, but as the width is smaller the dot patterns of its scale space map merge faster than those of the former. Figure 6(d) shows an END model with both the included angles equal to $\pi/4$ and Figure 6(e) shows another END model with both the included angles equal to $3\pi/4$. The scale space map of either of these models is similar to that of the END model in Figure 6(a) Figure 6(f) shows an END model with one angle equals to $\pi/4$ and the other $\pi/2$. The scale space map initially shows two dot patterns located at the angular points of the model. As the iteration proceeds the dot patterns attract each other but they do not merge. The dot patterns at the weaker corner (angle = $\pi/2$) terminates whereas the other persists. Figure 6(g) shows an END model with one angle $\pi/4$ and the other $3\pi/4$. Figure 6(h) shows another END model with one angle $\pi/2$ and the other $3\pi/4$. The scale space map of either of these models is similar to that of the model in Figure 6(f). Figure 7 shows STAIR models with different included angles and their scale space map. Figure 7(a) shows a STAIR model with both the included the angles are equal to $\pi/2$.

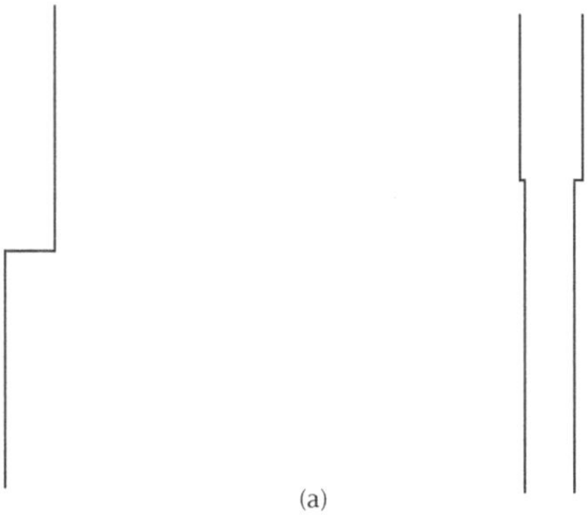

(a)

FIGURE 7 *(Continued)*

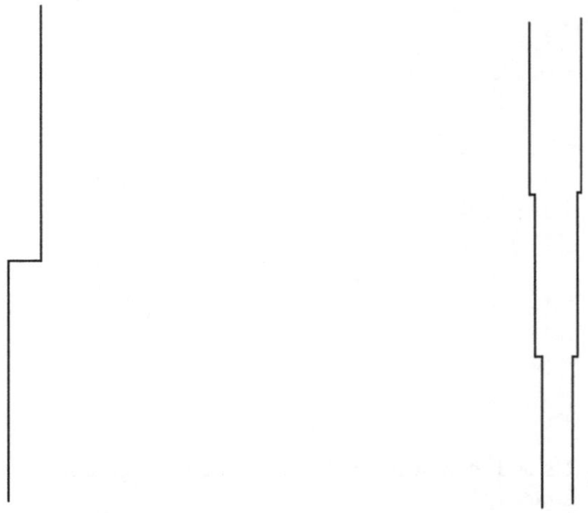

(b)

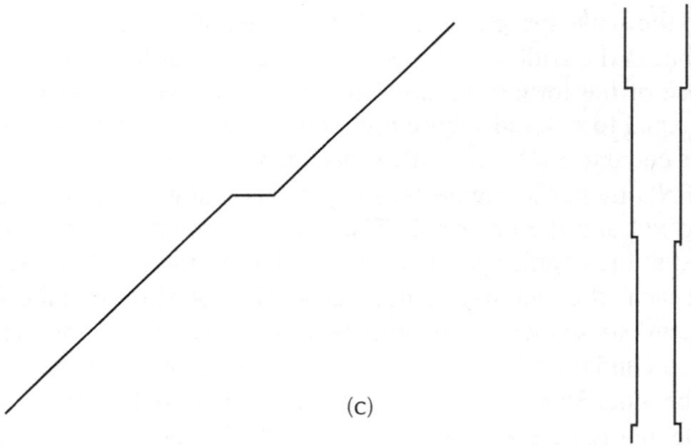

(c)

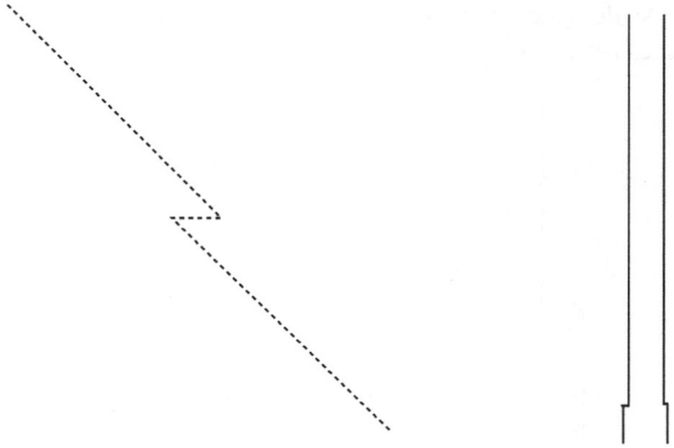

(d)

FIGURE 7 *(Continued)*

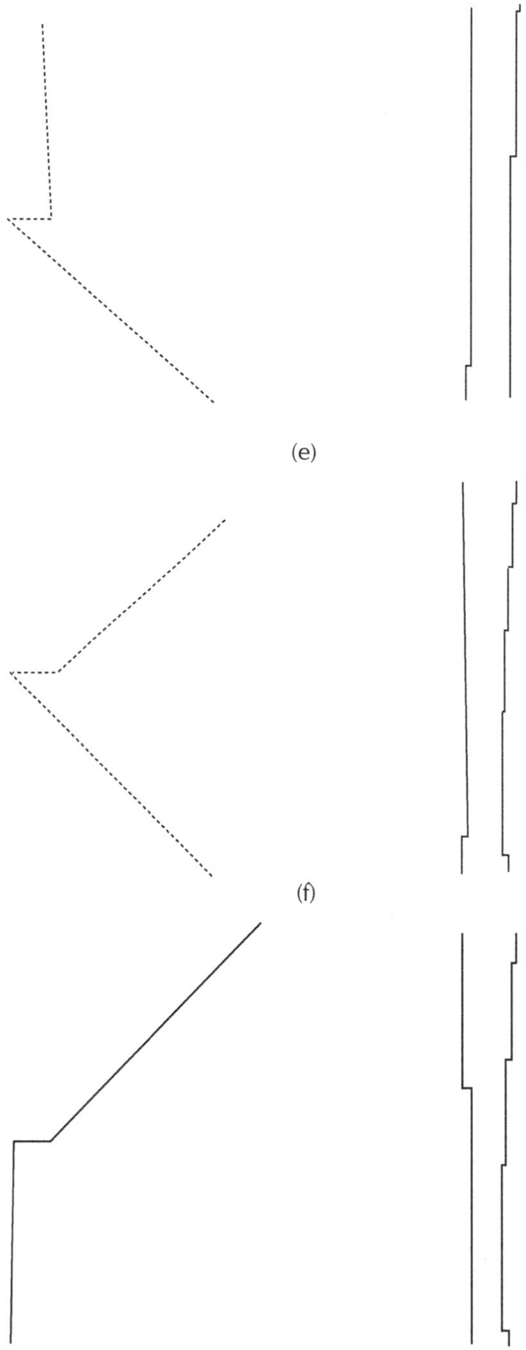

FIGURE 7 STAIR models and their scale space map. The left figure is the model and the right figure is its scale space map.

The scale space map consists of two persistent dot patterns located at the angular points of the model. The dot patterns repel each other as the iteration scale increases. Figure 7 (b) shows another STAIR model with both angles equal to $\pi/2$ but with width half that of the model in Figure 7(a). The scale space map of the model is almost similar to that of the model in Figure 7(a) but as the width of this model is smaller the dot patterns repel earlier. Figure 7(c) shows a STAIR model with both the included angles equal to $3\pi/4$. The scale space map consists of two persistent dot patterns each located at the angular points of the model.

The dot patterns initially attract each other but subsequently repel as the iteration proceeds. Figure 7(d) shows another STAIR model with both angles equal to $\pi/4$. The scale space map consists of two persistent dot patterns located at the angular points of the model. The dot patterns initially attract each other but remain separated without being merged as the iteration proceeds. Figure 7(e) shows a STAIR model with one angle equals to $\pi/4$ and the other $\pi/2$, Figure 7(f) shows a STAIR model with one angle equals to $\pi/4$ and the other $3\pi/4$ and Figure 7(g) shows a STAIR model with one angle equals to $\pi/2$ and the other $3\pi/4$. The scale space map of each of these models consists of two persistent dot patterns located at the angular points of the model. The dot patterns initially attract each other for once only but subsequently repel as the iteration proceeds.

We conclude that the scale space map of a Γ model consists of a single persistent dot pattern which may either be stationary or may exhibit movement as the iteration scale increases. The scale space map of an END model initially consists of two dot patterns located at the angular points of the model. If both the included angles be equal then as the iteration proceeds, the two dot patterns attract each other and merge to become a single dot pattern which may be stationary. If the angles be different then as the iteration proceeds though the dot patterns initially attract each other but the one at the weaker corner terminates and the other persists. The scale space map of a STAIR model consists of two persistent dot patterns located at the angular points of the model. The dot patterns may either repel each other or may initially attract each other for once only but subsequently repel. The dot patterns do not merge nor do they disappear. The models that have been analyzed are single and double corner models.

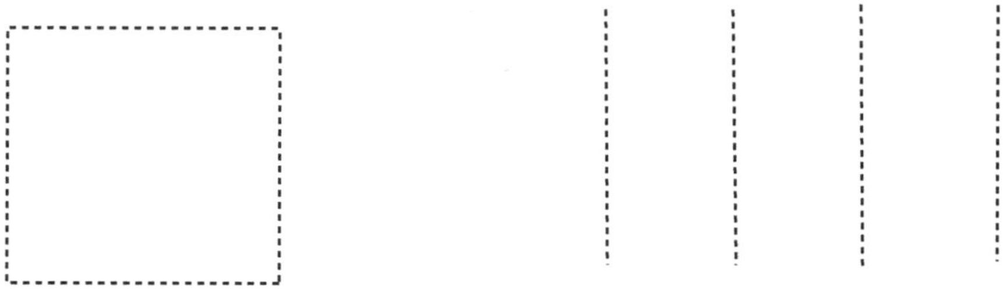

FIGURE 8 *(Continued)*

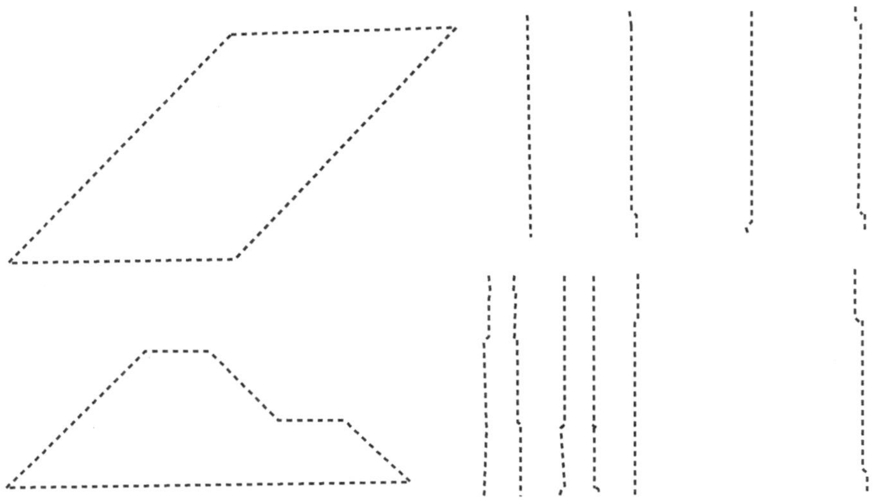

FIGURE 8 A number of polygons and their scale space map.

There is no explicit analysis of models that consist of more than two corners. In order to give the reader a perspective of the scale space behavior of models that consist of more than two corners, a number of polygons and their scale space map are shown in Figure 8. These maps are in conformity with the scale space behavior of the models that have been analyzed.

13.9 TREE ORGANIZATION AND CORNER DETECTION

To detect corners on digital curves, a tree representation similar to that in [187] and [119] is constructed from the scale space map of a curve. The tree representation is made taking into account three features of dot patterns namely, persistency, sign and movement. The persistency and movement are determined by tracking the dot patterns. The sign of dot pattern is the sign of the extreme curvature generating the dot pattern.

If a dot pattern originating from the finest possible scale survives at least 50% of the iteration scale [119] then a vertical line is drawn from the finest scale location of the dot pattern to the largest iteration scale that it survives. This vertical line which does not split as the iteration scale decreases is regarded as a tree which has a single root and has neither a branch nor a leaf. If a dot pattern does not survive at least 50% of the iteration scale then we look for the sign and the movement of the dot pattern. If a non-surviving dot pattern (which does not survive at least 50% of the iteration scale) has the same sign as that of a nearest persistent dot pattern either onto its left or right and if the non-surviving dot pattern and the persistent dot pattern is found to attract each other without being merged to become a single dot pattern then the non-surviving and the persistent dot pattern together are indicative of an END model present on a curve. In this case, a vertical line starting from the finest scale location to its terminating scale is drawn and this vertical line is joined to the persistent vertical line by a horizontal line so as to help comparison between the length of the non-surviving and the persistent line at the subsequent stage of corner detection. If two dot patterns having the same sign attract each other and subsequently merge to a single dot pattern which reaches at 50% of the iteration scale then corresponding to each dot pattern that initially appears a vertical line starting from the originating scale to the merging scale, is drawn. Another vertical line is drawn corresponding to the merged dot pattern. This vertical line starts from the merging scale and ends at the largest scale survived by the merged dot pattern. The three vertical lines are joined at the merging scale by horizontal lines. The three vertical lines form a tree whose root is the vertical line corresponding to the merged dot pattern and the other two vertical lines are its leaves. If a non-surviving dot pattern having sign opposite to that of its nearest persistent dot pattern is found to repel and repelled by the persistent dot pattern or they are initially found to attract each other for once only and subsequently repel then the two dot patterns together are indicative of a STAIR model present on a curve. In this case, the non-surviving dot pattern is replaced by a vertical line originating from the finest scale location of the non-surviving dot pattern and ending at the termination scale is drawn and this non-surviving vertical line is joined with the persistent vertical line by a horizontal line. If a non-surviving dot pattern (which does not survive at least 50% of the iteration scale) does not interact with its nearest persistent dot pattern (which persists at least 50% of the iteration scale) then the non-surviving dot pattern is discarded from the organization. The dot patterns whose movements are not in conformity with the scale space behavior of different corner models, the dot patterns whose curvature measure in the threshold (0.0001) and the dot patterns which survive only a single scale are all discarded from the organization. A non-surviving dot pattern having the same sign as that of a nearest persistent dot pattern cannot repel each other. A non-surviving dot pattern with such behavior is discarded from the organization.

FIGURE 9 *(Continued)*

FIGURE 9 Tree organization of the scale space map of two END models. The left figure is the scale space map and the right figure is its tree organization.

Corner detection is made on the basis of the persistency of the roots and branches/leaves of the tree organization. A persistent root which does not have a branch/leaf corresponds to a corner point located at the finest scale location of the root. If a root has branches and leaves then the length of the root is first compared with that of the branches. If the length of the root is greater than that of the branches then the root only corresponds to a corner point, located at the point of orthogonal projection of the root on the *x*-axis. If the length of the root does not exceed the length of the branches, then each branch corresponds to a corner point located at the point of orthogonal projection of the branches provided that the length of the leaves originating from a branch fall the length of the branch, otherwise the leaves originating from the branches gives the location of the corner points.

13.10 SPACE REQUIREMENTS AND COMPUTATIONAL LOAD

In the Rattarangsi-Chin algorithm [119] as the varying window size is used the space requirements by the Gaussian filter coefficients is different for different window size. The space requirement (s) is exactly equal to the window size (w). The window size is varied in arithmetic progression whose first term is 3, common difference is 2 and the last term is n – 1, when n is even and n, when n is odd. The number of terms of the progression is n/2 – 1, when n is even and (n – 1)/2, when n is odd. So the total space requirements by the filter coefficients is

$\sum s = (n-2)(n+2)/4$, when n is even

$= (n-1)(n+3)/4$, when n is odd.

In the present approach as the constant window size w = 3 is used, the space requirements by the filter coefficients is small and finite (= 3). This shows that the present approach reduces the space requirements considerably. In the Rattarangsi-Chin algorithm the space requirements is $O(n^2)$ whereas, in the present approach the space requirements does not depend on the data size, it is small and finite. To determine the computational load we note that the smoothing process in [119] and in the present approach are parallel in nature. So in order to compare the computational load of either of the smoothing processes, it is sufficient to determine the same at each point. In the Rattarangsi-Chin smoothing process, for a window size of $w = 3$, the number of multiplications (m) and additions (a) required are $m = 3$ and $a = 2$, for a window size of $w = 5$, $m = 5$ and $a = 4$, for a window size of $w = 7$, $m = 7$ and $a = 6$ and so on for higher window size. For a window size of $w = 2j + 1$ which should not exceed the length of the curve, the number of multiplications required is $m = 2j + 1$ and the number of additions is a = $2j$. This shows that the number multiplications as well as additions each form an arithmetic progression. For a digital curve with n points the number of terms in each series is $(n - 1)/2$, when n is odd and $n/2 - 1$, when n is even. So the total number of multiplications and additions required are:

$$\sum m = (n-1)(n+3)/4, \quad \text{when n is odd}$$

$$= (n-2)(n+2)/4, \quad \text{when n is even}$$

$$\sum a = (n-1)(n+1)/4, \quad \text{when n is odd}$$

$$= n(n-2)/4, \quad \text{when n is even}$$

So the total number of arithmetic operations required at the smoothing stage is $(n^2 + n - 2)/2$, when n is odd and it is $(n^2 - n - 2)/2$, when n is even. So the com putational load of the smoothing process in [119] is $O(n^2)$ at each point. In the present approach at each iteration and at each point, the number of multiplications and additions required are $m = 3$ and $a = 2$. It has already been shown that the number of iterations to be performed on a digital curve is $(n - 1)/2$, when n is odd and it is $n/2 - 1$, when n is even. So the total number of arithmetic operations required by the iterative smoothing process is $(5n - 5)/2$, when n is odd and it is $(5n - 10)/2$, when n is even. So the computational load of the iterative smoothing process is $O(n)$ at each point.

13.11 EXPERIMENTAL RESULTS

The corner detector that has been developed in this chapter is applied on the same digital curves. The corner points are indicated by bold solid circles on each curve. These are shown in Figure 10 through Figure 13. These figures also show the corner points as obtained by the Rattarangsi-Chin algorithm. From these results it is found that the present corner detector compares favorably with the Rattarangsi-Chin algorithm.

To show the robustness of the corner detector to noise, white Gaussian noise is added to the leaf-shaped curve making the noise level σ vary from 0.5 to 2.5 in an interval of 0.5. The noisy curves are shown in Figure 14 indicating the corner points on them by bold solid circles. From these figures it is evident as the noise level increases no additional corner point is detected. The location of the corner points on the noisy curves is comparable to that on the original image.

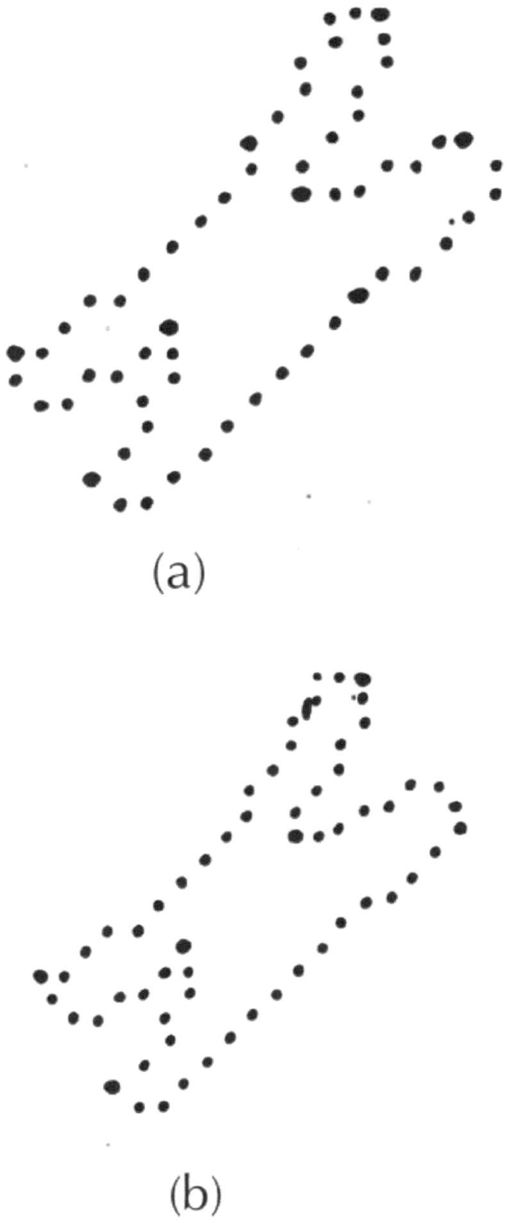

FIGURE 10 The chromosome shaped curve. The corners are indicated by bold solid circles. (a) Present method, (b) Rattarangsi-Chin method.

(a)

(b)

FIGURE 11 The figure-8 curve, the corners are indicated by bold solid circles. (a) Present method, (b) Rattarangsi-Chin method.

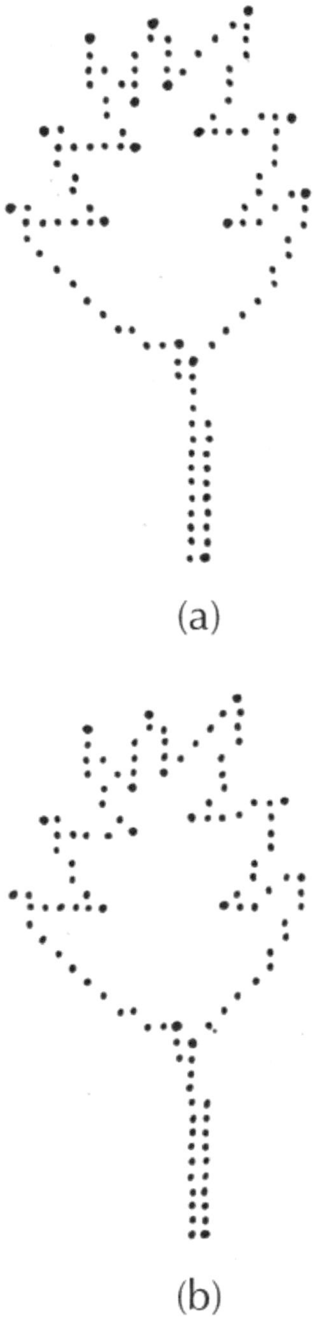

FIGURE 12 The leaf-shaped curve, the corners are indicated by bold solid circles. (a) Present method, (b) Rattarangsi-Chin method.

(a)

(b)

FIGURE 13 A curve with four semi circles, the corners are indicated by bold solid circles. (a) Present method, (b) Rattarangsi-Chin method.

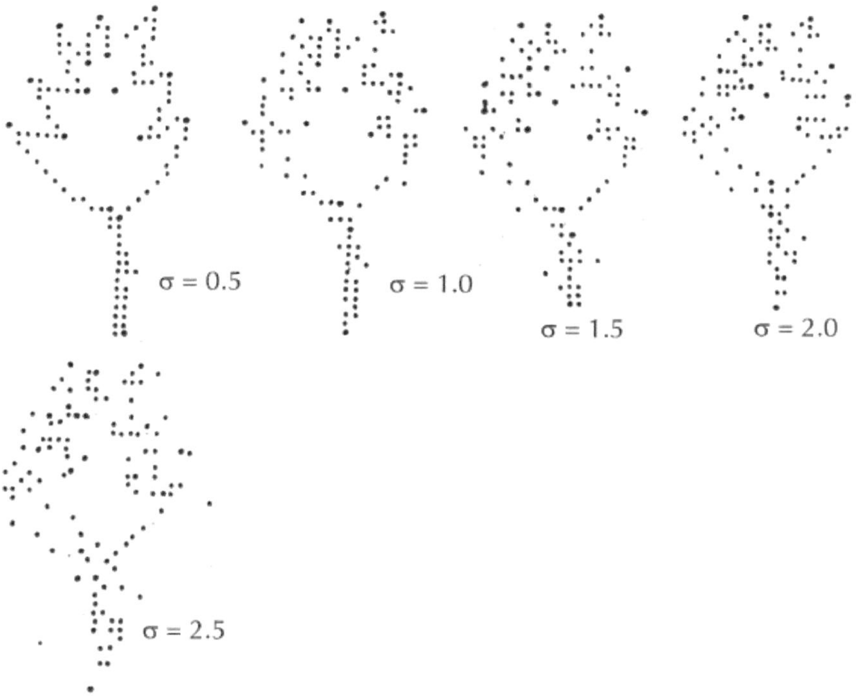

FIGURE 14 The noisy leaf at varying noise levels, the corners are indicated by bold solid circles.

KEYWORDS

- **Constant window Size**
- **Dot patterns**
- **Gaussian filter coefficients**
- **Scale space behavior**
- **Scale space map**

14 Corner Detection Using Bessel Function as Smoothing Kernel

CONTENTS

14.1 INTRODUCTION

In this chapter we use Bessel function as a smoothing kernel for scale space analysis and corner detection on digital curves. The scale space behavior of different corner models such as Γ model, END model and STAIR model is analyzed. The scale space map of a digital curve is converted into a tree representation and corners are detected and located in a process of interpreting the tree. The numerical problems that arise in the implementation are addressed and feasible solutions are proposed.

It has been proved in [13] that among a large number of kernels the Gaussian kernel has scale space property, which means that as one moves from coarser scale to finer scale new features may be introduced, but the existing ones never disappear and as one moves from finer scale to coarser ones no new feature is introduced in the smoothed curve. But this kernel is applicable to continuous inputs (signals and curves) only. In order that, this kernel can be applied on digital inputs it has to be discretized so that scale space property is preserved.

Lindeberg [79], on the other hand, has developed a discrete scale space theory and based on it he has shown that the digital filter coefficients derived from Gaussian distribution used in reference [119] do satisfy scale space property. But with these coefficients (digital Gaussian) it is not possible to blur an input by arbitrary amount. The parameter σ in the Gaussian distribution is responsible for blurring. The higher the value of σ is, the greater is the amount of blurring. Each window length in the digital Gaussian filter coefficients correspond to a specific value of σ. If σ_w and σ_{w+2} are the values of the blurring parameter for the window length w and $w + 2$ of the digital Gaussian filter coefficients then it is not possible to blur an input with a value of σ such that $\sigma_w < \sigma < \sigma_{w+2}$.

But the scale space theory demands that the smoothing kernel should not only satisfy the scale space property, but it should also be capable of blurring the input by arbitrary amount. This is possible only if the blurring parameter is continuous in nature. So in order to smooth digital inputs, discretization of Gaussian kernel so that it satisfies scale space property is not enough. The blurring parameter of the discrete scale space kernel should be continuous (a kernel is called scale space kernel if it satisfies scale space property and has the ability to blur an input with an arbitrary amount of smoothing).

In an attempt to derive a scale space kernel that meets the essence of scale space theory Lindeberg [79], instead of discrediting Gaussian kernel, has developed a genuinely discrete scale space theory for scale space analysis of discrete signals and curves with continuous scale parameter. Initially he has derived a two kernel (a smoothing kernel with two filter coefficients) with scale space property and then obtained a generalized binomial kernel with finite support (finite number of filter coefficients). Using semi-group and symmetric property he has suggested the form of a discrete scale space kernel with continuous scale (blurring) parameter and concluded that for discrete signals the most reasonable discrete scale-space kernel with a continuous scale parameter t is:

$$T(n, t) = \exp(-t)\, I_n(t),\ t > 0 \tag{14.1}$$

where $I_n(t)$ is the modified Bessel function of integer order. It is necessary that the scale parameter be continuous so that the inputs can be blurred with an arbitrary amount of smoothing and you are not locked to a fixed predetermined levels of scale that make event tracing difficult [53]. We address the numerical problems that arise in the implementation of the kernel $T(n, t)$ and propose feasible solutions.

A 2D digital curve is converted into two 1D signals each of which is convolved with the kernel $T(n, t)$. The extreme curvature points are located at different values of t creating the scale space map. The map is then converted into a tree representation and corners are detected and located in a process of interpreting the tree. Experiments are carried out on a number of digital curves and the corner detector is found to operate successfully even in presence of noise.

14.2 EVALUATING *T(N, T)*

The smoothing kernel in Equation (14.1) is to be evaluated for different values of n and t so that it can be convolved with a digital curve. The function subprograms generating the values of $I_0(t)$, $I_1(t)$ and $I_n(t)$, $n > 2$ are available in Numerical Recipes [117]. But these functions involve $\exp(t)$ for $t \geq 3.75$ as a product term which increases catastrophically for increasing value of t and hence cannot be used to evaluate $T(n, t)$. But we find that $T(n, t)$ involves $\exp(-t)$. So evaluation of $T(n, t)$ can be carried out by evaluating $I_0(t)$ and $I_1(t)$ for $t \geq 3.75$ after omitting the exp(t) term from them. The value of $T(0, t)$ and $T(1, t)$ for $0 < t < 3.75$ are generated by multiplying $I_0(t)$ and $I_1(t)$, $0 < t < 3.75$ by exp($-t$). The Appendix A shows the function subprograms generating $T(0, t)$, $T(1. t)$ and $T(n, t)$ for $n > 2$. These subprograms are the modified versions of those available in Numerical Recipes [117]. The modification has been done so as to avoid arithmetic overflow. The parameter "iacc" involved in the function subprograms is set to 400 instead of its original value 40 so as to attain the desired degree of accuracy. The final results are insensitive to values of "iacc" greater than 400.

14.3 KERNEL WITH FINITE SUPPORT

The smoothing kernel defined in (14.1) has an infinite support. Consequently the convolution of a signal with this kernel requires the kernel to be converted into one with finite support. Since, we are dealing with 2D digital curves hence the signals are x_i denoting the x-coordinate and y_i denoting the y-coordinate of the ith point of the curve. Assuming that the curve is open and has infinite number of points, the convolution of the coordinates with the smoothing kernel is defined by:

$$X_i(t) = \sum_{n=-\infty}^{n=\infty} T(n;t)x_{i-n} \text{ and } Y_i(t) = \sum_{n=-\infty}^{n=\infty} T(n;t)y_{i-n} \tag{14.2}$$

where, $X_i(t)$ and $Y_i(t)$ are the smoothed coordinates. At the implementation level, the infinite sum on the right hand sides of the last expressions is to be approximated by a finite one. For this the infinite sum is truncated for some large value of n (N say) producing the following expressions:

$$X_i(t) = \sum_{n=-N}^{n=N} T(n;t)x_{i-n} \text{ and } Y_i(t) = \sum_{n=-N}^{n=N} T(n;t)y_{i-n} \tag{14.3}$$

The value N of n is so chosen that the truncation error does not exceed an error limit ε. Assuming that x_i and y_i are bounded so that $\max(|x_i|) = K_1$, $\max(|y_i|) = K_2$ and $K = \max(K_1, K_2)$ we obtain the sufficiency condition for error tolerance as:

$$2K \sum_{n=N+1}^{\infty} T(n;t) \leq \varepsilon \tag{14.4}$$

So, the number of filter coefficients that should be generated within the error limit ε is given by the maximum value of n satisfying:

$$\sum_{n=-N}^{N} T(n;t) \geq 1 - \varepsilon / K \tag{14.5}$$

Now we need to select the value of ε so as to determine the value of N. This problem has not been addressed by Lindeberg [79]. It is possible to select K in such a way that for a given ε the value of ε/K is less than 10.0e-6 and *vice-versa*. We note that both

$I_0(t)$ and $I_1(t)$ are polynomial and the minimum absolute error in approximating $I_0(t)$ and I_1(t) is of the order of 10e-7 [1]. This is why the sum of $T(n; t)$ as n varies from –N to N, can never be correct beyond the sixth decimal place and for a given ε, K may be so selected that the desired accuracy may not be attained. To overcome this difficulty instead of truncating the infinite convolution $X_i(t)$ and $Y_i(t)$ as done in [35] we truncate the infinite sum of $T(n,t)$. The number of filter coefficients to be evaluated from $T(n,t)$ is determined by N where:

$$\sum_{n=-N}^{N} T(n;t) \geq 1 - \varepsilon`\varepsilon` > 0 \tag{14.6}$$

In contrast to [79] it may be noted here that the number of filter coefficients to be evaluated is independent of input data size [79] and depends on t only. For a given t, the same and equal number of filter coefficients operates on the curve. Since the maximum of the absolute error in approximating $I_0(t)$ and $I_1(t)$ is of the order of 10e-6 hence ε is selected as ε = 10e-6. One may select larger values of ε but it will make the results var(t) = t and the infinite sum on $T(n,$ t) = 1 more inaccurate and it may also cause overshoots in the scale space map.

So we use the expression (14.3) to convolve the coordinates x_i and y_i and use the condition (14.6) to determine the number of filter coefficients to be considered. It may be noted here that if the number of filter coefficients is N then the window length of the smoothing kernel is $2N + 1$ because $T(n, t) = T(-n, t)$.

14.4 SELECTING DISCRETE VALUES OF T

In order that a signal can be defocused with an arbitrary amount of blurring it is necessary that the smoothing parameter t be continuous. But at level of implementation the parameter t should be discretized because it is impossible to generate the smoothed curve at all values of t. The central idea involving the continuous scale parameter is that it is possible to smooth a curve with arbitrary amount of blurring instead of fixed predetermined levels of smoothing. Smoothing with a continuous scale parameter provides a theoretical framework in which the degree of smoothing can be varied arbitrarily. This is the advantage of discrete smoothing kernel with continuous scale parameter over digital Gaussian kernel.

But discretization of a continuous parameter is a non-trivial problem. This problem has not been addressed by Lindeberg [79]. We address the problem of selecting discrete values of the continuous parameter t in real situation. For this we discretized t at an interval of 0.01 starting with an initial value of 0.01. The initial value and the step size of the discrete levels could be some other values except that both should be sufficiently small so that t can be looked upon as a continuous parameter. To discretize t, we start with its initial value and compute the condition (14.6) to determine the number of filter coefficients (N). Then t is incremented with a step size of 0.01 and the condition (14.6) is evaluated to find out whether the number of filter coefficients has increased. The least value of t for which the number of filter coefficients increases is the next discrete value of t and the new value of N gives the number of filter coefficients at this value of t. As long as the increase in the value of t does not affect the number of filter coefficients the t value is incremented with a step size of 0.01. So, the discrete values of t are determined in such a way that for two successive least discrete values t_1 and t_2 of t with $t_2 > t_1$ if the number of filter coefficients satisfying the condition (14.6) be N_1 and N_2 then $N_2 > N_1$. Following this rule, the parameter t is discretized from its minimum value to the maximum allowable value.

14.5 TERMINAL VALUE OF T

As already stated if the number of filter coefficients is N then the window length of the smoothing kernel is $2N + 1$. Suppose that the value of t for this window length is t_{max}. If the curve is open then the value of N should be such that $2N + 1$ do not cross over the end points of the curve. If the curve is closed then $2N + 1$ should be such that aliasing effect is avoided. In other words t should be increased until aliasing occurs. If a closed curve consists of m points then to avoid aliasing effect N should be such that $m = 2N + 1$. So the maximum number of filter coefficients that should be evaluated is:

$$N_{max} = m/2 - 1$$

$$= (m - 1)/2 \quad (14.7)$$

and the terminal value of t (t_{max}) is the least value of t for which the number of filter coefficients is N_{max}.

14.6 SCALE SPACE MAP

To compute the curvature at a point of a digital curve we use the first and second order finite differences of the smoothed coordinates $X_i(t)$ and $Y_i(t)$ and the curvature measure following Rattarangsi and Chin [119]. The curvature measure at the ith point and for a given t is given by:

$$\kappa_i(t) = \Delta X_i(t)\, \Delta^2 Y_i(t) - \Delta Y_i(t)\, \Delta^2 X_i(t) \quad (14.8).$$

The digital curve is convolved with the kernel $T(n;t)$ and the local extrema of curvature are located at sampled values of t. The values of i for which $|\kappa_i(t)|$ exceeds $|\kappa_{i-1}(t)|$ and $|\kappa_{i+1}(t)|$ give the location of the extreme curvature points.

The t values are sampled using the increasing N criterion. In order to construct the scale space map, along the x-axis (horizontal), the ordinal number i of the points of the curve is shown and along the y-axis (vertical) N is shown. The local maxima of absolute curvature at varying N are plotted as points on the xy half-plane. The image showing the location of the maxima of absolute curvature at varying N is the scale space map of the curve. The map consists of an aggregate of dot patterns, some of which grow reaching the maximum allowable value of N that is N_{max}, some others terminate as N increases. A pair of dot patterns may grow and merge to a single dot pattern which may reach N_{max}. The dot patterns that persist are indicative of those maxima that are detected at all levels of details, fine as well as coarse. The dot patterns that appear temporarily are indicative of those maxima that are detected at the fine scales but disappear as the degree of smoothing increases. A number of digital curves and their scale space maps are shown in Figure 1 through Figure 4.

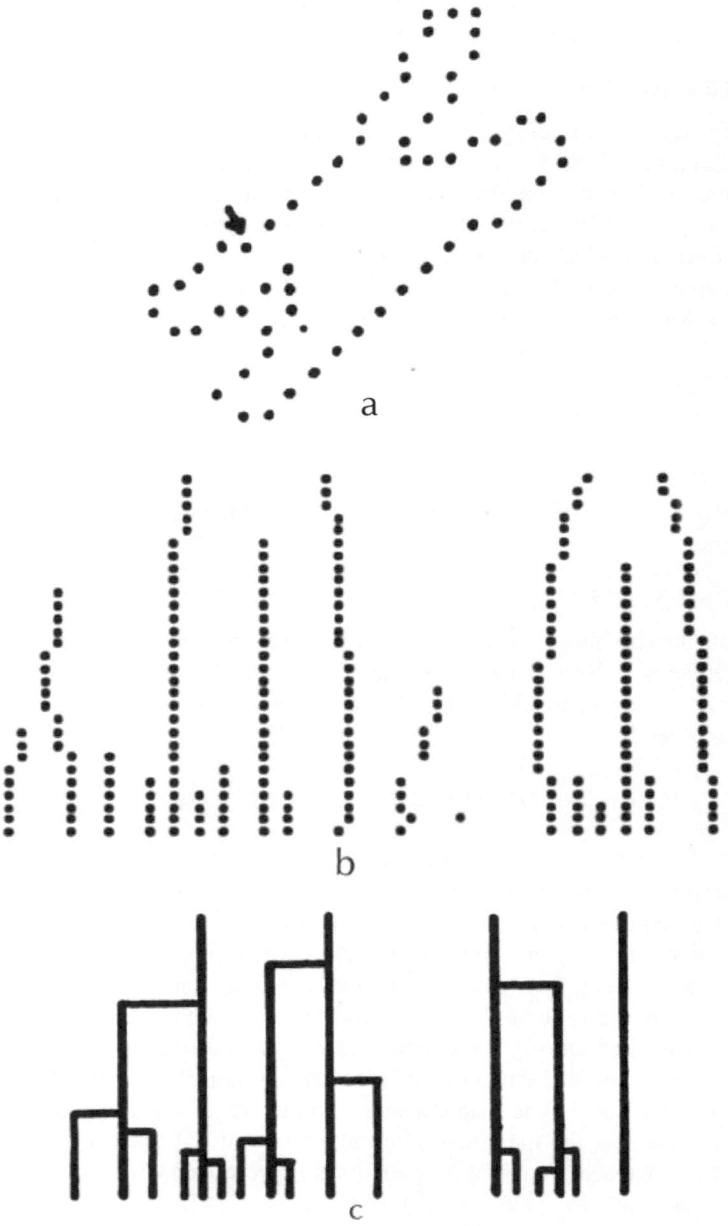

FIGURE 1 (a) Chromosome shaped curve, (b) scale space map, and (c) tree representation.

FIGURE 2 (a) Figure-8 curve, (b) scale space map, and (c) tree representation.

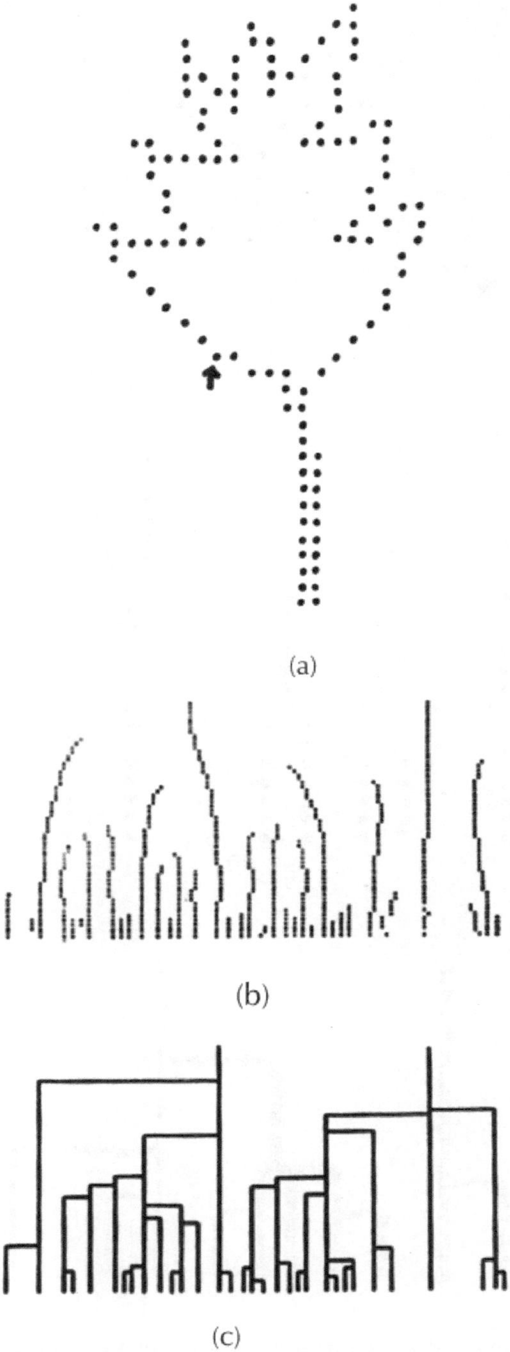

FIGURE 3 (a) Leaf-shaped curve, (b) scale space map, and (c) tree representation.

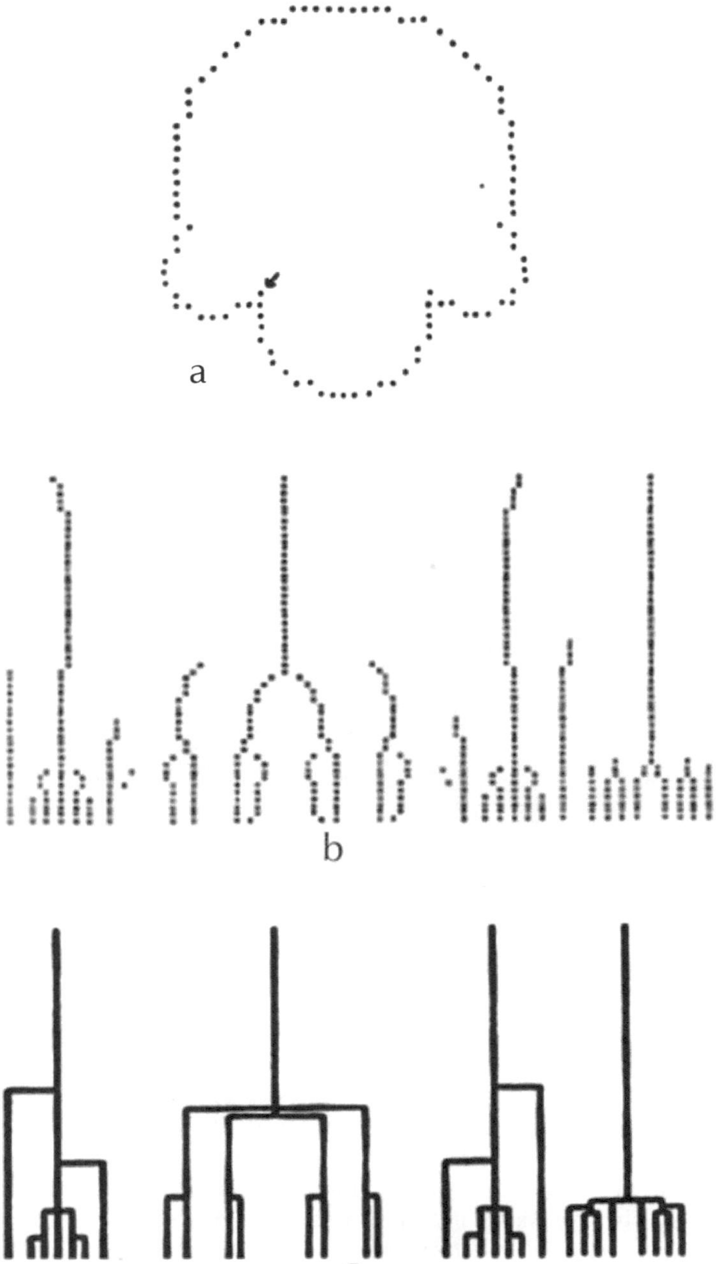

FIGURE 4 (a) Curve with four semi circles, (b) scale space map, and (c) tree representation.

14.7 EXPERIMENTAL RESULTS

The scale space behavior of corner models such as Γ model, END model and STAIR model when convolved with the discrete scale space kernel $T(n, t)$ are shown in Figure 5 through Figure 7. The models being open curves the smoothing should be performed up to a value of t for which N does not go beyond the end points of the model.

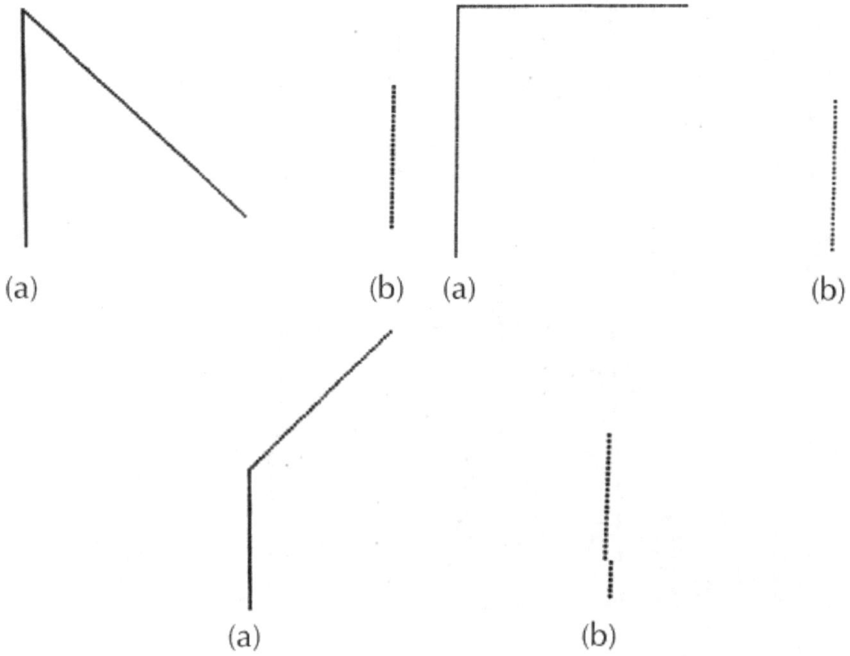

FIGURE 5 Scale space map of Γ models (a) the model and (b) the scale space map.

We have applied our corner detector on a number of digital curves. To detect corners the scale space map of the digital curve is converted into a tree representation and the corners are detected and located in a process of interpreting the tree. The methodology is similar. The Figure 1 through Figure 4 show the tree representation of the scale space map of a number of digital curves. The bold lines in the figures indicate the limbs of a tree. The Figure 8 through Figure 11 shows the corners on four digital curves together with the results of the Rattanragsi-Chin algorithm [119]. The corners are indicated by bold solid circles. We observe that the corner detector developed to detects the least number of corner points on circular objects (Figure 11).

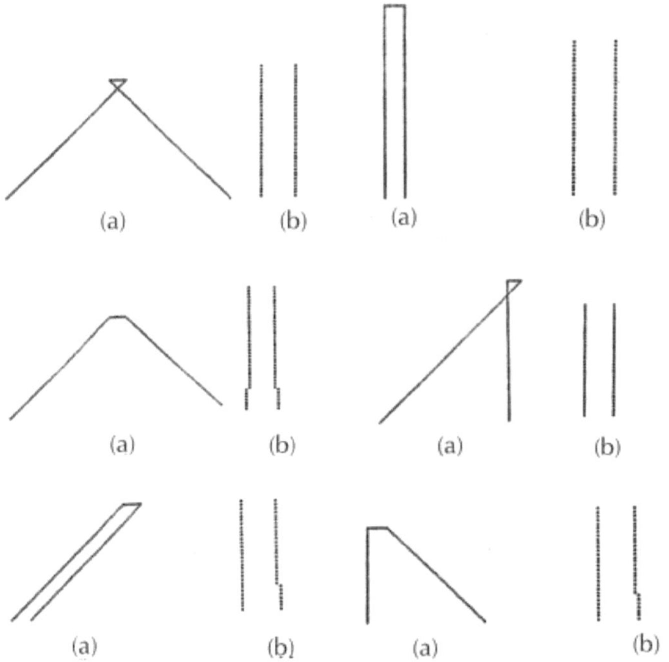

FIGURE 6 Scale space map of END models (a) the model and (b) the scale space map.

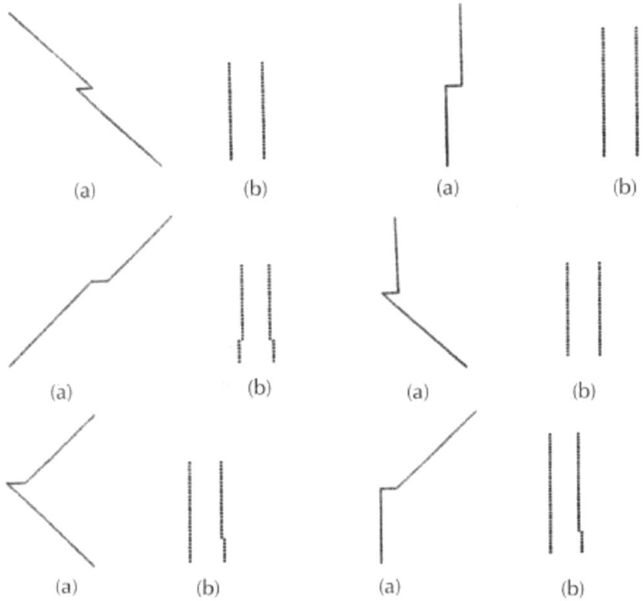

FIGURE 7 Scale space map of STAIR models (a) the model and (b) the scale space map.

a

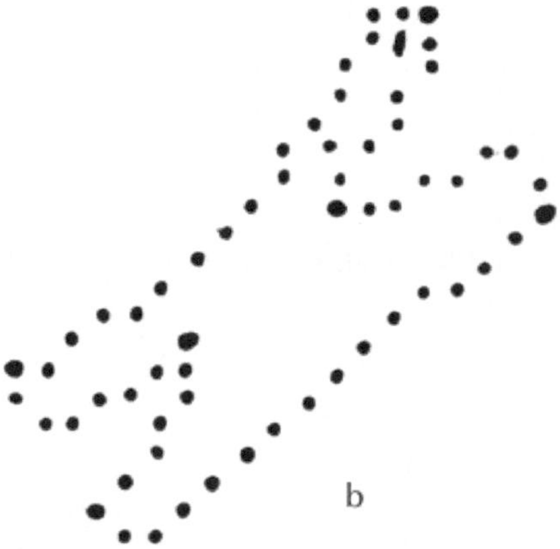

b

FIGURE 8 Chromosome shaped curve. The bold solid circles indicate corners. (a) Present method and (b) Rattarangsi-Chin method.

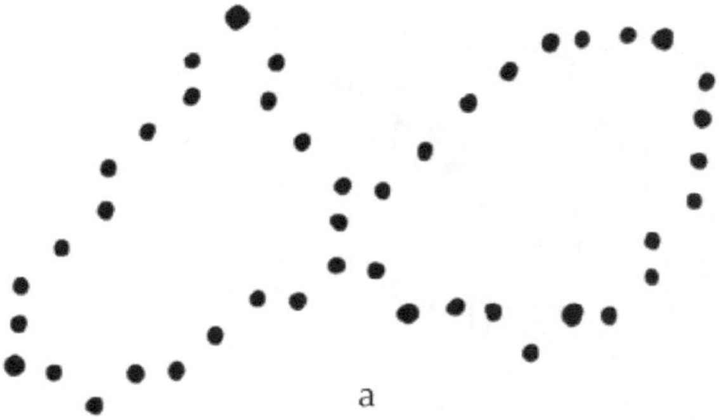

a

FIGURE 9 *(Continued)*

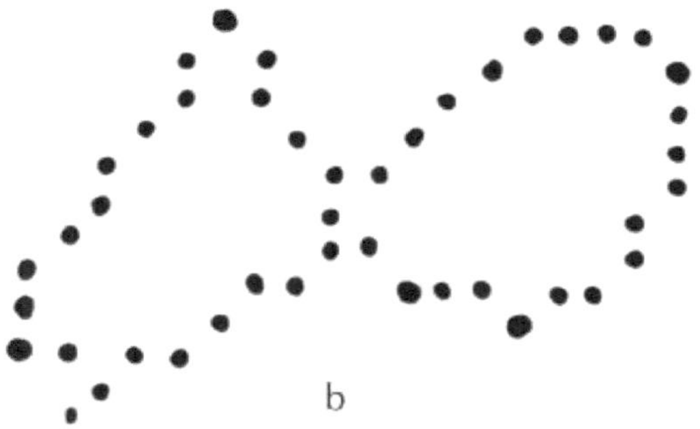

b

FIGURE 9 Figure-8 curve, the bold solid circles indicate corners. (a) Present method and (b) Rattarangsi-Chin method.

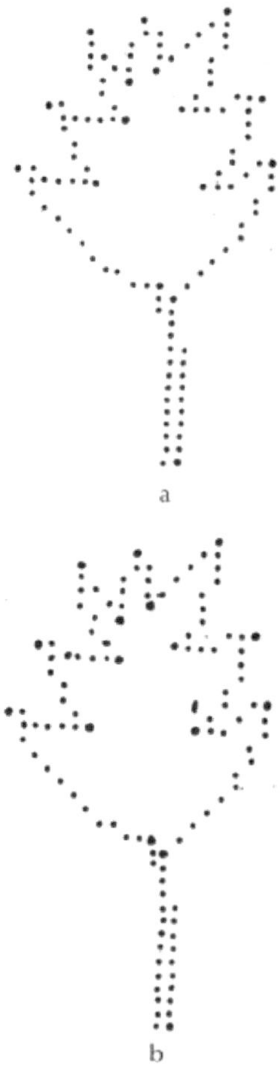

FIGURE 10 Leaf shaped curve, the bold solid circles indicate corners. (a) Present method, (b) Rattarangsi-Chin method.

a

b

FIGURE 11 Curve four semi circles, the bold solid circles indicate corners. (a) Present method and (b) Rattarangsi-Chin method.

The corner detector developed here is also robust to noise. To establish the robustness of the corner detector we have applied noise on the chromosome curve with different levels of noise varying from noise level $\sigma = 0.5$ to $\sigma = 2.5$ at an interval of 0.5. The noisy chromosomes are shown in Figure 12. These figures also indicate the corner points on the noisy curve with the help of bold solid circles. It may be observed that even in presence of noise the number of corners detected is the same as the number of corners on the original chromosome curve and the locations of comer points on the noisy curves are comparable to those on the original image.

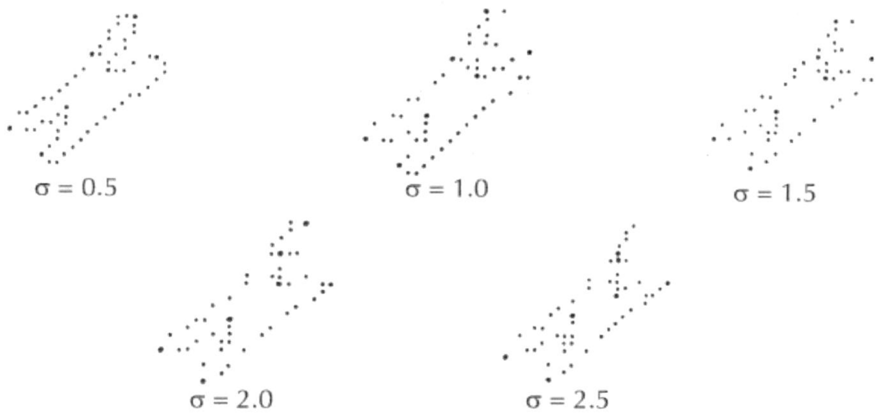

FIGURE 12 Chromosome at different noise levels, the bold solid circles indicate corners.

KEYWORDS

- **Bessel function**
- **Corner detector**
- **Dot patterns**
- **Scale space map**
- **Tree representation**

15 Adaptive Smoothing Using Convolution with Gaussian Kernel

CONTENTS

15.1 INTRODUCTION

In the existing multiscale smoothing (scale space) technique, each point of a curve is subjected to various degrees of smoothing. At each level of smoothing the extreme curvature points/zero crossings are detected. In this approach, it is essential to smooth a curve at all levels of detail, to construct a scale space map, to convert the map into a tree representation and to interpret the tree to detect and locate the corners. In this chapter, we suggest a method [130] for determining the degree of smoothing in the various regions of a curve based on the local nature of the curve.

15.2 CURVATURE VARIANCE AS A MEASURE OF ROUGHNESS

The more noise a curve has, the more smoothing is required before the actual curvature can be determined. The noise present on a curve is detected by the rapid variation along the curve over a comparatively small region. This rapid variation exhibits itself in the form of roughness on the curve. At rough regions there is a high variation in curvature and *vice versa*. So, we propose to measure the variation in the roughness of a curve using variance of curvature. In general, different regions of the curve have different roughness and so different amount of smoothing is required at different regions of a curve. We use Gaussian filter for smoothing the curve. Since the window size of the Gaussian filter determines the amount of smoothing the filter can impart, we determine the actual window size the filter should have at each point of the curve. In an attempt to find out the actual window size, we make use of the variance of curvature. As the window size of the filter is increased the variability in curvature measured by the variance of curvature is supposed to decrease, since the larger window size of filter imparts higher degree of smoothing to the curve. We start with the smallest window size and continually increment the window size until the variance of curvature starts decreasing. Since, the degree of smoothing applied to a region is determined based on the roughness of the curve in the region. So, this approach is known as adaptive smoothing. This method avoids the construction of the complete scale space map. We also avoid tree representation and interpretation. So, instead of smoothing a curve at all levels of detail, we determine the level of smoothing required to be imparted to each point based on the roughness of the curve in the neighborhood of the point. The roughness of a curve generally varies from point to point. So, the degree of smoothing is determined adaptively based on the nature of the curve in the neighborhood of a point. Each point will usually require a different degree of smoothing. The central problem is to determine the roughness of a curve in the neighborhood of a point. We propose to take the variance of curvature in the neighborhood of point as a measure of roughness. The higher the variance is, the rougher the curve is in the neighborhood of a point. In order to determine curvature it is necessary to smooth the curve. The curve is smoothed using the digital Gaussian filter coefficients. The digital Gaussian filter coefficients for window size w = 3 are given by:

$$c_{-1} = 0{:}2236;\ c_0 = 0{:}5477 \text{ and } c_1 = 0{:}2236 \tag{15.1}$$

The digital Gaussian filter coefficients for window size higher than w = 3 are obtained by repeated convolution of these coefficients with themselves. We take three points, namely, $(x_{i\pm1}, y_{i\pm1})$ and (x_i, y_i) and using:

$$X_i = c_{-1}x_{i-1} + c_0x_i + c_1x_{i+1} \text{ and } Y_i = c_{-1}y_{i-1} + c_0y_i + c_1y_{i+1} \tag{15.2}$$

We get the smooth coordinates (X_i, Y_i) corresponding to the input point (x_i, y_i). The same technique can be applied to find (x_{i-1}, y_{i-1}) and (x_{i+1}, y_{i+1}). We calculate the curvature at the point (x_i, y_i) using the Formula (15.2) after replacing (x, y) by (X, Y). We propose to denote the curvature at the ith point for window size w by $\kappa_{i,w.}$ We compute the curvature value at three points (x_i, y_i), $(x_{i\pm1}, y_{i\pm1})$ and compute the variance of these curvatures using

$$(\sigma_{i,w})^2 = \Sigma(\kappa_{i,w})^2/w - \Sigma(\kappa_{i,w}/w)^2 \tag{15.3}$$

The smoothed coordinates X_i and Y_i with window size w = 5 are given by:

$$X_i = c_{-2}x_{i-2} + c_{-1}x_{i-1} + c_0x_i + c_1x_{i+1} + c_2x_{i+2};$$

$$Y_i = c_{-2}y_{i-2} + c_{-1}y_{i-1} + c_0y_i + c_1y_{i+1} + c_2y_{i+2}; \tag{15.4}$$

The curvature and its variance are computed using Formula (15.2) and (15.3) with w = 5.

If this variance does not exceed the variance with w = 3 then the curvature at the ith point is the curvature value with w = 5, otherwise the window size has to be increased to 7. The window size is incremented in this way and for each window size w, the Gaussian filter coefficients are computed by convolving the filter coefficients are for window size w − 2 with those for w = 3. These filter coefficients are used to compute the smoothed coordinates X and Y using the formula:

$$X_i = \sum_{j=-(w-1)/2}^{j=(w+1)/2} c_j x_{i+j} \quad Y_i = \sum_{j=-(w-1)/2}^{j=(w+1)/2} c_j y_{i+j} \tag{15.5}$$

The curvature value at the point (x_i; y_i) using

$$(\sigma_{i,w})^2 = \sum_{-(w-1)/2}^{(w-1)/2} (\kappa_{i,w})^2 / w - \sum_{-(w-1)/2}^{(w-1)/2} (k_{i,w} / w)^2 \tag{15.6}$$

As it has been already stated, the greater the window size is, the more smoothing the Gaussian filter imparts to the curve and the less smooth the input curve is (i.e. the rougher the curve is), the higher is the value of variance of curvature. This is why when it is found that the value of curvature variance for window size w falls down the value of curvature variance for window size $w - 2$, then the curve has attained the necessary amount of smoothing in the neighborhood of a point and the window size w is taken as the required window size at the point. Thus, if $(\sigma_{i,w})^2 \leq (\sigma_{i,w-2})^2$ then the curvature at the point (x_i, y_i) is $\kappa_{i,w}$. On the other hand, if $(\sigma_{i,w})^2 > (\sigma_{i,w-2})^2$, the window size should be increased by 2 and the variance of curvature at window size w is to be compared with the variance of curvature at window size $w - 2$. The maximum value of w should never exceed the length of the curve in order to avoid aliasing.

The procedure determines the amount of smoothing required in the neighborhood of each point adaptively based on the roughness of the curve. The curvature values determined by smoothing are used to detect corners on the digital curve. The local maxima of the curvature values are the corner points provided the curvature value is in threshold. If the curvature at (x_i; y_i) determined by adaptive smoothing exceeds the curvature at the points (x_{i-1}, y_{i-1}) and (x_{i+1}, y_{i+1}) then the point p_i(x_i, y_i) is a corner point. The close corners are removed by merging technique.

15.3 EXPERIMENTAL RESULTS AND DISCUSSION

The procedure has been applied on four digital curves namely, a leaf shaped curve (Figure .1), a chromosome shaped curve (Figure 2), a figure-8 curve (Figure 3), and a curve with four semi circles (Figure 4). Solid circles on each curve have indicated the corners. These figures also show corner points obtained by the Rattarangsi–Chin algorithm [119]. As seen from these results, the present procedure, in most of the cases, detects more corner points than the Rattarangsi–Chin algorithm and it does not detect false corners.

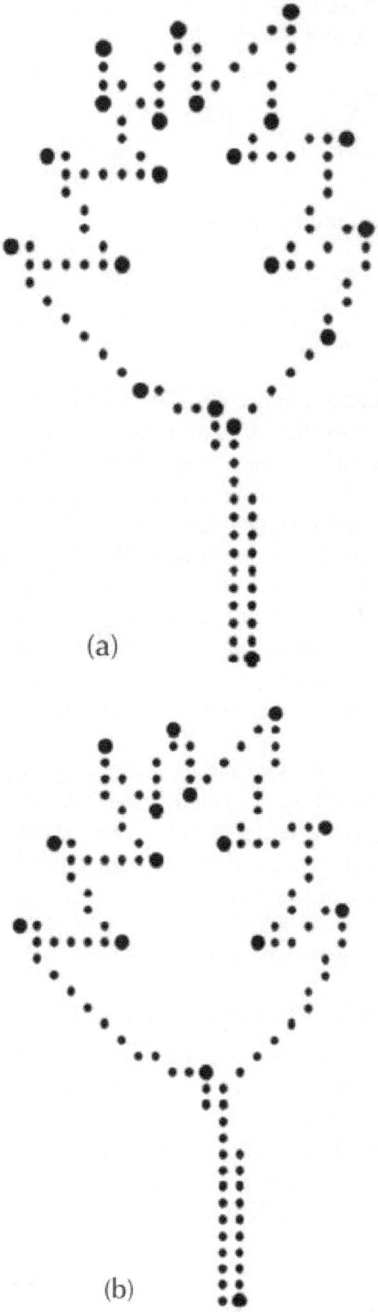

FIGURE 1 A leaf shaped curve. Bold solid circles indicate the corners. (a) Proposed method and (b) Rattarangsi-Chin method.

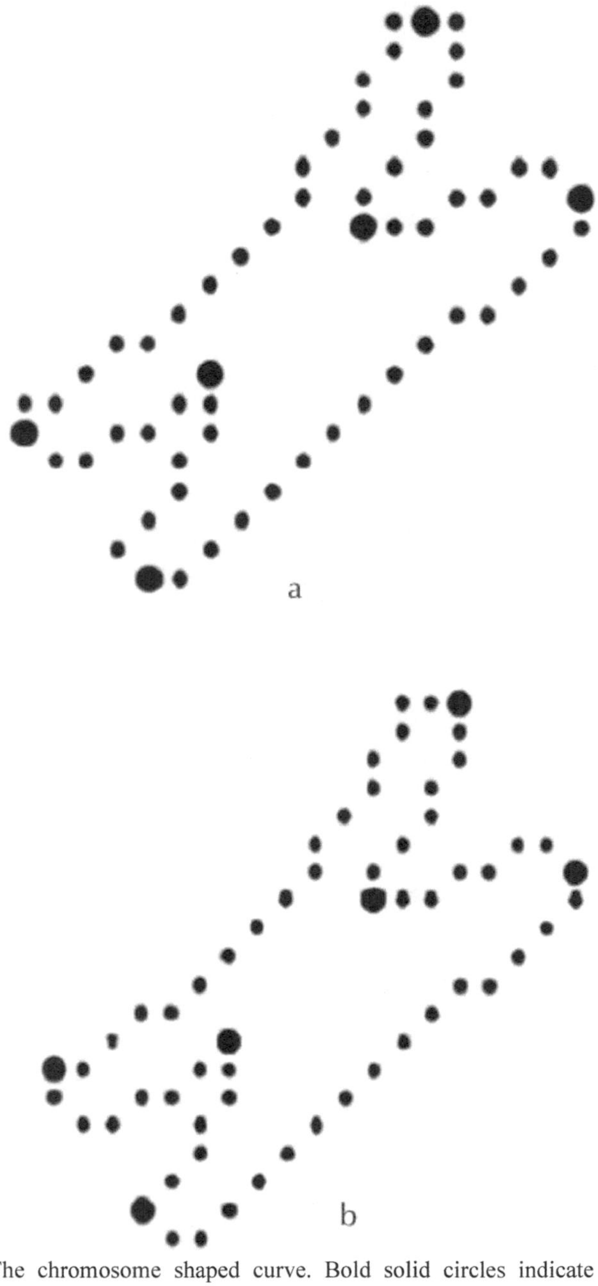

FIGURE 2 The chromosome shaped curve. Bold solid circles indicate the corners. (a) Proposed method and (b) Rattarangsi-Chin method.

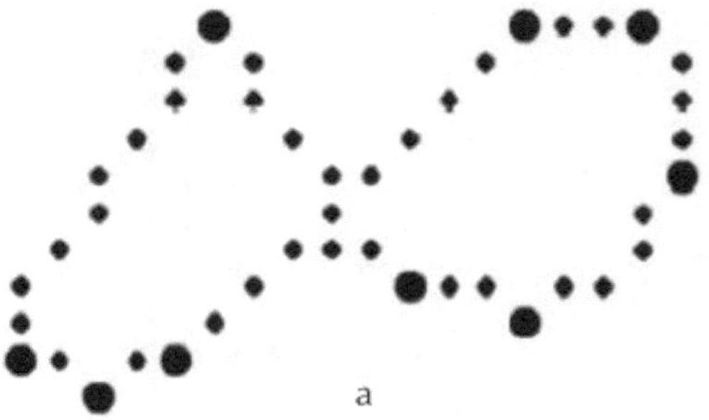

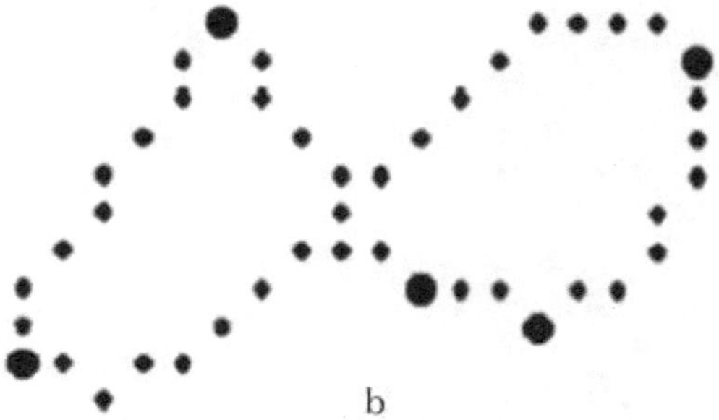

FIGURE 3 The Figure-8 curve. Bold solid circles indicate the corners. (a) Proposed method and (b) Rattarangsi-Chin method.

FIGURE 3 *(Continued)*

FIGURE 4 A curve with four semi circles. Bold solid circles indicate the corners. (a) Proposed method and (b) Rattarangsi-Chin method.

This is evident from the fact that, if we try to make a representation of the curve using a suitable function (straight line/circles) then the approximation error in the algorithm [119] is higher than that given by the present method. As we see some new corner points have been found in Figure 1, Figure 3, and Figure 4. The Rattarangsi-Chin algorithm performs excessive smoothing and this is why these corner points were not detected by this method. We note that in Rattarangsi-Chin algorithm the curve is smoothed by varying the window size from its minimum value of w = 3 to its maximum value which is equal to the length of the curve. But in the present technique, the window size is increased as long as the variance of curvature increases. Whenever, the variance of curvature at window size w falls down the variance of curvature at window size w − 2, we get the required window size for smoothing at a point. At any point of the curves that have been used for experiment, the window size is always lowest its maximum possible value. Thus, in the present technique the computation involved in smoothing the curve is lower than that involved in the Rattarangsi-Chin algorithm. The additional computation involved in the present method is the computation of variance of curvature. But this is the cost of adaptive smoothing. This cost is compensated by the improved experimental results found in the present technique. The corner detector is also found to be robust with respect to Gaussian noise $\sigma < 2$. For higher values of σ the performance of the corner detector degrades.

KEYWORDS

- **Adaptive smoothing**
- **Gaussian digital filter coefficients**
- **Multi scale smoothing technique**
- **Merging technique**
- **Rattarangsi-Chin algorithm**

16 Application of Polygonal Approximation for Pattern Classification and Object Recognition

CONTENTS

16.1 INTRODUCTION

This part of the book essentially handles the applications of polygonal approximations in structural pattern classifications and 2D occluded scene recognitions. The content of Chapter 17 deals with dissimilarity measure between two polygons and classification of structural patterns based on smoothed versions of polygons. Usually two geometrical measures are proposed in literature [67] to quantify the dissimilarity between two irregular polygons. These measures capture the intuitive notion of the dissimilarity between shapes and are related to the minimum value of the intersecting area of the polygons on superposing one on the other in various configurations. These measures are edge based and vertex based dissimilarity, but they are computationally heavy. A more easily computable measure of dissimilarity [67], referred to as the minimum integral square error between the polygons, is discussed in Chapter 17. Based on this latter measure classifications of structural patterns are performed. Experimental results involving the classification of the noisy boundaries of the four Great Lakes, Erie, Huron, Michigan, and Superior, using this integral square error measure, are presented. But at the time of classification, we consider the smoothed versions of polygons. Hence, in Chapter 17 we consider a scale preserving smoothing algorithm for polygons. The input to the algorithm is a polygon η and the output is its smoothed version η_ε. η_ε, which contains all the scale information that η contains, is called the linear minimum perimeter polygon (LMPP) of η within a tolerance of ε. The quantity ε controls the degree of smoothness and approximates η to η_ε. From the LMPP a representation for a polygon approximating η can be procured, which is invariant to scale and translation changes. Examples of smoothing maps are presented [68].

Many computer vision systems model objects using polygons. If objects occlude each other or do not appear entirely in view, we need to match a polygon scene fragment with the polygon representation of the model objects. Hence, in Chapter 18 we present a way to match polygon fragments. Using polygon moments and cross moments we compute a dissimilarity measure between two fragments. If the dissimilarity measure is less than a small number, then we can preliminarily conclude similar fragments. Otherwise, the polygon fragments are dissimilar and are not considered any further. In computing the dissimilarity measure, we also find a coordinate transform that maps one fragment to another. By using this coordinate transform, we test whether the scene polygon fragments really belong to an occluded object. These polygon moments can be computed by using just the end points of the line segments [73].

In Chapter 19, we consider a computer vision algorithm that recognize and locate partially occluded objects [74, 131]. The scene may contain unknown objects that may touch or overlap giving rise to partial occlusion. The algorithm is based on hypothesis generation and verification paradigm. The paradigm iteratively generates and tests hypotheses for compatibility with the scene until it identifies all the scene objects. Polygon representations of the object's boundary guide the hypothesis generation scheme. Choosing the polygon representation turns out to have powerful consequences in all phases of hypothesis generation and verification. Special vertices of the polygon called "corners" help to detect and locate the model in the scene. Polygon moment calculations lead to estimates of the dissimilarity between scene and model corners, and determine the model corner location in the scene. Extraction of the largest set of mutually compatible matches forms a model hypothesis. Using a coordinate transform that maps the model on to the scene, the hypothesis gives the proposed model's location and orientation. Hypothesis verification requires checking for region consistency. Experimental results give examples of all phases of recognizing and locating the objects.

In Chapter 20, we present a new approach to recognize and locate partially occluded rigid objects from a given scene and generate a belief about the scene using assumption based truth maintenance (ATM) system. The ATM system is basically a tool for belief revision. It explores multiple potential solutions and can work out efficiently with inconsistent information. In practice, sometimes occlusion of objects in a 2D scene may occur due to the presence of objects which are not described in our primary knowledge base and which may appear to be an object, in addition to the model objects of our primary knowledge base. Hence, after detection of such an event, question of revising belief about the scene may arise to establish a new belief. The present approach to recognize and locate an occluded scene is completely different from the existing paradigm based on the concept of hypothesis generation and verification [66].

To tackle the problem of occluded object recognition under uncertainty in Chapter 21 we consider a new interpretation of the multidimensional fuzzy reasoning. Subsequently, we realize that new interpretation through back propagation type neural network for recognizing occluded object based on voting scheme. At the learning stage of the neural network, fuzzy linguistic statements are used. Once learned, the non-fuzzy features of an occluded object can be classified. At the time of classification of the non-fuzzy features of an occluded object we use the concept of fuzzy singleton. An effective approach to recognize an unknown scene which consist a set of occluded objects is to detect a number of significant (local) features on the boundary of the unknown scene. Thus the major problems fall into the selection of the appropriate set of features (local) for representing the object in the training stage, as well as in the detection of these features in the recognition process. The features should be invariant to scale, orientation and minor distortions in boundary shape. The performance of the proposed scheme is tested through several experimental studies [133].

KEYWORDS

- **Hypothesis generation**
- **Linear minimum perimeter polygon**
- **Object recognition**
- **Pattern classification**
- **Polygon fragments**

17 Polygonal Dissimilarity and Scale Preserving Smoothing

CONTENTS

17.1 INTRODUCTION

This chapter deals with the measure of dissimilarity between two irregular polygons. Polygons are normalized to have either the same area or the same perimeter. The dissimilarity measure is used for classification of a closed boundary into one of a finite number of classes. The results of classification of closed boundaries are used in the area of automatic recognition of characters, industrial parts, maps, and airplanes. The suggested references are not exhaustive. For a more detail review of these areas interested readers are referred to [49, 103, 137].

Initially, we consider the techniques used to compare two irregular polygons. The first approach extracts a set of global numerical features from the polygons. The polygons are then compared using a norm in the numerical feature space. Some popular feature sets are the Fourier descriptors and their variants [54, 112, 135, 181, 183, 199], moments [3, 38, 64], chords [159], and the circular autogressive coefficients [69]. Some of these features are invariant to certain elementary transformations. Furthermore, they can often be used to reconstruct the original polygons. Another approach [2, 25, 42, 46, 49] reduces each polygon into a linear string of symbols. The distance between the respective strings, such as their correlation [42], the Levenshtein metric between the two strings, can be used as dissimilarity between the two polygons. Some of these distance measures are sensitive to noise [30]. There are other techniques for classification, example, the syntactic approach [31, 49, 105, 107, 176, 191, 195], the relaxation approach [30, 144] which are not based on dissimilarity measure between polygons.

The geometrical dissimilar measures between two polygons do not extract any features from the polygons. If the geometric resemblance between the pair of polygons is high, then the dissimilarity measure has low numerical values. Such dissimilarity measures are usually either edge based or vertex based. In case of edge based dissimilarity measure the two polygons η and μ are geometrically superimposed so that the ith edge of η falls alongside the jth of μ, coinciding their midpoints. In this configuration the non-overlapping area between the polygons, that is $E_{i,j}(\eta, \mu)$ is defined as the sum of the areas of the two polygons minus twice their overlapping area. The minimum of $E_{i,j}(\eta, \mu)$ over all i and j is defined as E(η, μ) which represents the dissimilarity between the two polygons. Similarly we can determine the vertex based dissimilarity measure. In case of vertex based dissimilarity measure to determine the non-overlapping area between the polygons, that is $V_{i,j}(\eta, \mu)$ the polygons are superimposed with the ith vertex of one coinciding with the jth vertex of the other and their angular bisectors fall alongside each other.

The method of superposition ensures the dissimilarity between the overall shapes of the two polygons. Hence, both E(η, μ) and V(η, μ) are small if and only if the geometrical resemblance between η and μ is high. Both the edge based dissimilarity E(η, μ) and the vertex based dissimilarity V(η, μ) between η and μ are zero if and only if η is a rotated, scaled, or translated version of μ.

But the dissimilarity measure based on the non-overlapping area measure is computationally heavy. Hence, in the next section we consider, another geometrical measure which behaves like the non-overlapping area, but is easier to compute. This measure is termed as the minimum integral error between the polygons. Using the integral error measure as a criterion, structural pattern classification can be performed. Actual classification of structural patterns of polygons is based on smoothed polygons.

Past decades considerable interests are shown for automatic recognition of structural pattern. In this context polygons have played a major role since the outer boundary of an object without holes can be approximated as a polygon. The advantages of such representations can be found, in [103, chapter VII]. Since the time required for processing the polygon is dependent on the number of edges it possesses, the polygons are usually approximated using a smoothing technique, such as the split and merge technique or the linear scan technique. These techniques and their variants have been well described in [103, pp. 161–184]. But none of these techniques preserves all the scale information contained in the unsmoothed boundary. To justify this statement, let us consider η and t which are two unsmoothed polygons with t being a scaled version of η. Here the scaling factor is $k > 0$. Let η^* and t^* be the corresponding smoothed versions, which are obtained using any of the algorithms known in the literature. Even though t is a scaled version of η, none of the currently available techniques can guarantee that η^* is a scaled version of t^*.

In this chapter, we propose a smoothing scheme which can indeed guarantee the preservation of scale information. The input to the scheme is a polygon η and its output is η_e, the smoothed version of η, referred to as linear minimum perimeter polygon (LMPP) of η within the tolerancee. The quantitye, $0 \leq e \leq 1$, is termed as the tolerance factor. The value $e = 0$ yields η_e identical to η and as e increases η_e approximates η more and more crudely. Within reasonable limits of e, η_e indeed preserves the scale information in η and yields e as a single control parameter by which the smoothing can be controlled.

A natural consequence of this technique is a representation for a smoothed version of a shape, which is invariant to changes in scaling and the translation of the coordinate system in which the original shape is drawn.

17.2 INTEGRAL ERROR DISSIMILARITY MEASURES [67]

Let us consider two polygons η and μ which are normalized to have the same unity perimeter. The two polygons are superimposed so that the ith edge of η falls alongside the jth edge of μ. The midpoints of the two edges coincide. This midpoint is considered as origin of a new rectangular coordinate system. Here the abscissa lies on these common edges. Now both the polygons are traversed in a clockwise direction starting from the origin. Let $E_i(\alpha)$ be the unique point reached on η after a traversal of length α, and let $F_j(\alpha)$ be the unique point reached on μ after a traversal of α, $\alpha \in [0, 1]$. To measure the point wise dissimilarity between the points $E_i(\alpha)$ and $F_j(\alpha)$ any norm in R^2 can be used. We consider the square error $\|E_i(\alpha) - F_j(\alpha)\|^2$ as the measure of dissimilarity. The cumulative effect of this local dissimilarity is obtained by integrating the quantity over α in the interval [0, 1]. Thus, we obtain the integral square error between the two polygons relative to their $i - j$ edges as:

$$D_{i,j}(\eta, \mu) = \int_0^1 \left\| E_i(\alpha) - F_j(\alpha) \right\|^2 d\alpha \ .$$

We minimize the quantity over all possible values of i and j. We call it the minimum integral square error $D(\eta, \mu)$

$$D(\eta, \mu) = \min_{i=1,\ldots\ldots} \int_0^1 \left\| E_i(\alpha) - F_j(\alpha) \right\|^2 d\alpha \, .$$

Remark 17.1

In the definition of $D_{i,j}(\eta, \mu)$, the square of the L_2 norm is used to measure the point wise dissimilarity between $E_i(\alpha)$ and $F_j(\alpha)$.

Remark 17.2

The edge based minimum integral square error between η and μ given by $D(\eta, \mu)$ is a pseudometric.

For the detail study on the capability of $\|E_i(\alpha) - F_j(\alpha)\|^2$ to capture the local dissimilarity between $F_j(\alpha)$ and $E_i(\alpha)$ for various shapes are available in [98].

Example 17.1 [67]

Let us consider η and μ as the square and rhombus respectively. Here the minor angle of the rhombus is Φ. Using geometrical and trigonometrical properties we get,

$D(\eta, \mu) = (10/192)\,(1 - \sin \Phi)$
which is a monotonically decreasing function of Φ. In this case, $D(\eta, \mu)$ is a monotonically increasing function of $E(\eta, \mu)$ [98]. Thus, we can demonstrate the utility of the integral $\int_0^1 \left\| E_i(\alpha) - F_j(\alpha) \right\|^2 d\alpha$ dα to quantify the overall dissimilarity.

Example 17.2 [67]

Let us consider a hexagon and a heptagon as shown in Figure 1. Figure 1 shows the superposition of the two polygons for computing $D_{1,1}$. The edges of η and μ are shown in Figure1 are represented by the Arabic and Roman numerical respectively. The functions $\|E_1(\alpha) - F_5(\alpha)\|^2$ and $\|E_1(\alpha) - F_1(\alpha)\|^2$ are plotted against α in Figure 2. Figure 2 shows the capability of $\|E_i(\alpha) - F_j(\alpha)\|^2$ to capture the point wise dissimilarity between the polygon η and μ are shown in Figure 1. By inspection we get $D(\eta, \mu) = D_{1,1}(\eta, \mu)$.

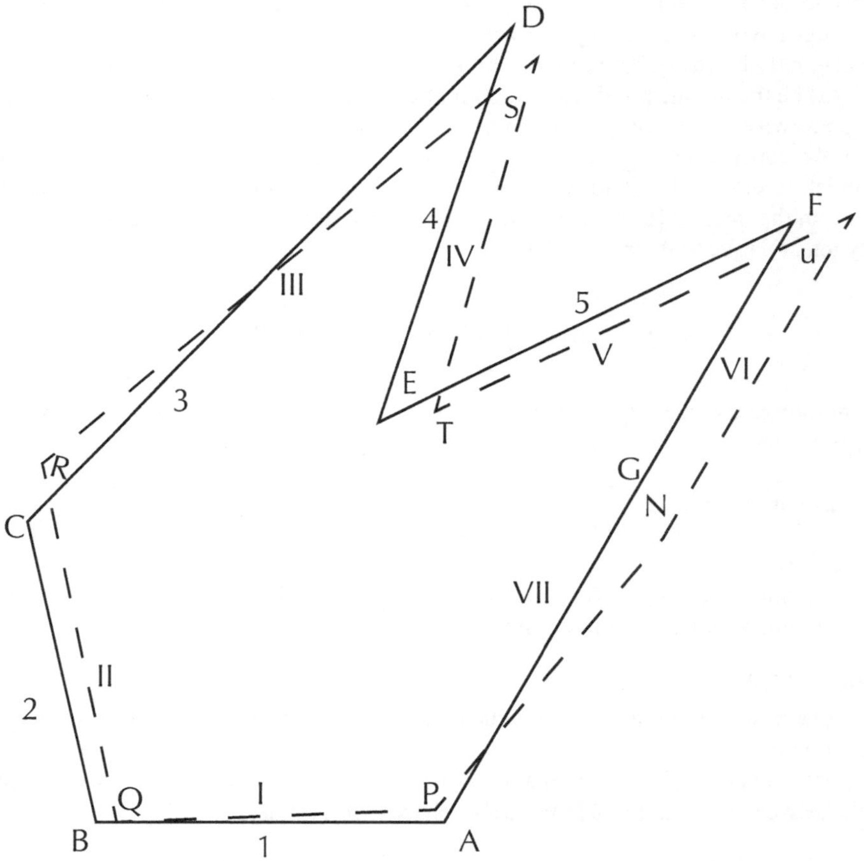

FIGURE 1 Superimpose of hexagon η, ABCDEF, and the heptagon μ, PQRSTUW [67].

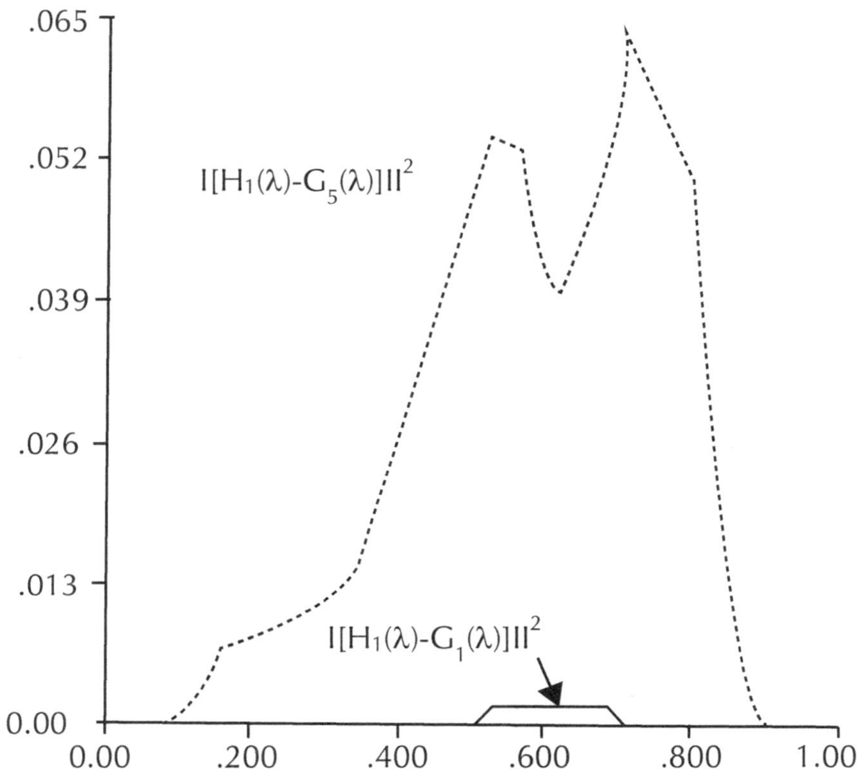

FIGURE 2 The point wise dissimilarity between the polygons η and μ of **Figure 1** [67].

17.3 COMPUTATION OF DISSIMILARITY MEASURES [67]

To compute $D_{i,j}(\eta, \mu)$ for any two indexes i and j, the functions $E_i(\alpha)$ and $F_i(\alpha)$ are evaluated at K points. The integer K is determined by δ the step size of α. It is independent of the number of edges in both η and μ. The integral:

$$\int_0^1 \left\| E_i(\alpha) - F_j(\alpha) \right\|^2 d\alpha$$

is approximated by the sum

$$\sum_{k=0}^{K-1} \left\| E_i(\alpha) - F_j(\alpha) \right\|^2 \delta$$

where
$K\delta = 1$.

Remark 17.3

The expression requires, the computation of K values of $E_i(\alpha)$ and $F_i(\alpha)$, K squared errors and 2K additions.

Remark 17.4

If η and μ have N and M edges, for evaluating $D(\eta, \mu)$ requires the computation of all the MN integrals $D_{i,j}(\eta, \mu)$. However, the same minimum can effectively be obtained by computing $D_{i,j}(\eta, \mu)$ for a subset of the MN integrals.

Let O_η and Q_μ be the centers of gravity of η and μ respectively. If the ith edge of η falls alongside the jth edge of μ, $\delta^{i,j}$ represents the distance between O_η and Q_μ. From Figure 3 we observe that $D_{i,j}(\eta, \mu)$ attains its minimum when i = p and j = q. Therefore, then $\delta^{p,q}$ becomes relatively small. Conversely, if $\delta^{r,s}$ is relatively large, it is unlikely that $D_{r,s}(\eta, \mu)$ is the minimum integral. Thus, using the quantities $\delta^{i,j}$ a proper subset of λ of the set $\{1, N\} \times \{1, M\}$ can be chosen to compute $D_{i,j}(\eta, \mu)$ only for those configurations $(i, j) \in \lambda$ for evaluating $D(\eta, \mu)$. The pair (i, j) is said to be a promising pair if $\delta^{i,j}$ is less than or equal to a predefined threshold, say δ^*. The set of all the promising pairs is defined as:

$$\lambda = [(i, j)|(i, j) \in (1, N), \times (1, M), \delta^{i,j} \leq \delta^*].$$

Thus the required estimate of D is,

$$(\lambda, \mu) = [D_{i,j}(\lambda, \mu)]. \tag{17.1}$$

Remark 17.5

The positions of O_η and Q_μ relative to the vertices of η and μ respectively, computed once only and are stored in polar representations. Subsequently, to test if any pair (i, j) is in λ, the locations of O_η and Q_μ in this configuration can be obtained by one linear transformation.

Remark 17.6

To obtain a suitable value for δ^*, in supervised approach to pattern recognition where a set of training samples for each class is given, calculate the value of $\delta^{i,j}$ for the optimum integral of each class. Collect such values for various classes and obtain a conservative estimate for δ^* for each class.

Remark 17.7

If value of δ^* is underestimated, the (η, μ) in (17.1) becomes close to $D(\eta, \mu)$, and thus becomes a good approximation of $D(\eta, \mu)$.

Remark 17.8

An alternative approach is to order the pairs (i, j) in the set $\{1, N\} \times \{1, M\}$ in ascending values of $\delta^{i,j}$. The integral $D_{i,j}(\eta, \mu)$ is computed for the first 1 pairs. A heuristic approach to estimate the number of elements in λ for the polygons under consideration is to make $\#\lambda = M + N$. In such a case, out of the MN distinct integrals, only M + N of them are considered.

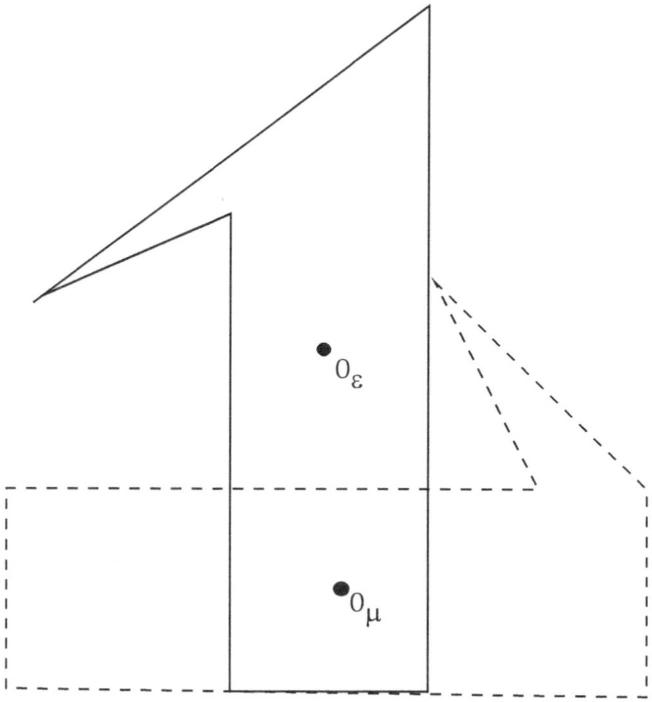

FIGURE 3 The distance between the centers of gravity O_η and Q_μ of η and μ respectively represents the degree of dissimilarity between two particular configurations [67].

Remark 17.9

In practice, all the integrals corresponding to the pairs in λ need not be calculated. We are interested only in the one that yields the minimum value. Initially, D(η, μ) is assigned the value of the first integral computed. If, in the subsequent computation of any $D_{i,j}(\eta, \mu)$, the value of the integral exceeds the currently stored value of D(η, μ), the computation of that integral is terminated. If, however, the integral is evaluated to its completion, the value of D(η, μ) is updated as the minimum of its current value and the latest $D_{i,j}(\eta, \mu)$. This reduces the computation time considerably.

For details computation time, interested readers are referred to [98]. One important area of application of dissimilarity measure the classification of structural patterns. Consider a pattern classification problem with J classes. Let η_i be the ideal polygon associated with the ith class. Let μ' be the test boundary to be classified, approximated by the polygon μ. The classification of μ can be performed by computing $D(\eta_i, \mu)$ for i = 1, J and the pattern assigned to the class which minimizes this dissimilarity measure. Since the η_i's and μ usually have a large number of edges, the computation of D(η, μ) is time consuming. Hence, it is better to smooth the polygons under consideration, before evaluating the dissimilarity measure. The smoothed version of any polygon τ is a polygon τ^* which approximates it according to some criterion and which possesses fewer edges. The actual smoothing process can be performed by using any of the techniques known in the literature, such as the split and merge technique. We have chosen to smooth the polygons using a variant of its minimum perimeter polygon (MPP) [94]. This variant, termed as linear minimum perimeter polygon (LMPP), has some interesting scale preserving properties and is described in [98]. The actual classification as shown in section 17.5 is achieved by computing the dissimilarities between the smoothed versions of the η_i's and the smoothed version of μ. In the next section we discuss on smoothing of polygons.

17.4 SMOOTHING OF POLYGON [68]

In this section, we first consider MPP. Then we define the LMPP and demonstrate its properties. Subsequently, we discuss with examples the use of the LMPP in smoothing maps.

Definition 17.1

Let η be any polygon specified by an ordered sequence of points in the plane. $\eta = \{P_i/i = 1, N\}$. Let δ_i be a prespecified circular or polygonal constraint domain in the neighborhood of P_i and let η' be any polygon specified by the sequence $\eta' = \{S_i/i = 1, N, S_i \in \delta_i\}$. The η' is any approximation of η in which the point P_i is perturbed to a new location S_i within the domain δ_i. The polygon η^* which satisfies η' and which has the minimum perimeter is called the MPP of η.

Sklansky et al. [156, 157] and Montanari [94] have suggested algorithms for computing the MPP if the disjoint domains δ_i are polygonal or circular, respectively. If the constraint domains are all disjoint circles with radius g, the MPP η^* satisfies the following:

$$\eta^* = \text{argument}\ \min_{\eta'} [f(\eta›)] \tag{17.2}$$

where

$\eta› = \{S_i/i = 1, N; \|P_i - S_i\| \leq \gamma\}$

and

$f(\eta') = \sum_{i=1}^{N} \|S_i - S_{i+1}\|$

with $S_{N+1} = S_1$. A vertex S_i of η^* is called an active vertex if it lies on the boundary of the constraint disk. In such a case

$$\|P_i - S_i\| = \gamma.$$

Montanari proved [94] that the MPP η^* possesses the following properties.

Property 17.1

η^* can be completely and uniquely specified by the sequence of active vertices.

Property 17.2

Let S_j and S_k be the first active vertices on either side of a vertex P_i of η. If the line joining S_j and S_k intersects the circular constraint disk around P_i, the vertex S_i is not an active vertex. In such a case, the edges of the MPP through S_i are collinear and can therefore be merged. This is depicted in Figure 4.

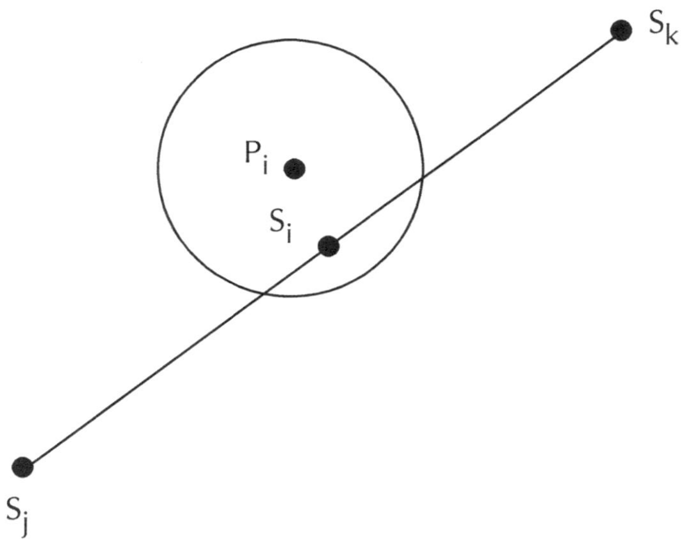

FIGURE 4 The line joining S_j and S_k intersects the constraint disk around P_i.

Property 17.3

If the line joining S_j and S_k lies outside the constraint disk around P_i, the vertex S_i becomes active. It lies on the boundary of the disk at the point where the bisector of the angle $S_j S_i S_k$ is normal to the boundary (see Figure 5).

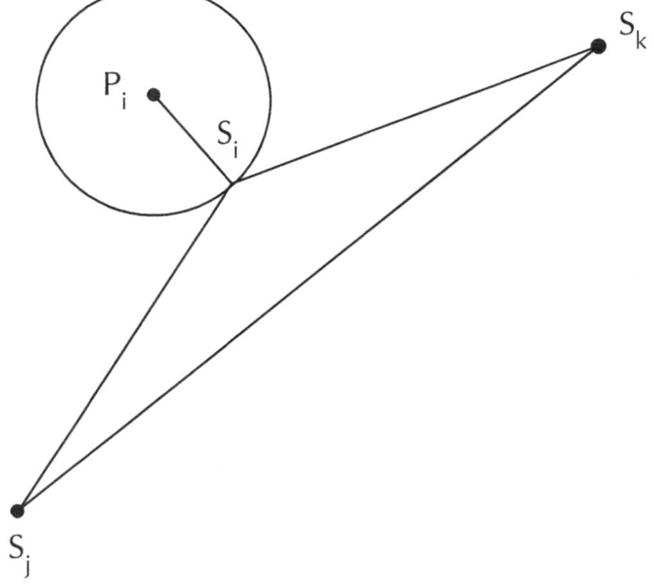

FIGURE 5 The line joining S_j and S_k lies outside the constraint disk around P_i.

17.5 LINEAR MINIMUM PERIMETER POLYGON [68]

Kashyap and Oommen extend the results of Montanari to formulate the LMPP.

Definition 17.2

The LMPP of a polygon η is its MPP obtained by making the radii of the constraint disks directly proportional to the perimeter of η. The constant of proportionality e is called the tolerance factor, and the LMPP of η obtained using a tolerance factor of η is given by η_e.

17.5.1 Statement of the Problem

Let $\eta = \{P_i / i = 1, N\}$. η_ε is obtained as:

η_ε = argument $\min_{\eta'}$ [f(η›)] (17.3)

with

$\eta› = \{S_i / i = 1, N; \|S_i - P_i\| \le \varepsilon L\}$

where

$$L = \sum_{i=1}^{N} \left\| P_i - P_{i+1} \right\|, \; P_{N+1} = P_1 \tag{17.4}$$

and

$$f(\eta') = \sum_{i=1}^{N} \left\| S_i - S_{i+1} \right\|, \; S_{N+1} = S_1 \, .$$

Remark 17.10

η_e equals η if and only if e is identically zero.

We state the following result [68] to demonstrate the scale preserving property of the LMPP.

Theorem 17.1 [68]

Let η be any closed boundary and, within the tolerance e, η_e be its LMPP. Let t be a scaled version of η and the scale factor is $k > 0$. The LMPP of t within the same tolerance e, that is τ_e, is a scaled version of η_e, where the scale factor is same as *k*, and the constraint disks are all disjoint.

Proof

Let $\eta = \{P_i / i= 1, N\}$. The perimeter of η that is $|\eta| = L$. Let its LMPP be $\eta_\varepsilon = \{S_i / i = 1, N\}$, with perimeter $|\eta'|$. Let $\tau = \{T_i / i= 1, N\}$be the scaled version of η. Therefore, by hypothesis $|\tau| = kL$. Hence the constraint disks around T_i must have radius εkL.

We construct the polygon $\tau^* = \{V_i / i = 1, N\}$, using η, η_e and the following rules.

Rule 1

If S_i is nonactive render V_i nonactive.

Rule 2

If S_i is active, V_i is the unique point on the boundary around T_i with the angular equality, $P_{i-1}P_iS_i = T_{i-1}T_iV_i$, as shown in Figure 6.

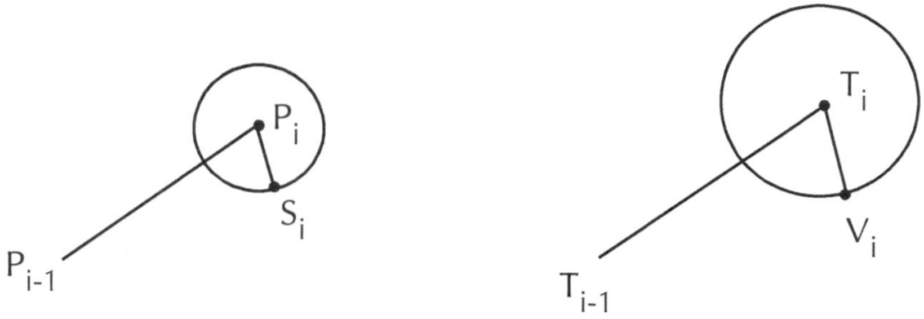

FIGURE 6 Representation of angular equality $P_{i-1}P_iS_i = T_{i-1}T_iV_i$.

By construction t* is a scaled version of η_e. The scaling factor is exactly k. Thus $|\tau|^* = k|\eta_\varepsilon|$.

Lemma 17.1 [68]

t* is the LMPP of t within the tolerance e.

Proof

Let $\tau^+ = \{W_i / i = 1, \ldots\ldots\ldots, N\}$ be any polygon satisfying $||W_i - T_i||$ £ εkL for $i = 1, \ldots., N$ and it is distinct from τ^*. Now we have to prove by contradiction that, $|t^+| > |t^*|$.

Suppose $|t^+|$ £ $k|\eta_\varepsilon|$. Using the polygons t and t$^+$ construct a polygon η^+ inside η with perimeter $1/k$ $[|t^+|]$. Thus $|\eta^+|$ £ $|\eta_\varepsilon|$ which implies that η^+ is the LMPP of η; but it is impossible, because by definition, η_e is the unique [94] LMPP of η. This in turn implies that t* is the LMPP of t within the tolerance e.

Remarks 17.11

The uniqueness of the MPP requires that the constraint disks are disjoint. Therefore, η preserves the scale information in η if and only if:

$< \min\ [||P_j - P_{j+1}||]/2\ \sum_{i=1}^{N} \|P_i - P_{i+1}\|$

where $$P_{N+1} = P_1 \qquad (17.5)$$

Remark 17.12

In many applications the range of e is not adequate because of the fact that two neighboring pixels may be adjacent vertices of a polygon which represents a boundary [47, 103]. In such cases, further smoothing can be achieved by preprocessing η to ensure that the constraint disks are disjoint. If P_k and P_{k+1} are two vertices in η obeying $\|P_k - P_{k+1}\| \le \varepsilon L$, we approximate η by merging P_k and P_{k+1} to a single point. The LMPP η is computed using the approximated version of η as shown in Figure 7.

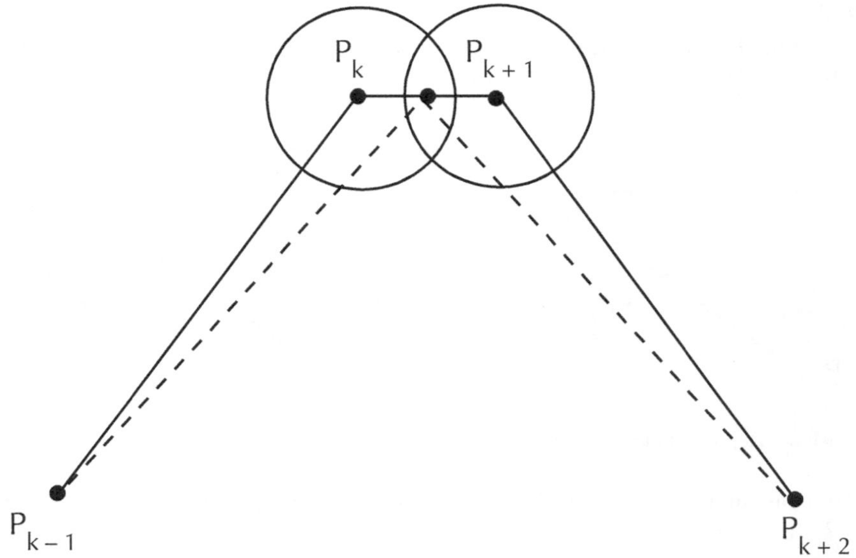

FIGURE 7 Approximation of η.

Remark 17.13

A consequence of Theorem 17.1 is that if e satisfies Equation (17.5), the LMPP provides a representation for a smoothed version of η which is invariant to scaling and translation of the original figure η.

The use of this representation in pattern classification is discussed in earlier section. For smoothing, we introduce an error norm which is proportional to the perimeter of the polygon to be smoothed. We exploit the properties of the active and nonactive vertices of the MPP and the uniqueness of the MPP. The error norm introduced here is augmented by the latter properties of the MPP to yield the scale preserving property of the LMPP claimed in Theorem 17.1. The LMPP merges all the edges which are "almost collinear". Thus, the number of edges is drastically reduced. Further, the vertices of LMPP fall on the boundaries of the constraint disks. Therefore, the vertices which primarily distinguish the shape of the polygon are preserved, as in Figure 5. Example 17.3 illustrates this phenomenon.

Example 17.3 [68]

Let us consider the quadrilateral ABCD of Figure 8. Let it be represented by τ and let L be the radii of constraint disks around the vertices of the quadrilateral ABCD, that is $L = \|AB\| + \|BC\| + \|CD\| + \|DA\|$.

Let the triangle abc be the LMPP t of τ. Let the angles of the triangle abc be α_a, α_b, and α_c, at the vertices a, b, and c, respectively, and let the lengths of the edges opposite to a, b, and c be l_a, l_b, and l_c, respectively. Therefore, the sequence of angles and lengths of the normalized LMPP is the ordered sequence $\{(\alpha_a, l_a/\Sigma), (\alpha_b, l_b/\Sigma), (\alpha_c, l_c/\Sigma)\}$, where å $= l_a + l_b + l_c$.

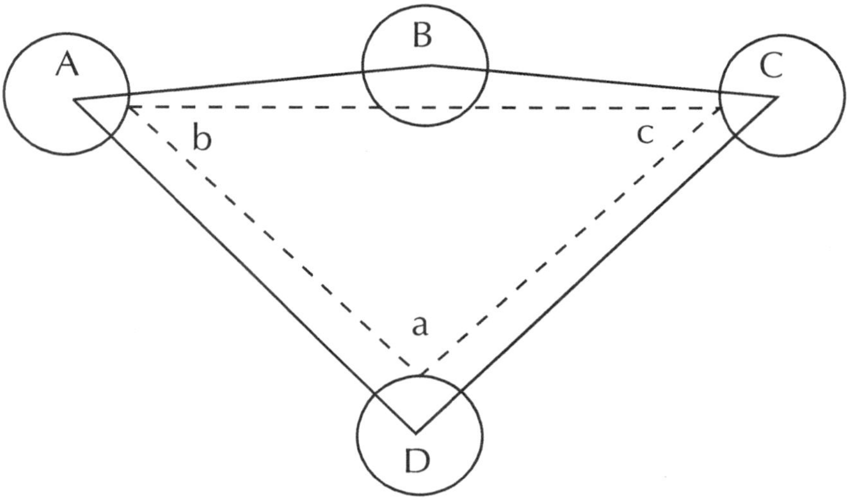

FIGURE 8 Representation of the quadrilateral ABCD and the triangle abc [68].

17.6 EXPERIMENTAL RESULT

To demonstrate the effectiveness of scale preserving approach to polygonal approximation we consider the Great Lakes of North America obtained from the National Geographic Magazine collection drawn at a scale of 32 mi/in. They were appropriately scaled to fit an 8×11 in frame. These pictures are eventually represented by an array of 90×90 pixel. A simple boundary tracking algorithm with constant thresholds is applied to extract the boundaries of the lakes. The boundary of a map is used as the original polygon and the LMPP is constructed as its smoothed version. The effect of smoothing using various values of ϵ is shown for Lake Erie in Figure 9(a)–(e). The original map is shown in Figure 9(a). The smoothed versions of 17.9(a) for ϵ equals of 0.001, 0.002, 0.005, and 0.01 are shown in Figure 9(b)–(e), respectively. It can be seen that a value of $\epsilon = 0.005$ very much acceptable from application point of view because, it preserves the principal vertices of η, and reduces the number of edges by about 60%. The approximation can be increased in terms of crudeness if the value of ϵ increases. For instance Figure 9(e) is a very crude approximation of the original map of Figure 9(a) for $\epsilon = 0.01$.

A scale preserving smoothing technique is valuable in the area of cartography. Maps of islands and lakes can be drawn to great precision provided a small scale factor is used. However, if a miniature map of an original has to be obtained, a fair amount of smoothing has to be done.

In this case study, we have presented a smoothing technique which has some interesting scale preserving properties. The smoothed version, known as the LMPP, is a method of approximating η by η_e where the latter has a number of edges less than or equal to the former.

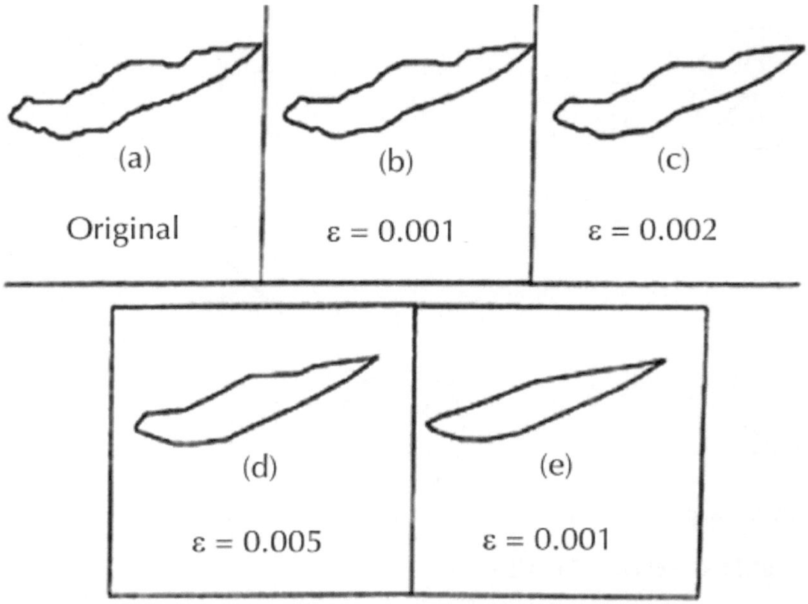

FIGURE 9 Examples of smoothing the boundary of Lake Erie. (a) Is the original and (b)–(e) are the smoothed version of the original map [68].

A pattern classification experiment is performed using the boundaries of the four Great Lakes, Erie, Huron, Michigan, and Superior. The maps of the lakes are obtained from [81] and are drawn in a similar fashion.

17.6.1 Generating Noisy Boundaries

To provide test boundaries for the classification process, unsmoothed boundaries are noisily degraded as follows. Let P_i be any arbitrary point on the boundary, whose adjacent vertices are P_{i-1} and P_{i+1}. Let the centroid of the triangle P_{i-1}, P_i, P_{i+1} is the point T_i. Then, the point P_i is noisily displaced to the point S_i which is on the line joining P_i and T_i.

$$S_i = P_i + \alpha(T_i - P_i)$$

where, α is a Gaussian random variable of mean zero and variance σ^2. If $0 \le \alpha \le 1$, the point S_i is a point in between P_i and T_i. If α is negative or greater than unity, the point S_i lies on the infinite line joining P_i and T_i, but is not in between them. This noise generating mechanism ensures a global and a local deformation on the boundary. The ideal boundaries and some typical noisy boundaries of the four Great Lakes are given in Figure 10 with $\sigma^2 = 6.25$. Note that the distortions in the shape are significant.

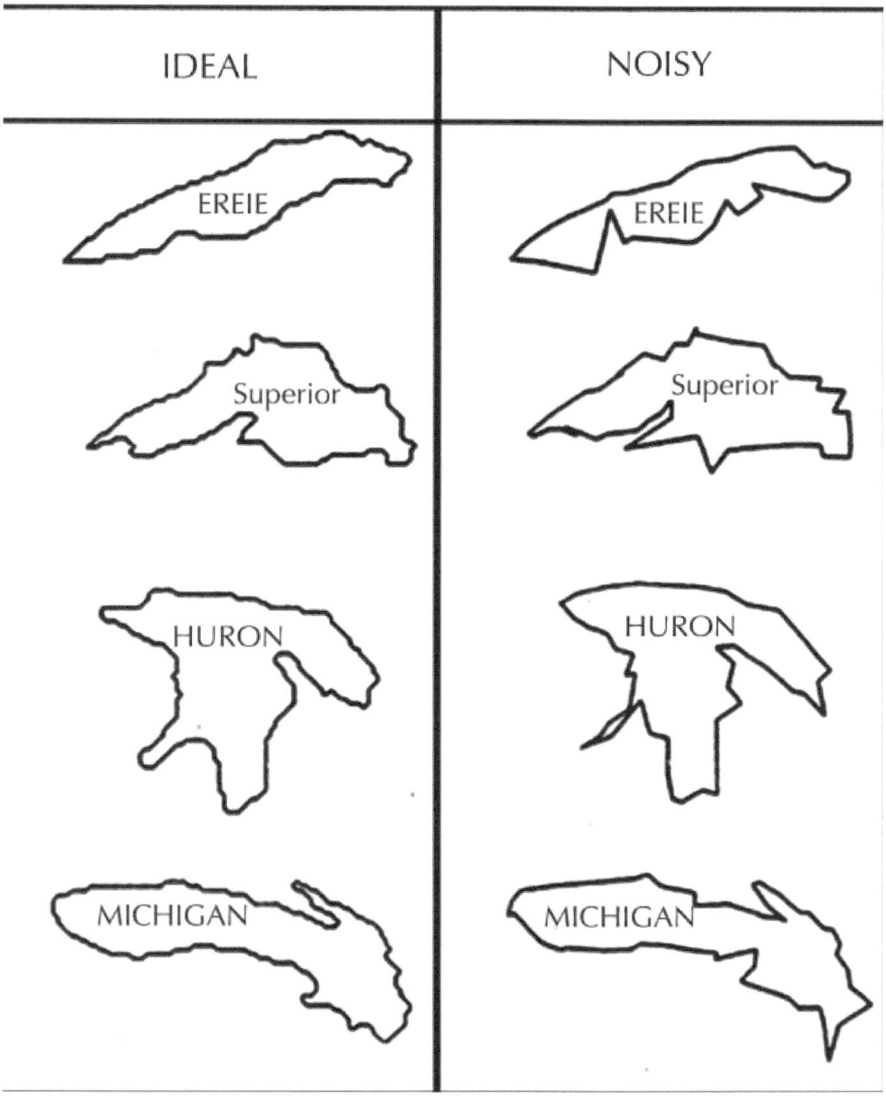

FIGURE 10 Ideal and noisy boundaries of the four Great Lakes [67].

17.6.2 Classification Results

The reference boundaries and the noisy boundaries are smoothed. These smoothed versions are compared using the minimum integral square error measure. The test boundary is assigned to the class which minimized this dissimilarity. One hundred noisy boundaries are tested using 20 different maps and five different variance levels ranging from 1.0 to 9.0. Out of the 100, 97 of them were correctly classified. The details of the classification are presented in Table 1 [67]. The third column of the Table 1 gives the total number boundaries tested. The last column of the Table 1 gives the class to which the given boundary is wrongly classified. The total classification score is 97%.

TABLE 1 Classification of Lakes [67]

Lake	Variance	Total Number	Number Correct	Mis. Class
Erie	1.00	5	5	---
	2.25	5	5	---
	4.00	5	5	---
	6.25	5	5	---
	9.00	5	5	---
Huron	1.00	5	5	---
	2.25	5	5	---
	4.00	5	5	---
	6.25	5	5	---
	9.00	5	5	---
Michigan	1.00	5	5	---
	2.25	5	5	Superior
	4.00	5	5	---
	6.25	5	5	---
	9.00	5	5	---
Superior	1.00	5	5	---
	2.25	5	5	---
	4.00	5	5	---
	6.25	5	4	Michigan
	9.00	5	5	---

Figure 11 [67] demonstrates how the measure $D(\cdot, \cdot)$ is able to discriminate between two classes which are structurally distinct. The value of D(ideal Erie, μ) is plotted against the corresponding value of D(ideal Huron, μ) for the various noisy boundaries of both Lakes Erie and Huron. From the figure, we can observe that as the two lakes are structurally very much dissimilar, the values of the dissimilarity measures are also widely different. The clustering of the noisy boundaries obviously separates them into the two classes. These results justify the use of the geometrical dissimilarity measures for the structural classification of closed boundaries.

In this case study we have considered the problem of quantifying the dissimilarity between two irregular polygons η and μ. A measure D(η, μ), termed as the minimum integral square error between η and μ, is proposed. The computation of D(η, μ) is presented, and its use in pattern classification is verified in experiments involving the four Great Lakes Erie, Huron, Michigan, and Superior.

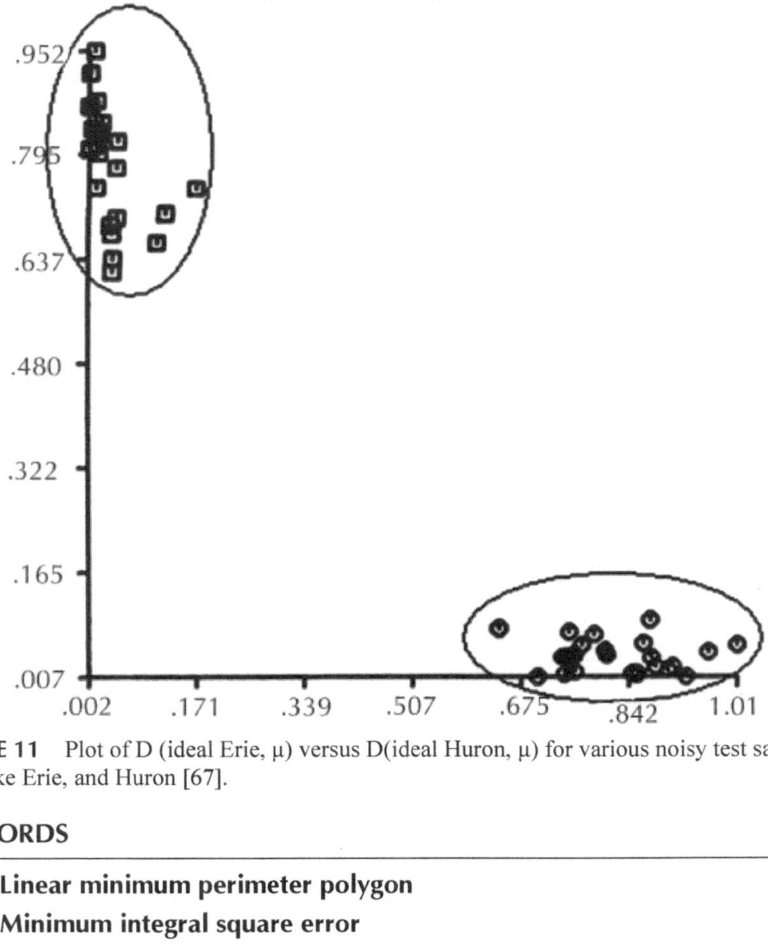

FIGURE 11 Plot of D (ideal Erie, μ) versus D(ideal Huron, μ) for various noisy test samples, μ of Lake Erie, and Huron [67].

KEYWORDS

- **Linear minimum perimeter polygon**
- **Minimum integral square error**
- **Minimum perimeter polygon**
- **Polygonal dissimilarity**
- **Tolerance factor**

18 Matching Polygon Fragments

CONTENTS

18.1 INTRODUCTION

Polygonal approximation of model objects that may appear in an image is used in many tasks of computer vision and graphical system. The object may consist of curves and line segments. Errors between the polygon representation and a curve can be reduced to an arbitrarily small value by using more and more vertices to obtain a better piecewise approximation of the curve. The basic aim of polygon approximation is to control the huge data explosion which is inherent in vision and graphical systems. Representing the collinear boundary points of a curve by line segments, we can drastically reduce the boundary representation from several 1,000 points to a very less number of points. Thus, we can save the computational time and memory requirement to store, display, and manipulate the objects.

In model based computer vision system to perform the task of object recognition, we need to compare the stored model polygons with polygons extracted from the scene. The resulting matches are used to recognize and locate the objects in the scene.

Sometimes the objects are either occluded or partially visible in the scene. Under such situation, we match only the visible parts of the polygons and subsequently take suitable hypothesis based on the partial match between model object and scene to identify and locate the objects in the scene. Hence, matching polygon fragments has an important role in the task of occluded object recognition. The rest of this chapter, we discuss the matching of polygon fragments [73].

18.2 REVIEW OF THE STATE OF ART

The matching of polygon fragments is performed either through the measure of similarity or through the measure of dissimilarity between the fragments. The best coordinate transform places one fragment on to the other. The coordinate transform helps locate a model object in the scene and thus, generates a hypothesized scene consists of model objects.

There are several methods for matching polygon fragments. Sometimes, depending upon the problem at hand local features such as vertex position, interior angle at the vertex, line segments slope, or line segment length are used [73] for matching polygonal fragments. In this case, the object appears at a standard orientation and position then a matching scheme may use all the mentioned features, otherwise for objects that may appear in different positions and orientations, only the features like vertex angle and line length are used. Based on such features the dissimilarity measure is represented as:

$$d = \sum_{i=1}^{n} \omega_i \left(Mf_i - Sf_i \right)$$

where, Mf_i and Sf_i denote the respective model and scheme features and ω_i represents a weight for feature f_i. A larger weight assigns more importance to the feature.

To determine the coordinate transform for the match we may align the bisectors of the vertices. We may also find the locus of coordinate transforms that align the two unequal line segments. We save the result of the best coordinate transform for other compatible evidence [161]. But due to noise and suboptimal polygon approximation, we may obtain bad segmentations which cause non-unique placement. Thus, make the approach unsuitable for these cases. In [98, 100] quantizing the fragment into n points using a small step is adopted. The matching algorithm then aligns the polygon fragments and pairs the matching points on the fragments. Finally, the matching algorithm sums the squared errors between each pair of points. The sum of squared errors represents the non-overlapping area between the fragments. A large non-overlapping area indicates two dissimilar polygons, while a small non-overlapping area indicates two similar polygons. Aligning the two fragments requires a coordinate transform that maps one fragment onto the other. Oommen [98] aligns the complete polygons using edges or vertices. He measures the dissimilarity between whole polygons. For example, designate two polygons by η and μ. To calculate the non-overlapping area between η and μ, Oommen normalizes the two polygons with respect to perimeter and aligned them at an edge or vertex. An edged based similarity measure superimposes η on top of μ, so that the jth edge of η aligns with the ith edge of μ, and the midpoints of the two edges also coincide. The vertex based similarity measure superimposes η and μ. Thus, the jth vertex of η aligns with the ith vertex of μ, and the bisector of the two vertices also coincides. But for partially occluded objects the normalization technique is not suitable. Parui and Majumder [100] aligns the two fragments by finding the coordinate transform that minimizes the least squared error between pairs of matching points. The quantizing of the polygon fragment into n points underutilizes the merit of the polygon approximation. The polygons represent groups of collinear points. The quantization reverts back to representing the line segments with large number of points. After deriving the coordinate transform and least squares error we find them to contain sums of the n quantized points from both fragments. We call these sums polygon moments and polygon cross moments. We demonstrate how to compute them using the end points of the line segment. This allows us to compute the dissimilarity between two polygon fragments by using just the coordinates of the end points of the line segment. Polygon moments can be used to compute local features that describe the polygon fragment's shape. Combination of polygon moments which is invariant to rotation and translation represent the local features. We achieve the initial matching between two polygon fragments by comparing the invariant moments which have better discriminating power than the local properties like angle and line length.

18.3 DISSIMILARITY BETWEEN TWO POLYGON FRAGMENTS [73]

Koch and Kashyap extends [73] Ommen's polygon dissimilarity measure and Parui's polygon fragment dissimilarity measure. We assume that the two polygon fragments have equal length but can have a different number of segments with varying lengths. One fragment is the model fragment and other fragment is the scene fragment. We compute the coordinate transform that aligns the two fragments. We then minimize the squared error. The minimal squared error measures the dissimilarity between the two fragments. We derive the coordinate transform by quantizing the two fragments into n points. The coordinate transform and minimal squared error calculation involve sums of the n models and scene coordinate points. We compute these sums using the coordinated of the line segment's end points.

18.3.1 Least Squares Error between Two Polygon Fragments

We assume that both fragments have equal total length. But they may have line segments differing in number and in length. To measure the dissimilarity between the n points of two polygon fragments, L_1 and L_2, we consider the following expression.

$$D(L1, L2) = \frac{1}{2}\sum_{i=1}^{n}\left[(L_1x_i - L_2x_i)^2 + (L_1y_1 - L_2y_i)^2\right]. \tag{18.1}$$

$D(L_1, L_2)$ represents the distance between two fragments L_1 and L_2.

Let $L_{2\tau}$ represents the rotated and translated version of L_2 where θ represents the rotation and (T_x, T_y) represents the translation. Therefore, the dissimilarity between two fragments, L_1 and L_2 is:

$$J(L_1, L_2) = \min_\tau D(L_1, L_{2\tau}). \tag{18.2}$$

Thus, to calculate J for any pair of model fragment, M and scene fragment, S, find the coordinate transform, τ, that minimizes $D(S, M_\tau)$. For i = 1, n (Mx_i, My_i) represent n points along a model fragment M. The coordinate transform, τ, determines the n points along M_τ which is obtained by a rotation, θ and a translation, (T_x, T_y), to each point, (Mx_i, My_i) of the model fragment M, that is:

$$M_\tau x_i = M\, x_i \cos(\theta) - M\, y_i \sin(\theta) + T_x, \text{ and}$$
$$M_\tau y_i = M\, x_i \sin(\theta) + M\, y_i \cos(\theta) + T_y \text{ for } i = 1, n.$$

To determine J(S, M) the following expression is minimized over τ:

$$D(S, M_\tau) = \frac{1}{2}\sum_{i=1}^{n}\left[(Sx_i - M_\tau x_i)^2 + (Sy_i - M_\tau y_i)^2\right]. \tag{18.3}$$

First the minimum of θ, that is θ^* is estimated. The coordinate systems are changed without loss of generality so that

$$\frac{1}{n}\sum_{i=1}^{n} Sx_i = 0 \text{ and } \frac{1}{n}\sum_{i=1}^{n} Mx_i = 0.$$

Using θ^* we then find the best estimate for the translation, (T^*_x, T^*_y), by finding the (T_x, T_y) that minimizes (18.3). The following represents the best estimate of (θ, T_x, T_y) in terms of the n scene coordinate points and n model coordinate points:

$$\theta^* = \tan^{-1}(V'_2 / V'_1) \tag{18.4}$$

where

$$V'_2 = \frac{1}{n}\sum_{i=1}^{n} Sy_i Mx_i - \frac{1}{n}\sum_{i=1}^{n} Sy_i \ \frac{1}{n}\sum_{i=1}^{n} Mx_i - \frac{1}{n}\sum_{i=1}^{n} Sx_i My_i + \frac{1}{n}\sum_{i=1}^{n} Sx_i \ \frac{1}{n}\sum_{i=1}^{n} My_i$$

and

$$V'_1 = \frac{1}{n}\sum_{i=1}^{n} Sx_i Mx_i - \frac{1}{n}\sum_{i=1}^{n} Sx_i \ \frac{1}{n}\sum_{i=1}^{n} Mx_i + \frac{1}{n}\sum_{i=1}^{n} Sy_i My_i + \frac{1}{n}\sum_{i=1}^{n} Sy_i \ \frac{1}{n}\sum_{i=1}^{n} My_i.$$

$$T^*_x = \frac{1}{n}\sum_{i=1}^{n} Sx_i - \cos(\theta)\frac{1}{n}\sum_{i=1}^{n} Mx_i + \sin(\theta)\frac{1}{n}\sum_{i=1}^{n} My_i. \tag{18.5}$$

$$T^*_y = \frac{1}{n}\sum_{i=1}^{n} Sy_i - \sin(\theta)\frac{1}{n}\sum_{i=1}^{n} Mx_i - \cos(\theta)\frac{1}{n}\sum_{i=1}^{n} My_i. \tag{18.6}$$

Parui and Majumder [100] obtain similar results for the best estimate of θ. The minimum square error J(S, M) is obtained from the optimal coordinate transform parameters. Thus, we obtain the squared error in terms of the n scene points the n model points and the optimal coordinate transform parameters

$$J = V_3 - T^2_x - T^2_y - 2\cos(\theta)\, V_1 - 2\sin(\theta)\, V_2 \tag{18.7}$$

where

$$V_1 = \frac{1}{n}\sum_{i=1}^{n} Sx_i Mx_i + \frac{1}{n}\sum_{i=1}^{n} Sy_i My_i,\ V_2 = \frac{1}{n}\sum_{i=1}^{n} Sy_i Mx_i - \frac{1}{n}\sum_{i=1}^{n} Sx_i My_i,$$

and

$$V_3 = \frac{1}{n}\sum_{i=1}^{n} Sx_i^{\,2} + \frac{1}{n}\sum_{i=1}^{n} Mx_i^{\,2} + \frac{1}{n}\sum_{i=1}^{n} Sy_i^{\,2} + \frac{1}{n}\sum_{i=1}^{n} My_i^{\,2}.$$

The dissimilarity between the two polygonal fragments is obtained from J(S, M). In course of calculating J, we can find the coordinate transform that takes the model fragment onto the scene fragment. The polygon moments of the model fragment is represented as, (18.4)–(18.7)

$$\frac{1}{n}\sum_{i=1}^{n} Mx_i, \quad \frac{1}{n}\sum_{i=1}^{n} My_i, \quad \frac{1}{n}\sum_{i=1}^{n} Mx_i^{\,2}, \text{ and } \frac{1}{n}\sum_{i=1}^{n} My_i^{\,2}.$$

The terms for the model fragment appear in Equations (18.4)–(18.7). Similar terms exist for the scene fragment. The polygon cross moments are represented as,

$$\frac{1}{n}\sum_{i=1}^{n} Sx_iMx_i, \frac{1}{n}\sum_{i=1}^{n} Sx_iMy_i, \frac{1}{n}\sum_{i=1}^{n} Sy_iMx_i, \text{ and } \frac{1}{n}\sum_{i=1}^{n} Sy_iMy_i.$$

The cross terms have coordinates from both fragments. The dissimilarity calculation between two fragments needs all these terms. From the boundary point data to the polygon approximation and then to quantize the polygon approximation wastes computer time. It does not utilize the merit of the polygon approximation. Next, we compute the polygon moments and polygon cross moments using the end points of the polygon line segments.

18.3.2 Polygon Moments and Polygon Cross Moments for Single Line Segments

In this section, we show how to compute the polygon moments and polygon cross moments for single line segments by using just the end points of the line segments. In the next section, an extension of these results gives the polygon moments and polygon crosses moments for groups of line segments or polygon fragments. This results in a computational savings when computing the moments for a polygon fragment [73].

Let

$$\eta_{rs} = \frac{1}{n}\sum_{i=0}^{n-1} (Au_i)^r (Bv_i)^s, \tag{18.8}$$

which represents the (r + s)th order moment between n u-coordinate points from line segment L_1 and n v-coordinate points from line segment L_2. This represents both polygon moments and polygon crosses moments. For polygon moments $L_1 = L_2$, and the u and v coordinates come from the same line segment. For polygon crosses moments $L_1 \neq L_2$. The u and v represent the same coordinates. The polygon cross moments between the x coordinates of line segments L_1 and the x coordinates of line segment L_2 is described by u = v = x. We find η_{rs} for any order, though the calculation of the dissimilarity requires the highest order of two. In case of polygon moment the restriction that the two line segments L_1 and L_2 have the same length does not create any problem.

But this restriction create problem for polygon cross moments, because they have coordinates from two different line segment. In case of polygon cross moments the two fragments either have the same total length but have different number of line segments or have the same number of line segments with vertices in different positions along the fragments. To overcome this problem, for model fragment, the fragment matching algorithm finds vertices that do not exist in the scene fragment and places them in the scene fragment. Whereas, for a scene fragment, the matching algorithm finds vertices in the scene fragment that do not exist in the model fragment and places them in the model fragment. This procedure, used to compute the cross moments, divides the model and scene fragment into the same number of equal length line segments.

Consider the Figure 1 The line segment L_1 have endpoints (L_1x_0, L_1y_0) and (L_1x_n, L_1y_n). u stands for either the x or y coordinate on the L_1 line segment. The line segment L_2 has endpoints (L_2x_0, L_2y_0) and (L_2x_n, L_2y_n). v stands for either the x or y coordinate on the L_2 line segment.

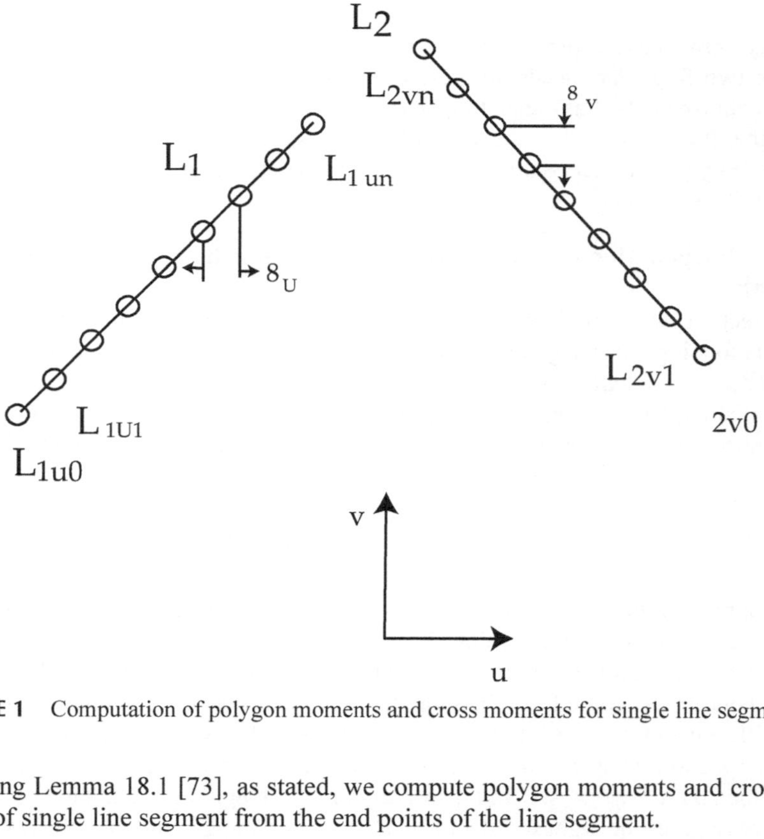

FIGURE 1 Computation of polygon moments and cross moments for single line segments.

Using Lemma 18.1 [73], as stated, we compute polygon moments and cross moments of single line segment from the end points of the line segment.

Lemma 18.1[73]

$$\eta_{rs} = \sum_{l=0}^{s}\sum_{j=0}^{r}\binom{r}{j}\binom{s}{l}(L_1u_0)^{r-j}(L_1u_n - L_1u_0)^j(L_2v_0)^{s-l}(L_2v_n - L_2v_0)^l\frac{1}{j+l+1} \quad (18.9)$$

where $\binom{R}{S}$ represents the combination of r things taken s at a time.

Proof [73] Line segments L_1 and L_2 are divided into n equally spaced points

$$\text{Let, } \Delta_u = \frac{L_1u_n - L_1u_0}{n} \text{ and } \Delta_v = \frac{L_2v_n - L_2v_0}{n}, \quad (18.10)$$

where Δ_u is the distance between the samples on line segment L_1 in the u direction, and Δ_v is the distance between the samples on line segment L_2 in the v direction.
The first point and the distance between the samples can express each of the n points

$$L_1u_i = L_1u_0 + i\,\Delta_u, \text{ and } L_2v_i = L_2v_0 + i\,\Delta_v \text{ for } i = 1, n\text{-}1 \quad (18.11)$$

From Equation (18.8) and Equation (18.11) we get

$$\eta_{rs} = \frac{1}{n}\sum_{i=0}^{n-1}(L_1u_0 + i\Delta_u)^r\,(L_2v_0 + i\Delta_v)^r.$$

Using the binomial expansion, we get,

$$\eta_{rs} = \sum_{j=0}^{r}\sum_{l=0}^{s}\binom{r}{j}\binom{s}{l}(L_1u_0)^{r-j}(L_1u_n - L_1u_0)^j(L_2v_0)^{s-l}(L_2v_n - L_2v_0)^l\,\delta$$

where, Δ_u and Δ_v are substituted from Equation (18.10) and where

$$\delta = \sum_{i=0}^{n-1}\frac{i^l}{n^{l+1}}, \text{ and } l = j + l..$$

To evaluate the term δ we consider the following identity from Knuth [72],

$$\sum_{i=0}^{n} i^l = \sum_{k=0}^{l} k!\left\{{l \atop k}\right\}\binom{n+1}{k+1} \quad (18.12)$$

where $\left\{{l \atop k}\right\}$ represents Stirling's numbers of the second kind. Thus, we set

$$\delta = \sum_{k=0}^{l} k!\left\{{l \atop k}\right\}\binom{n}{k+1}\frac{1}{n^{l+1}}.$$

Expanding and simplifying, we get

$$\delta = \sum_{k=0}^{l}\frac{1}{k+1}\left\{{l \atop k}\right\}\frac{n!}{(n-k-1)!}\frac{1}{n^{l+1}}.$$

For large n, all the terms in δ become zero except for k = l. Therefore, for large n, using the identity $\left\{ {l \atop k} \right\} = 1$, the k = l term for δ becomes 1/(l + 1). Thus,

$$\eta_{rs} = \sum_{j=0}^{r} \sum_{l=0}^{s} \binom{r}{j} \binom{s}{l} (L_1 u_0)^{r-j} (L_1 u_n - L_1 u_0)^j (L_2 v_0)^{s-l} (L_2 v_n - L_2 v_0)^l \frac{1}{l+1}.$$

Using l = j + l we get (18.9)

Using Lemma 18.1, we can calculate the polygon moments and polygon cross moments for single line segments from just their end points.

Example 18.1 Now we calculate the first polygon moment for the model fragment's y coordinates.

Let r = 1, s = 0, u = y, and L_1 = M. Therefore we get,

$$\frac{1}{n} \sum_{i=1}^{n} My_i = \eta_{10} = My_0 + \frac{My_n - My_0}{2} = \frac{My_0 + My_n}{2}.$$

Similarly, we can calculate the second order polygon crosses moment between the model fragment's y coordinates and the scene fragment's x coordinates.

Let r = l, s = l, u = x, v = y, L_1 = M, and L_2 = S. We get,

$$\frac{1}{n} \sum_{i=1}^{n} Sx_i My_i = \eta_{11}$$

$$= Mx_0 Sy_0 + \frac{(Mx_n - Mx_0) Sy_0}{2} + \frac{Mx_0 (Sy_n - Sy_0)}{2} + \frac{(Mx_n - Mx_0)(Sy_n - Sy_0)}{3}$$

$$= \frac{2Mx_0 Sy_0 + Mx_0 Sy_n + Mx_n Sy_0 + 2Mx_n Sy_n}{6}.$$

With the help of Lemma 18.1, the computation of moments becomes very efficient. In the next section, we combine the polygon moments to form the polygon moments for groups of line segments.

18.3.3 Polygon Moments and Polygon Cross Moments for Polygon Fragments

To calculate the fragment moments, we consider the polygon moments and polygon cross moments for line segment groups. Using the polygon moments for single line segments, we combine the polygon moments and polygon cross moments for single line segments to find the moments for groups of line segments. These moments represent the dissimilarity between two fragments.

Let

$$\eta^J_{rs} = \frac{1}{n} \sum_{i=0}^{n_J - 1} (L_1 u_i^J)^r (L_2 v_i^J)^s. \tag{18.13}$$

which represents the cross polygon moment between the u coordinates of fragment L_1's Jth line segment and the v coordinates of fragment L_2's Jth line segment. Here polygon moments represent a special case of the cross polygon moments. By assigning L_1 to L_2 we have the polygon moments for fragment L_1. Now we stipulate the following result [73]

Lemma 18.2

Let M_{rs} denote the r + s order moment between fragment L_1 and L_2. Then

$$M_{rs} = \sum_{J=1}^{m} \lambda_J \eta_{rs}^{J} / \sum_{J=1}^{m} \lambda_J \tag{18.14}$$

where λ_J represents the length of line segment J.

Proof [73]From Equation (18.8):

$$M_{rs} = \frac{\sum_{i=0}^{n_1-1}(L_1 u_i^1)^r (L_2 v_i^1)^s + \ldots\ldots\ldots + \sum_{i=0}^{n_m-1}(L_1 u_i^m)^r (L_2 v_i^m)^s}{n_1 + \ldots\ldots\ldots + n_m}$$

$$M_{rs} = \sum_{J=1}^{m} \sum_{i=0}^{n_J-1} (L_1 u_i^J)^r (L_2 v_i^J)^s / \sum_{J=1}^{m} n_J.$$

Using (18.13)

$$M_{rs} = \sum_{J=1}^{m} n_J \eta_{rs}^{J} / \sum_{J=1}^{m} n_J. \tag{18.15}$$

But

$$\lambda_J = n_J \varepsilon \tag{18.16}$$

where ε represents the distance between the samples for all line segments. Solving (18.16) for n_J and substituting in (18.15)

$$M_{rs} = \sum_{J=1}^{m} \frac{\lambda_J}{\varepsilon} \eta_{rs}^{J} / \sum_{J=1}^{m} \frac{\lambda_J}{\varepsilon}. \tag{18.17}$$

From (18.18) we can get (18.14).

If we consider the length of line segment as weight, then using the weighted sum of the moments for the individual line segments we can calculate polygon moments and polygon cross moments for groups of line segments or fragments. Thus, based on polygon moments for a model fragment and scene fragment, and the polygon cross moments we can generate a dissimilarity measure to decide the matching of two fragments.

18.3.4. Initial Matching of Fragments

Before computing the dissimilarity between a model fragment and a scene fragment, we consider invariant (under rotation and translation) moment features of fragments which are due to Hu [64]. These invariant moments, which is less accurate and less expensive than the actual dissimilarity measure of Equation (18.7) first prunes less promising matches between fragments. Subsequently, we apply dissimilarity measure of Equation (18.7) between the two fragments.

Hu [64] essentially consider combination of moments that have the rotation and translation invariant properties. The following gives the two low order invariant moments for a model fragment, M

$$\alpha_1 = \sigma_x + \sigma_y \quad \text{and} \quad \alpha_2 = \sqrt{(\sigma_x - \sigma_y)^2 + 4(\sigma_{xy})^2} \tag{18.18}$$

where

$$\sigma_x = \left(\frac{1}{n}\sum_{i=1}^{n} Mx_i^2 - \left(\frac{1}{n}\sum_{i=1}^{n} Mx_i\right)^2\right), \qquad \sigma_y = \left(\frac{1}{n}\sum_{i=1}^{n} My_i^2 - \left(\frac{1}{n}\sum_{i=1}^{n} My_i\right)^2\right)$$

and

$$\sigma_{xy} = \left(\frac{1}{n}\sum_{i=1}^{n} Mx_i My_i - \frac{1}{n}\sum_{i=1}^{n} Mx_i \frac{1}{n}\sum_{i=1}^{n} My_i\right).$$

The polygon moments of the model fragments produce the invariant moments, α_1 and α_2. Similar invariant moments can be obtained for the scene fragments. We compute the model fragment invariant moments during the learning phase of the objects and scene fragment invariant moments once during recognition phase. By comparing scene fragment invariant moments with model fragment invariant moments, using a threshold, an initial decision on the similarity between two polygon fragments can be obtained.

18.4 EXPERIMENTAL RESULTS

In this section, we state results using some 2D industrial objects. The objects are obtained from high contrast images. A simple thresholding algorithm separate the objects from the background and a boundary following routine extract the boundaries of the objects [73].

Case Study I Matching Two Fragments

Figure 2 shows the three model objects. The stars represent the vertices found by a split and merge polygon approximation algorithm.

From the end points of the polygon fragments of model object-2 of Figure 2 the polygon moments and cross moments are computed. Using the moments, we get the dissimilarity measure, J, and matching coordinate transform, (θ, T_x, T_y). Matching the polygon fragments gives a J = 0.35, and $(\theta, T_x, T_y) = (300.5, -60.0, 140.8)$. Thus, in Figure 3 we obtain translated and rotated version of model object in the occluded scene.

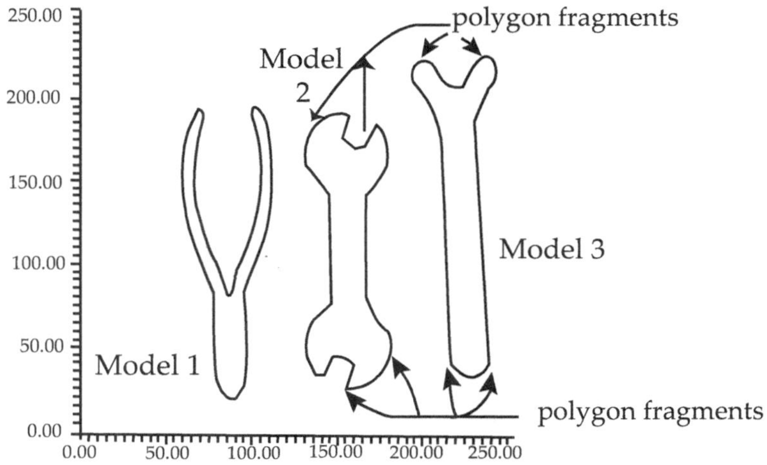

FIGURE 2 Polygon approximation of model objects.

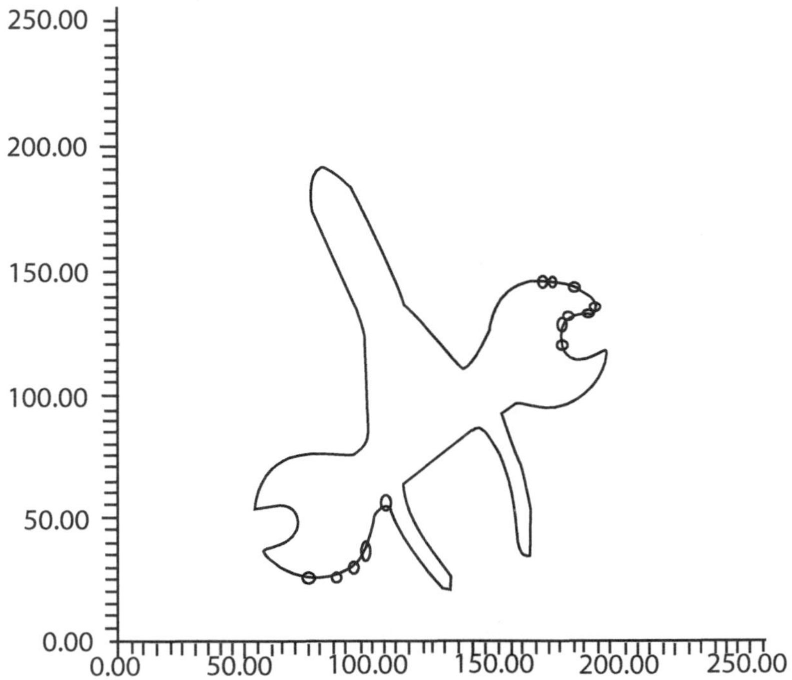

FIGURE 3 Two polygon fragments of the model object 2 (wrench) that correspond to rotated and translated versions of the fragments in Figure 2.

Case Study II Matching Discontinuous Fragments by Combination of Moments

According to Lemma 18.2 fragments can be combined to create a higher level feature by combining the polygon moments. Using these combined moments we can compute a new dissimilarity measure and coordinate transform to bring these higher level features into alignment. The results of the matching give a J = 0.45, and (θ, T_x, T_y) = (310.2°, –62.3, 150.5). Note that the value of the dissimilarity has gone up since more information is being used, and from Figure 4 we see that both model fragments are placed on occluded scene.

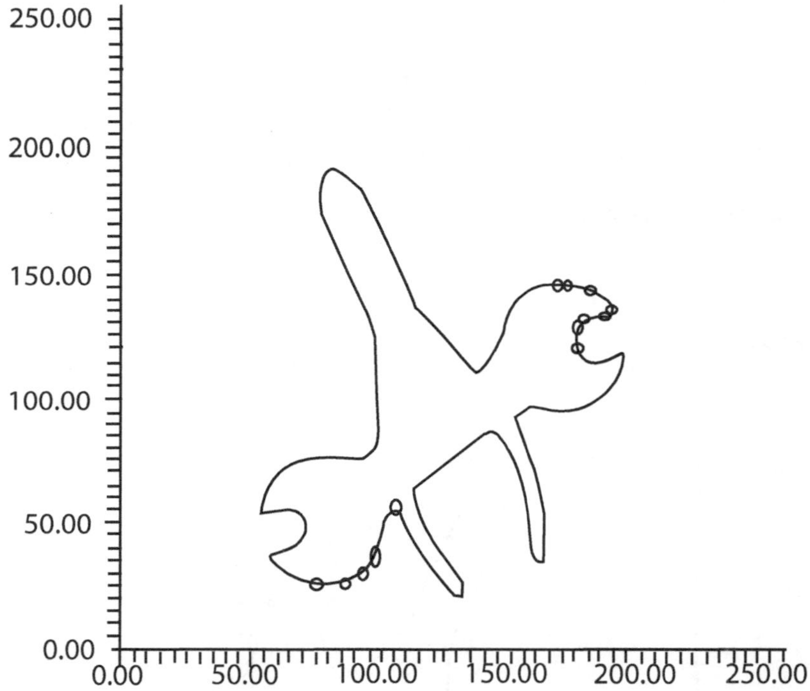

FIGURE 4 Placement of model 2 of Figure 2 on the occluded scene.

Case Study III Matching discontinuous Fragments with Small Distortions of Objects

Our method is immune to small distortions in the objects due to noise and camera tilt. Combining the polygon fragments to create two higher level features we find J = 4.25, and (θ, T_x, T_y) = (36.52, 137.3, –26.0). Figure 5 shows the result of applying the transform of the polygon fragments of the model 3 of Figure 2 on the occluded scene.

In this section, we develop a dissimilarity measure, J, to decide if two polygon fragments match. A J greater than a threshold, T_J, indicates two dissimilar fragments, and J less than T_J results in two similar fragments. This decision rule determines if a scene fragment matches a model fragment.

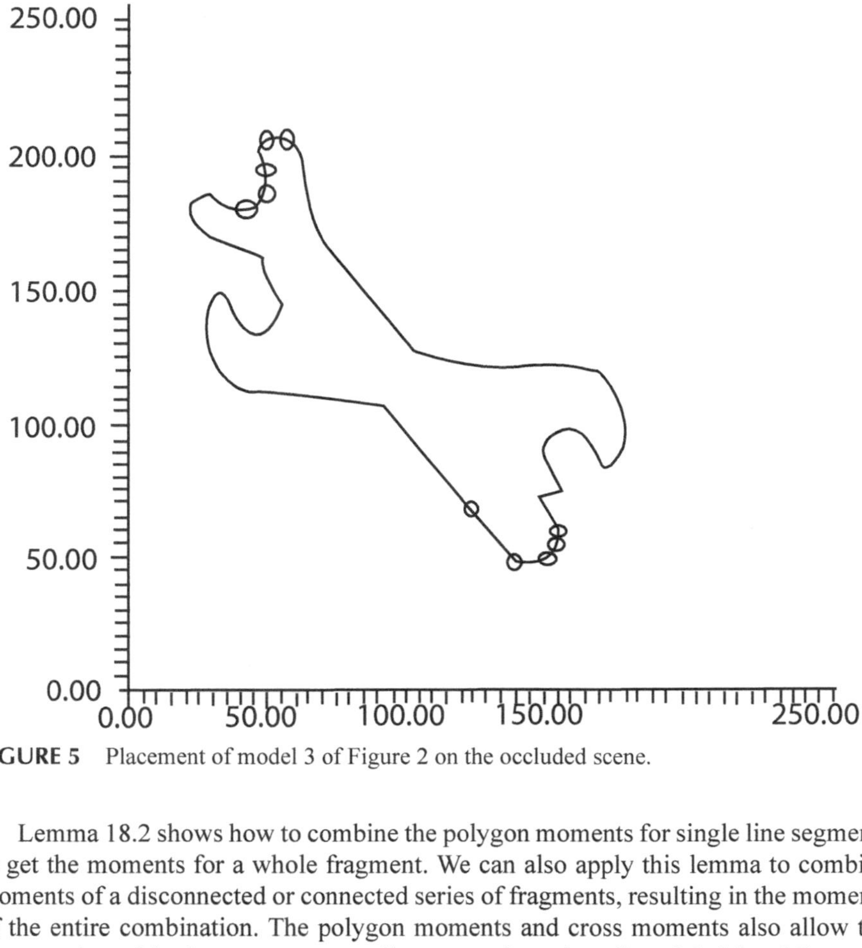

FIGURE 5 Placement of model 3 of Figure 2 on the occluded scene.

Lemma 18.2 shows how to combine the polygon moments for single line segments to get the moments for a whole fragment. We can also apply this lemma to combine moments of a disconnected or connected series of fragments, resulting in the moments of the entire combination. The polygon moments and cross moments also allow the computation of the least squares coordinate transform that aligns Model and Scene.

The certain combinations of these polygon moments lead to invariant moments. These invariant moments allow us to do preliminary matching between two fragments. We find these invariant moments to be more descriptive than other local features.

KEYWORDS

- **Dissimilarity**
- **Fragment matching algorithm**
- **Least squares error**
- **Polygon approximation**
- **Polygon fragments**

19 Polygonal Approximation to Recognize and Locate Partially Occluded Objects Hypothesis Generation and Verification Paradigm

CONTENTS

19.1 INTRODUCTION

The task of robots in automatic assembly and inspection requires object selection and determination of its orientation in an intelligent manner. Usually assembly machines expect uniform parts in precise locations and use elaborate jigs. But correct placement of objects and its ideal viewing under noisy and occluded situation present difficulties and decrease system flexibility. Multiple objects under occlusion cause significant problems in identifying and locating them in the workspace of the robot. Further the enormous amount of data makes the exhaustive search for template matching impractical. To reduce the data using features like area, perimeter, length, and breadth can be applied. Position and orientation of the object do not affect these features. A decision tree can determine the objects present by indicating which measurements to make based on the values of the other measurements. The measurements quickly narrow down the possible objects that might appear in the scene.

Unfortunately, the features like area, perimeter and so on work well, if objects are entirely visible. If object M1 is placed on top of M2, object Ml make M2's global features unrecognizable. By utilizing features such as protrusions and notches, a computer vision system can recognize objects with missing parts. Using these types of features, object characterization and recognition happen even in the presence of spurious, missing, or distorted features.

This chapter is mainly concerned with the application of polygonal approximation of objects to recognize and locate partially occluded objects. Figure 1 illustrates a typical image under consideration. Multiple objects may appear in the scene and they can touch or overlap. Under such circumstances, global features such as area and perimeter do not work in recognizing the objects in the scene.

19.2 ASSUMPTIONS OF VISION SYSTEM

We use computer vision to recognize and locate objects that may touch or overlap. We assume that, if an object appears in the scene, then the object belongs to a set called M. Models of objects shown a priori to the vision system, under ideal viewing conditions and with the entire object visible, comprise the model set M. In this context our basic philosophy of designing a vision system is that we can recognize the objects which we have seen earlier. "Seen earlier" basically forms the data base of our model based vision system. We choose polygons to represent the models of the objects. The polygon representation of an object reduces the number of its boundary points. Thus, we can reduce the large amount of data initially collected as an image of the object. The use of polygons to represent the objects adds the assumption that the boundary of the object alone distinguishes the object from other objects. Because of the 2D nature of polygons, we deal with rigid 2D planar objects whose position and orientation uncertainties lie in a plane parallel to the image plane. The assumption of rigid objects significantly reduces the ambiguous and inconsistent matches that arise in identifying and locating the objects.

The assumptions may limit our applications, but many stamped, cast, or rotationally symmetric objects, such as cogs and shafts, fit to the assumptions. We can extend the present concept of vision system from the 2D objects to 3D objects.

19.3 MAJOR ISSUES OF VISION SYSTEM

"Seen earlier" as stated basically represents knowledge about the objects to generate a hypothesis about an object in the scene. This knowledge is represented as a feature database containing distinguishing characteristics of the objects. To recognize occluded objects, instead of global features, we should consider local features such as corners, protrusions, holes, lines, texture, and curves and so on.

The hypothesis generation scheme considers the following three basic problems:

1. Features in the scene may not appear in the model.
2. Features in the model may not appear in the scene.
3. Each feature has an uncertainty associated with its numerical properties (position, location, length, etc).

Allowing multiple objects to appear in a scene generates the problem of spurious features and as a result the hypothesis generation scheme does not know which features to use or which features to reject. For instance, in Figure 1(c), the features from the unknown object which is placed on top of pliers or cutter may not exist in the pliers or cutter. Also two objects which overlap or touch each other generate a feature which does not belongs to neither object. In Figure 1(c) a feature extracted from the point where the boundary of the pliers or cutter meets the boundary of the unknown object may not belong to either pliers or cutter or the unknown object.

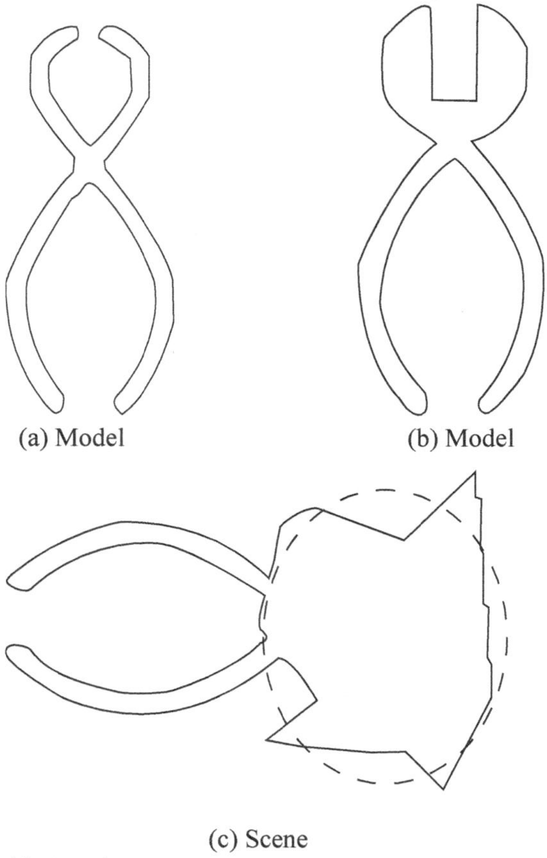

FIGURE 1 Model objects and scene.

Further the touch or overlap of multiple objects causes problem of missing features in the scene. An occluding object could cover up a feature, and a touching object could change a feature making the feature go undetected. In Figure 1(c), the unknown object occludes features belonging to the pliers or cutter. The hypothesis generation scheme has to deal with incomplete scene information.

The feature extractor causes noisy feature problem. Noise enters the scene and distorts the information supplied to the feature extractor. Usually, we do not expect a perfectly segmented scene. Always there exists segmentation error due to position of lights, surface material, quantization, reflections, and shadows. This causes imperfect and undetected features, and contributes to the first two problems by missing or adding features.

The spurious feature problem and the noise problem, or the missing feature problem and the noise problem, creates two separate problems. An algorithm of time complexity $O(n^2)$ can successfully solve the matching problem by using constraints. Here n represents the number of features. Considering all the problems simultaneously leads to an exponential time complexity algorithm. We may use heuristic measures and apply constraints to reduce the time complexity.

19.4 LITERATURE SURVEY

Global techniques are unable to recognize objects that are only partially visible. Global features computed for part of an object have in general no relation to those computed for the entire object. Hence, to avoid this problem, we consider local features, which depend only on portions of objects to perform recognition.

Wallace et al. [182] recognizes aircraft silhouettes using local shape descriptors. Here the basic assumption is that only one object appears in the image and that it is completely visible. The boundary of the object is traced to determine the peaks in the boundary's curvature function along with valleys of low curvature. Thus, peaks and valleys are separated. The local shape descriptors are obtained in terms of the arc length between peaks and the angle change between valleys. Objects are recognized by matching the list of lengths and angles of unknown objects with those of each model object. A distance measure is used between descriptor lists. Thus, the unknown object is recognized by the closest match. Though the local shape descriptor is used the recognition system does not handle the case of overlapping objects. Tejwani and Jones [171] describe a similar system that is able to recognize partial shapes.

Perkins [110, 111] describes a system to recognize overlapping objects. First edge points are detected in a gray-scale image which are linked and stored as chain codes. These edges are approximated by straight lines and circular arcs by fitting lines to the chain code data in **θ-s** (angle-arc length) space. In the learning stage, the system is shown each possible object. The system stores the detected curves as object models. At the time of recognizing objects, image curves are matched against model curves. Potential matches are checked using cross-correlation in **θ-s** space. Matches are verified by computing a transformation from model coordinates to image coordinates. Checking of edges in the expected directions at a list of points spaced along the model's perimeter are also considered. McKee and Aggarwal [85] use the similar methods.

Turney et al. [179] recognize and locate partially occluded 2D objects using a sub-template based version of the Hough transform. Instead of using edge points as in the normal Hough transform, overlapping segments of boundary are used. This is a sub-template based approach. For each segment, a vector is stored pointing to the object's centroid. Whenever a sub-template matches in the image, an accumulator at the position pointed to by the sub-template's vector is incremented. After all the object's sub-templates are matched against the edges in the image, the accumulator with an appropriate large value is considered to be the position of the object. One problem with Hough transforms is that false peaks may be generated. Turney et al. reduce this problem by assigning each sub-template weight based on its saliency.

Segen [152] describes another Hough transform based technique to recognize objects. It is basically a search based method. Initially, the objects are allowed three degrees of freedom (rotation, x position, y position). This is reduced 1D at a time through a series of 1D Hough transforms. This method avoids the need for large multidimensional accumulator arrays.

Bolles and Cain [23] consider a hypothesis-verify type paradigm to handle occluded object recognition problem. They consider simple local features such as rectangular corners and holes to represent model objects and scene. A set of matches represents scene features similar to model features. The vision system transforms the set of matches into an association graph where each node of the graph corresponds to a match and each arc connects two structurally compatible matches. They consider the orientation difference and distance between two model features to measure the structure between the model features. The same approach is adopted for the scene features the structure between the two matched are obtained by comparison of the orientation difference and distance difference. The largest set of structurally compatible matches is obtained from the largest completely connected subgraph or maximal clique of the association graph. These matched provide a hypothesis about an object in the scene and determine a coordinate transform to place the model object onto the scene. The vision system performs an extensive model analysis to determine focus features and nearby local features and then reduces the exponential time complexity for maximal clique. The tests of hypotheses are done by checking object boundary consistency. The system applies the obtained coordinate transform to the model boundary. Image analysis takes samples perpendicular and on either side of the projected model boundary thus, the vision system determines that object region lies inside the boundary and background region lies outside the boundary. If background region lies inside the boundary, then the hypothesis is detrimentally affected indicating a place where the unmatched section does not exist in the scene. If object region appears both inside and outside the boundary, that does not affect the hypothesis either positively or negatively, but indicates a place where an object occludes the boundary.

Ayache [11] uses polygons to represent objects. The polygon line segments are considered as features. The longer line segments are promising features. These promising feature, when match with scene line segments, generate initial hypotheses. Each hypothesis is given quality score. A coordinate transform places the promising features onto the scene. Matching other model segments revises the quality score. The coordinate transform is updated using Kalman filter. The quality score essentially determines how much the model object resembles the scene. The inconsistent and ambiguous matches are reduced by deleting hypothesized objects that have a low quality score. Ayache [12] also extends the method to allow for recognition of objects in low contrast situations.

Koch and Kashyap [74] use a hypothesis-test paradigm to recognize partially occluded objects. Using a similarity measure to test hypotheses, like Ayache's quality score, produces unsatisfactory results for heavily occluded objects. A high quality threshold will miss heavily occluded objects even for scene features correctly identified and located. A lower quality score will allow false detections. Koch and Kashyap use polygons to model the objects. For local features, they use polygon fragments, since they find the line segment features require a reliance on a polygon approximation algorithm that gives unique placement and number of vertices. They match polygon fragments by computing a dissimilarity measure and a coordinate transform that takes one fragment to another. The coordinate transform decides the structural compatibility between two matches, and rejects all inconsistent interpretations. To find cliques we grow match clusters around a cluster center. This approach has a smaller time complexity than that of the maximal clique algorithm used by [74], but does not necessarily find a maximal clique.

They use a similarity measure to determine which hypotheses to try first. By trying hypotheses of the most visible objects first, a verified hypothesis can place constraints on the other hypotheses or remove hypotheses. We know about a scene, the more tightly we can constrain the other hypotheses of scene objects. They use hypothesis testing to reveal inconsistencies between the hypothesized object and knowledge about the scene. Instead of sampling the model boundary to test the hypothesis, they use the efficient polygon representation to determine what the scene would look like if the object does appear in the scene. This method easily rejects objects which is not present in the scene. A positive test result allows removal of matching scene features from consideration. A negative result removes the hypothesis from consideration. Continuous hypotheses generating and testing for feasibility eventually reveal all the objects that belong to the set of models.

A new method has been proposed to recognize and locate partially occluded 2D rigid objects of a given scene. For this purpose, we initially generate a set of local features of the shapes using the concept of differential geometry. A computer vision scheme based upon matching local features of the objects in a scene and the models which are considered as cognitive database is described using hypothesis generation and verification of features for the best possible recognition. The hypothesis generation and verification scheme is based on both the smoothed polygonal approximation and spline representation of the boundaries of the model objects and scene.

19.5 FEATURE EXTRACTION AND MATCHING

19.5.1 Polygon Approximation

To obtain the polygon approximation of the object, the vision system finds the boundary of the object. A simple thresholding algorithm is used to segment the object from the background and Montanari's Algorithm [94] is used to determine the boundary. Of course objects seen under poor lighting conditions require more sophisticated algorithms [12] since the polygon approximation algorithm requires a closed boundary from the segmentation algorithm.

A polygon approximation algorithm decreases the number of boundary points. It also smooths out the quantization noise and segmentation errors introduced into the object boundary. The fitting of a polygon to the object boundaries helps in determining features that describe the shape of the object. This representation handles the large amount of data initially collected as an image of the object by representing the object with a minimum number of its boundary points. Vertices of the polygon depict local features extracted from the models under ideal viewing conditions.

We use an algorithm similar to the split and merge algorithm. The algorithm does not necessarily lead to an optimal solution of having a polygon with the smallest number vertices, but it usually places vertices near points of high curvature.

Pavlidis and Horiwitz also proposed a similar algorithm and use least squares to compute the error of fit. Sometimes this leads to an unconnected boundary because the estimated line does not necessarily have to include any boundary points. We use the collinear line fit test suggested by Pavlidis [106]. The resulting polygon groups the boundary points into line segments and reduces the data needed to represent the object. The line segments and vertices provide useful features in describing the shape of the object.

19.5.2 Polygon Fragments

To describe shape, humans use such terms as elongated, sharp corner, protrusion, and notches. Brady and Asada [16] use smoothed local symmetries (SLS) to describe shape. The SLS description can represent both the local bounding contour and the region it encloses. The SLS may also determine how to divide the object into subshapes, making it attractive for partially occluded object recognition.

We use polygon fragments or corners centered at high curvature vertices to describe the object's shape. Most algorithms estimate the curvature by fitting a smooth curve to the boundary point data and then calculate the curvature. Davis [29] discusses the problems in determining the length of the line segments used to estimate the curvature. The fitting of long line segments produces a smoothing affect and missed vertices will result. A smaller size produces a bad estimation due to noise and quantization.

We use the polygon approximation to estimate the curvature since the algorithm places vertices on or near large curvature values. We describe how the local descriptions of the model and scene are built using the fundamental results of differential geometry for curves and surfaces. We then describe the matching process that identifies models in the scene description and estimate the corresponding coordinate transformation. The complexity measure of the proposed algorithm is also considered. Finally, we present some results obtained from a number of scenes of rigid objects.

19.5.3 Fundamental Results of Differential Geometry for Curves and Surfaces

The problem of space recognition is analogous to recognition of curves in space. Therefore, well known results and theorems of differential geometry can be utilized in shape analysis [63, 20].

Let c be a curve in Euclidian space R^n of class $C \geq 1$ whose domain is I_S where s is the parameter. Then a differential geometry theorem that is particularly useful is [63].

Theorem 19.1

Every regular curve $c{:}I_S \rightarrow R^n$ can be parameterized by its arc length.

Given a curve c we can construct many more curves which follow the same route but whose rate of motion is different.

Definition 19.1

Let I_S and I_Q be open intervals in R, $c{:}I_S \rightarrow R^n$, be a curve and let $Q{:}I_Q \rightarrow I_S$ be a differentiable function. Then the composite function $c.Q = C_*{:}I_Q \rightarrow R^n$ is a curve called the reparametrisation of c by Q, where C_* is the composite function of c and Q.

Let C = c(s) be the parametric representation of the curve under analysis with s as the natural parameter (i.e. |dc/ds| = 1). Then the vectors t, n, b satisfy the Serret-Frenet equations [20, 63]

$$\begin{bmatrix} \dot{t} \\ \dot{n} \\ \dot{b} \end{bmatrix} = \begin{bmatrix} 0 & k' & 0 \\ -k' & 0 & \tau \\ 0 & -= & 0 \end{bmatrix} \begin{bmatrix} t \\ n \\ b \end{bmatrix} \qquad (19.1)$$

where k' = curvature, τ = torsion, t = tangent, n = normal at a point on the curve, b = binormal at a point on the curve, and the dot denotes the derivatives with respect to the natural parameters.

The parameter k' and τ are geometric invariants and their existence and uniqueness are guaranteed by the following theorem [20, 63].

Theorem 19.2

Let k'(s) and τ(s) be arbitrary continuous functions for $a \leq s \leq b$. Then there exists, except for position in space, one and only one curve c for which k'(s) is the curvature, τ(s) is the torsion and s is the natural parameter.

This tells us that an arbitrary three-dimensional smooth curve shape is captured by two scalar functions, curvature and torsion.

When the smooth curve under analysis lies in a plane, the torsion τ is equal to zero and the binormal b is constant. Thus, for the 2D curves, Equation (19.1) reduces to:

$$\begin{bmatrix} \dot{t} \\ \dot{n} \end{bmatrix} = \begin{bmatrix} 0 & k' \\ -k' & 0 \end{bmatrix} \begin{bmatrix} t \\ n \end{bmatrix} \qquad (19.2)$$

Hence, in this case curvature alone specifies a 2D smooth curve shape.

Surfaces are a natural geometric generalization of curves. But in the present text we are only concerned with 2D curves. Before we conclude this section we state the following result which is useful for the computation of curvatures of 2D shapes.

Result 19.1

The curvature of two-dimensional curves in the X-Y plane with parametric equation x = x(s) and y = y(s) is given by:

$$k' = \sqrt{\left(\frac{d^2x}{ds^2}\right)^2 + \left(\frac{d^2y}{ds^2}\right)^2}.$$

Proof

The position vector of any point on the curve is p = x(s)i + y(s)j. Then,

$$t = \frac{dp}{ds} = \frac{dx}{ds}i + \frac{dy}{ds}j \text{ and}$$

$$\frac{dt}{ds} = \frac{d^2x}{ds^2}i + \frac{d^2y}{ds^2}j$$

Now from Equation (2) we know that,

$$\frac{dt}{ds} = k'n,$$

therefore,

$$k' = \left|\frac{dt}{ds}\right| = \sqrt{\left(\frac{d^2x}{ds^2}\right)^2 + \left(\frac{d^2y}{ds^2}\right)^2}$$

19.5.4 Curvature Computation and Corner Detection

In our vision scheme we assume that both object and scene can be exposed to either good lighting or poor lighting condition. But the key feature of the vision system is that models and scenes are represented the same way. In order to build a model or a scene description, the following sequence of operations applied to the picture of the isolated object or of the scene.

1. If the contrast is high enough (i.e. if the lighting conditions are perfectly controlled), threshold the image and obtain the binary picture [140, 154].
2. If the contrast is not high enough (general lighting condition or poor lighting condition) find the edges by combining gradient and second order derivative information [11]. For subpixel precision we may use zero crossing of the convolution with Laplacian of Gaussian masks [86].
3. Either present the boundary of the binary picture, obtained from step (1), by smooth polygons [94] or lists the connected border points, obtained through edge following of step (2), and approximate the connected components with rough polygons [94]. Finally, we obtain smooth polygons from the rough polygons using the method stated in [94].
4. Generate an appropriate coordinate frame for model objects and scene.

5. Record the coordinates of the end points of the line segments of the finally obtained smooth polygons which represent the model object and scene.
6. Pass periodic cubic splines through the coordinates of step (5). Record the individual polynomial segment, represented by the periodic cubic spline between the end points of the line segments. Splines are particularly meant for suitable representation of the closed smooth curves. In case of open curves we may use only cubic splines. At this stage, we believe in the fact that curvature computation for corner detection (the local description of the shape) requires some smoothing [16] which we achieve by fitting the spline function through the recorded coordinates step (5).
7. Compute the curvature in the following way:

   (i) Consider either any model object or scene.

   (ii) Instead of representing smooth curve (either of any model object or the scene) obtained at step (6), through the arbitrary points [Coordinates of the point (x_k, y_k) are obtained according to the coordinate frame of step (4).] (x_k, y_k) $(k = 1, n)$ we represent it parametrically by two functions,

$$x = x(s),\ y = y(s)$$

   of the curve parameters. This can be accomplished by choosing a set of strictly monotone value s,

$$S_1 < S_2 <. < S_n$$

   and assuming that the periodic cubic spline passes through each of the sets (s_k, x_k) and (s_k, y_k) $(k = 1, n)$. Herein, it is assumed that the value of the first and second derivatives and or and with respect to s and s_1 and s_n are given [160]. For particular purposes, it often suffices to use natural cubic splines and do without special boundary conditions. As s runs through interval (s_1, s_n), the ordered pairs [Coordinates of the points $(x(s_k), y(s_k))$ are obtained according to the coordinate frame of step (4)] $(x(s_k), y(s_k))$ are the corresponding points on the curve obtained at step (6).The ideal values for the quantities s_k would be accumulated arc lengths of the curve obtained at step (6). Though these quantities are unknown, the iterative process to determine them is as follows:

$$s_1 = 0, \quad s_k = s_{k-1} + d_{k-1} \quad (k = 2, n)$$

   where, d_k is any predetermined small value as per our need of the problem (see Figure 1). Now the problem is to find out the coordinates of the ordered pairs $(x(s_k), y(s_k))$ as s runs through the interval (s_1, s_n). This problem is solved at the next step.

   (iii) Consider the following expression of d_k

$$d_k = \sqrt{(\Delta x_k)^2 + (\Delta y_k)^2}$$

where

$$\Delta x_k = x_{k+1} - x_k \quad \text{and} \quad \Delta y_k = y_{k+1} - y_k.$$

If we start from any known recorded coordinate of step (5), then the initial coordinate (i.e. for $s_1 = 0$), (x_1, y_1) is known. Now to find out the coordinate of (x_{k+1}, y_{k+1}), (i.e. for

$s_{k+1} = s_k + d_k$, where d_k is any small known predetermined value, s_k is known and [For notational simplicity the recorded pair $(x(s_k), y(s_k))$ as s_k runs through s_1 to s_n is represented by (x_k, y_k) for any given s_k] (x_k, y_k) is known), we may use the expression for d_k and the expression for cubic spline which is recorded at step (6) and which passes through the end points of a particular line segment (for further details see Figure 3).

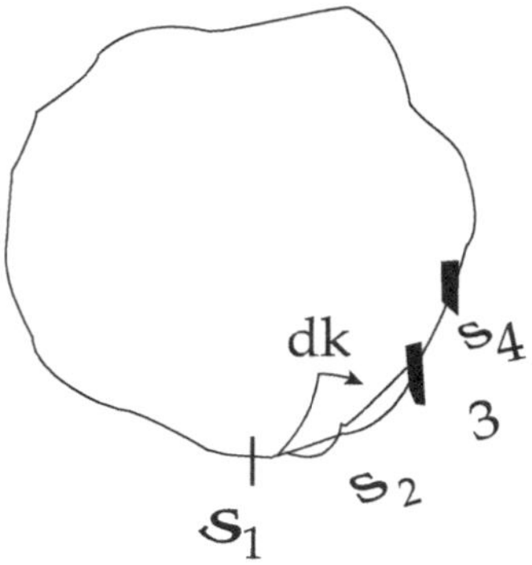

FIGURE 2 Curve representations through parametric spline.

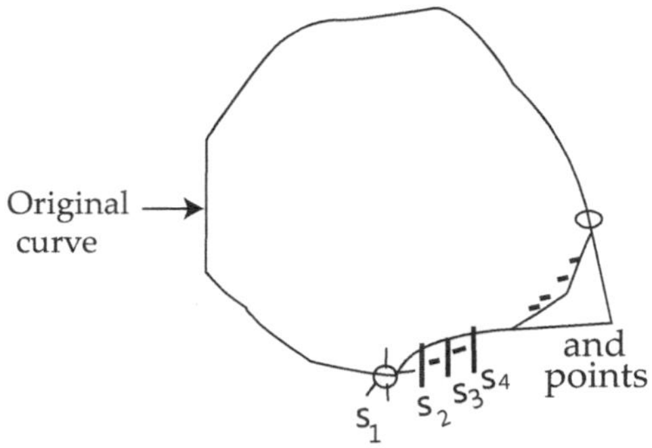

FIGURE 3 Spline representations of curves through the end points of smoothed polygons.

⊙ marks end points of the smooth polygons obtained at step (5). Dotted curves, between the two end points of the smooth polygons, are the recorded cubic splines of step (6). Coordinates of the end points are recorded at step (5). As s_k runs through s_1 to s_n, the coordinate $(x(s_k), y(s_k))$ on the curve (represented by dotted line), obtained at step (6) will be compute according to the method stated at step (7). The computation of (x_{k+1}, y_{k+1}) for s_{k+1} from the given (x_k, y_k) for s_k and d_k, requires the selection of the appropriate segment of the cubic spline. This selection can be done by comparing d_k with the Euclidean distance between (x_k, y_k) and the end point which is the immediate successor of (x_k, y_k). If the d_k is less than the said measure or same as the said measure we can select the same segment of the cubic spline as we did for (x_k, y_k). Otherwise we select another appropriate segment of the cubic spline.

We have actually applied BAIRSTOW'S method [131] for polynomials to determine (x_{k+1}, y_{k+1}). Thus, we determine the coordinate (x_k, y_k) as s_k runs through s_1 to s_n.

(iv) Apply the expression for curvature (see Result 1). The terms d^2x/ds^2 and d^2y/ds^2 at a particular coordinate (x_k, y_k) are numerically computed using the following expressions respectively:

$$x''_k = \frac{-x_{k+2} + 16x_{k+1} - 30x_k + 16x_{k-1} - x_{k-2}}{12d_k^2}$$

$$y''_k = \frac{-y_{k+2} + 16y_{k+1} - 30y_k + 16y_{k-1} - y_{k-2}}{12d_k^2}$$

Increment on arc length d_k is constant. If we choose d_k very small, then the computed coordinates $(x(s_k), y(s_k))$ will be very close to each other. Therefore, the numerical computation of d^2x/ds^2 and d^2y/ds^2 will be highly accurate and the computational error will be $o(d^4_k)$.

1. Corner points are detected by putting appropriate threshold(s) on the numerical values of the computed curvature. In our case, if the curvature exceeds 0.4 we declare this edge point to be a corner.

Thus, we can determine the local descriptions (in term s of the corner points) of the model objects and scene.

19.5.5 Matching Polygon Fragments

To form a hypothesis about the presence of a model in the scene, a matching algorithm matches model features to scene features. We use local features called corners or polygon fragments. A corner defines a group of line segments centered at a corner vertex. We discussed how to get the corner vertices from the polygon approximation.

To match polygon fragments or corners, we quantize both fragments into n points and determine the coordinate transform τ that minimizes the squared error between these points. The minimal square error J represents dissimilarity between the two polygon fragments. A J greater than a threshold T_J indicates two dissimilar corners, and J less than T_J results in two similar corners. This decision rule decides if a scene corner matches a model corner.

In calculating J and τ we consider the following terms for the model corner:

$$\frac{1}{n}\sum_{i=1}^{n} Mx_i, \frac{1}{n}\sum_{i=1}^{n} My_i, \frac{1}{n}\sum_{i=1}^{n} Mx_i^2, \text{ and } \frac{1}{n}\sum_{i=1}^{n} My_i^2$$

where (Mx_i, My_i), i = 1, n, represents n model points along a polygon fragment M. We call these terms the polygon moments of the model corner. Similar terms exist for the scene corner. We call the following terms the polygon cross moments:

$$\frac{1}{n}\sum_{i=1}^{n} Sx_i Mx_i, \frac{1}{n}\sum_{i=1}^{n} Sx_i My_i, \frac{1}{n}\sum_{i=1}^{n} Sy_i Mx_i, \text{ and } \frac{1}{n}\sum_{i=1}^{n} Sy_i My_i$$

where, (Sx_i, Sy_i), i = 1, n, represents n scene points along a scene fragment S. The cross terms have coordinates from both corners. To go from the boundary point data to the polygon approximation and then to quantize the polygon approximation wastes computer time and does not use the power of the polygon representation. We demonstrate how to compute these polygon moments and cross moments from just the end points of the polygon line segments. This avoids the expensive quantization step. We have also shown how to consolidate the moments of two unconnected polygon fragments to get the moments of the entire combination. Certain combinations of these polygon moments lead to invariant moments which are used as local features for model objects and scene.

19.6 MATCHING MODELS AND SCENE DESCRIPTION

To form a hypothesis about the presence of a model in the scene, a matching algorithm matches model features to scene features. We use local features called corners. We discussed how to get the corners from the smooth curves of the model objects and scene. To match corners we use the Euclidean distance between the curvatures of the model corner and scene corner. Instead of using curvature we may consider certain combination of the polygon moments as invariant local features. The model corner invariant moments can form a dictionary of model local features. Using the scene corner invariant moments, the matching algorithm can look up possible matches to the model corners through the dictionary. An Euclidean distance J greater than a threshold T_j indicates two dissimilar corners and less than T_j results in two similar corners. This decision rule decides, if a scene corner matches a model corner. After finding the match we compute the appropriate coordinate transformation [9, 161] T to place the coordinates of the matched point of the model on the corresponding point of the scene. T is represented by a parameter vector v = (cos θ, sin θ, t_x, t_y) where is the scaling factor, θ is the rotational parameter and t_x, t_y are translation parameters. If the model objects and scene are of same scale, is replaced by identity.

The model corners form a dictionary of model features which is defined earlier as our cognitive database. Using the scene corners the matching algorithm can look up possible matches to the model corners through this dictionary.

19.6.1 Hypothesis Generation

The hypothesis generation algorithm makes a list of scene corners and their matches for each possible model that could appear in the scene. To hypothesize an object in the scene, mutual compatibility constraints extract a set of consistent matches from a model's match list. First we discuss the mutual match compatibility [12, 73] between two matches followed by extracting a group of mutually compatible matches.

19.6.2 Mutual Compatibility and Multiple Matches

Suppose that two matches K_i for i = 1, 2 contain a scene corner S_i and a model corner M_i. The following four rules [73] test two matches K_1 and K_2 for mutual compatibility.

1. $S_1 \neq S_2$.
2. $M_1 \neq M_2$.
3. The two scene corners refer to the model corners belonging to the same model (M_1 and M_2 belong to the same model).
4. The structure between the two scene features equals the structure between the two model features.

The first three rules are self explanatory. The fourth rule comes from the assumption that rigid objects make up the model set. To have the same structure, two matches should have the same rotation and translation parameters (assuming scaling factor = 1). The very unlikely possibility that two matches have exactly the same coordinate transform parameters requires that the distance between them to be less than some small ε. That is:

$$\delta_\theta = |\theta_1 - \theta_2| < \varepsilon_\theta,$$
$$\delta_x = |t_{x1} - t_{x2}| < \varepsilon_t,$$
$$\delta_y = |t_{y1} - t_{y2}| < \varepsilon_t,$$

To compare two matches, we compare the rotational parameters first. If the difference between the rotation parameters has a value less than a certain threshold (ε_θ), then we fix the rotation parameter to the same value for both the matches and then compute the translational parameters. If the difference between the translational parameters has a value less than some threshold (ε_t), then the matches have compatible coordinate structure [73]. Otherwise the matches do not meet the structural compatibility requirements.

Suppose a group of model features supported one high level model feature and the same for the matching scene features. This group of mutually compatible matches would represent a matching of a high level model feature to a high level scene feature or the largest portion of the model identifiable in the scene. There are already plenty schemes [17, 23, 44, 56, 60, 73, 161] available for multiple compatible matches. But we apply the minimal spanning tree algorithm [155]. This algorithm has several benefits which are discussed. The algorithm starts with the smallest amount of data needed to form a solution and then adds consistent data instead of using all the data to form cluster around the match. The compatible matches satisfy the four compatibility conditions mentioned earlier. Conflicts can occur between pairs of matches which can be solved by taking the match with the highest structural compatibility. Thus, we find out the largest set of mutually compatible matches. The complexity of the algorithm is the order of o(nlogn) where n represents the number of points to be clustered. This technique can avoid stray matches by fixing the cluster centers. The matching procedure for finding the largest compatible match is discussed.

The storage requirements to realize this algorithm is not large. We have to store the corner points and the corresponding coordinates for model objects and scene. In addition to this, the spline functions between the two end points of the line segments of the smoothed polygon are also stored.

19.7 MATCHING PROCEDURE FOR COMPATIBLE MATCH

It is assume that the corner points of the model objects and scene are detected and marked by integers 1, 2, 3.

Consider the Table 1 which represents the Euclidean distance J between the curvature of every model corner and scene corner. A threshold T_j decides whether a scene corner matches a model corner. According to the numerical value of T_j a particular scene corner may be matched with more than one model corners and *vice versa*. This initial curvature matching between scene corner and model corner segregates the most dissimilar features (local features) between the scene and model objects. The symbol ψ indicates the similarity between the ith corner of the scene and the jth corner of the kth model. After the initial segregation of dissimilar features we decompose Table 1 into three parts, example Table 2, Table 3, and Table 4. On the individual tables (i.e. on Table 2, Table 3, and Table 4) we separately apply the algorithm for minimal spanning tree discussed in the text. Thus, we get the most compatible match between the scene corner and model corner. A node in a table would consist a match. An arc would exist between two nodes, if the two matches complied with the four compatibility constraints of the text. The decomposition of Table 1 into several parts (three parts in the present example) indicates that some of the recognition tasks as indicated by the network displayed in Figure 4 can be done simultaneously (i.e. parallel computation).

After determining the compatible structure on each table (i.e. on Table 2, Table 3, and Table 4) we compute the cluster centre for each compatible structure of each table. The cluster centre at each table is essentially formed by taking the average of all the rotational parameters θ for all compatible structures of each table. Thus in the example we get three cluster centres at three tables. The translation parameters (t_x, t_y) of all the compatible structures of each table are further modified to (t'_x, t'_y) due to the fixing up of all rotational parameters θ of each table to the cluster center θ', discussed earlier. Thus, we get the compatible structure of an arbitrary example as displayed in Figure 5.

TABLE 1 Scene-model objects array based on detected corner points.

	M_1						M_2							M_3				
Sccne Corners ↓ / Model Corners →	1	2	3	4	5	6	1	2	3	4	5	6	7	1	2	3	4	5
1	✓	✓	✓	✓	✓	ψ	ψ	✓	✓	✓	ψ	✓	✓	✓	✓	✓	✓	✓
2	ψ	✓	✓	ψ	✓	✓	✓	✓	✓	✓	✓	✓	✓	✓	✓	ψ	✓	✓
3	ψ	✓	✓	✓	ψ	✓	✓	✓	ψ	ψ	✓	✓	ψ	✓	✓	✓	✓	✓
4	✓	ψ	ψ	✓	✓	✓	✓	✓	✓	ψ	✓	✓	✓	✓	✓	ψ	ψ	ψ
5	✓	✓	ψ	ψ	✓	✓	✓	✓	ψ	✓	✓	✓	✓	✓	✓	✓	✓	✓
6	✓	✓	ψ	✓	✓	✓	ψ	✓	✓	✓	✓	ψ	✓	✓	ψ	ψ	✓	✓
7	✓	ψ	✓	✓	✓	ψ	✓	✓	✓	ψ	✓	✓	✓	✓	✓	✓	✓	✓
8	✓	✓	✓	✓	ψ	✓	✓	✓	ψ	✓	✓	✓	✓	ψ	✓	✓	✓	ψ
9	ψ	✓	✓	ψ	✓	✓	✓	✓	✓	✓	✓	✓	ψ	✓	ψ	✓	✓	✓
10	✓	✓	✓	ψ	✓	✓	✓	✓	ψ	ψ	✓	✓	✓	✓	✓	✓	ψ	✓
11	ψ	✓	✓	✓	ψ	✓	✓	✓	✓	✓	ψ	✓	✓	ψ	✓	ψ	✓	✓
12	✓	✓	✓	✓	✓	✓	ψ	✓	✓	ψ	✓	✓	ψ	✓	✓	✓	✓	✓
13	✓	✓	ψ	✓	✓	✓	✓	✓	ψ	✓	✓	✓	✓	✓	✓	✓	ψ	✓
14	✓	✓	✓	✓	✓	✓	✓	✓	ψ	✓	ψ	✓	✓	✓	✓	ψ	✓	✓

- Indicates the difference between the curvature of the model corner and scene corner. ᛘ Indicates the similarity between the curvature of the model corner and scene.

TABLE 2 Scene-M1 array.

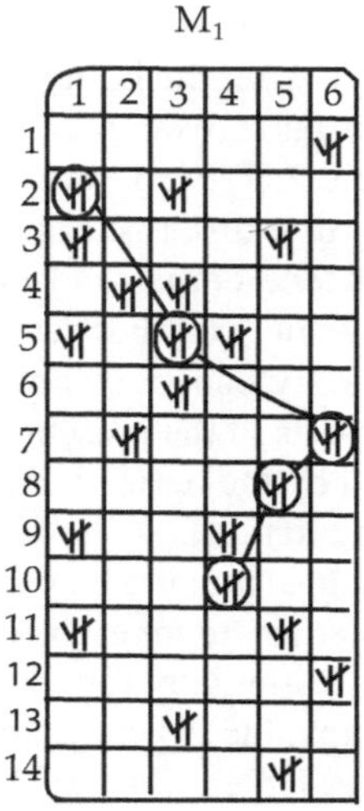

TABLE 3 Scene-M2 array

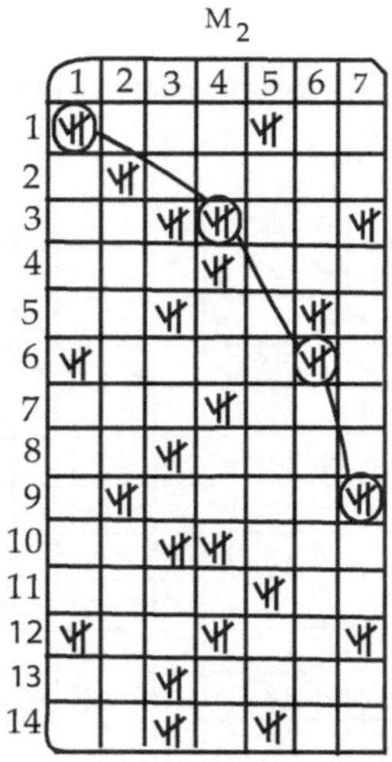

TABLE 4 Scene-M3 array

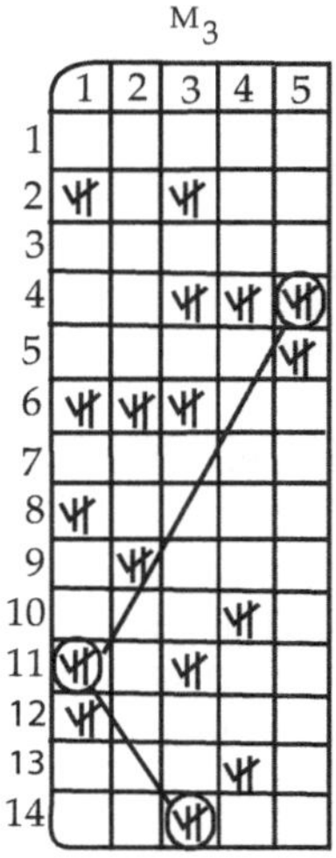

Indicates a node which would consist on a match. An arc between two satisfies the four comtibility constraints.

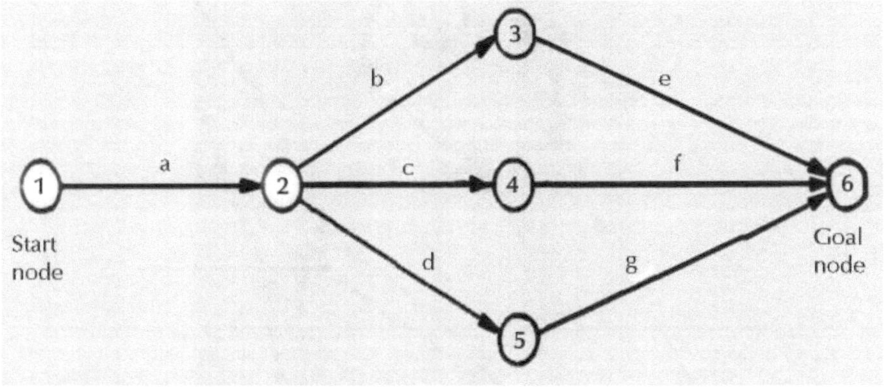

FIGURE 4 Different paths of computation for matching.

(a) Perform the task of curvature matching of Table 1.

(b) Apply minimum spanning tree of Table 2.

(c) Apply minimum spanning tree of Table 3.

(d) Apply minimum spanning tree of Table 4.

(e) Compute the cluster centre for Table 2 and the new t'_x and t'_y.

(f) Compute the cluster centre for Table 3 and the new t'_x and t'_y.

(g) Compute the cluster centre for Table 4 and the new t'_x and t'_y.

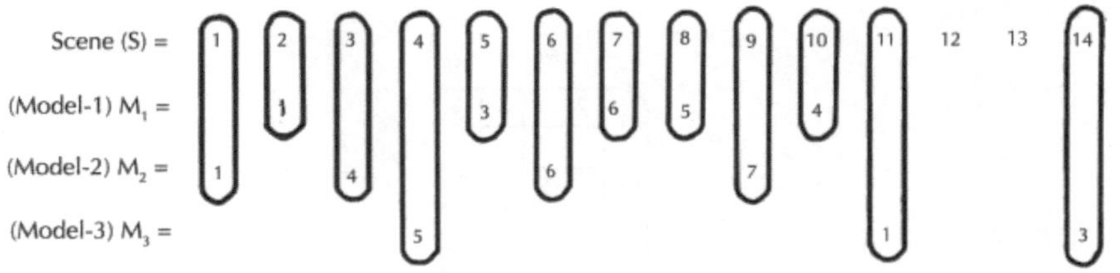

FIGURE 5 Compatible structure of match.

19.8 EXPERIMENTAL RESULTS AND DISCUSSION

Case Study 1

Figure 7 depicts the spline approximation of the boundary extracted from a image of a pliers on top of an adjustable wrench. The stars in the figure represent the corner points (local features). Using the corners of the wrench and pliers shown in Figure 6 as a model, the matching algorithm matches model features to scene features. A Euclidean measure J and an appropriate threshold T_J determine, if two corners match and result in a coordinate transform for the ultimate matching through structural compatibility using the minimal spanning tree algorithm.

FIGURE 6 Spline approximation of model objects.

FIGURE 7 Spline approximation of pliers overlapping wrench.

The following coordinate transform gives the location of the best possible wrench hypothesis:

$$\theta' = 95,\ T_x = 430 \text{ and } T_y = 35$$

where θ' is the cluster center, T_x and T_y are the average over t'_x and t'_y respectively. Similarly the following coordinate transform gives the best possible pliers hypothesis.

$$\theta' = 220,\ T_x = 135 \text{ and } T_y = 650$$

Figure 7 displays the result of taking these two hypotheses under consideration.

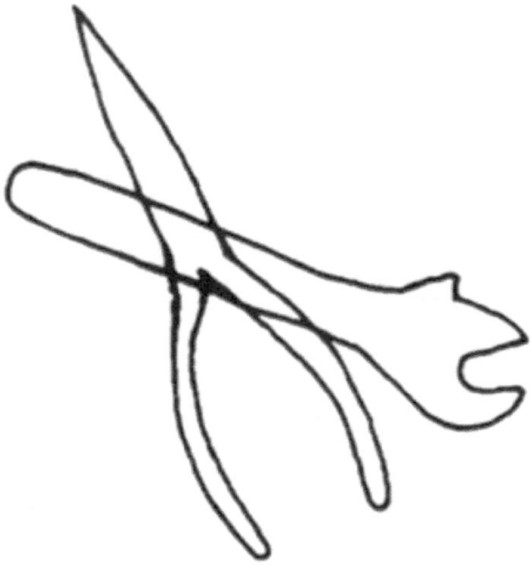

FIGURE 8 Location of wrench and pliers found in the scene.

Case Study 2

The feature extraction method produces the corners, represented by stars of the scene of Figure 9. The scene of Figure 10 has got two objects one place on top of other. Figure 9 represents the models and the corresponding corners represented by stars.

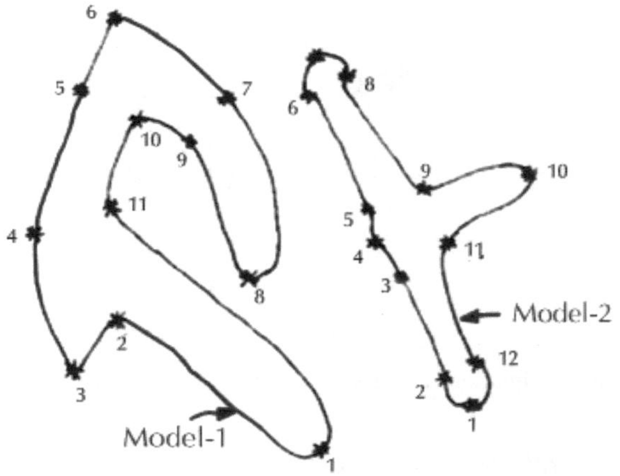

FIGURE 9 Spline approximation of model objects.

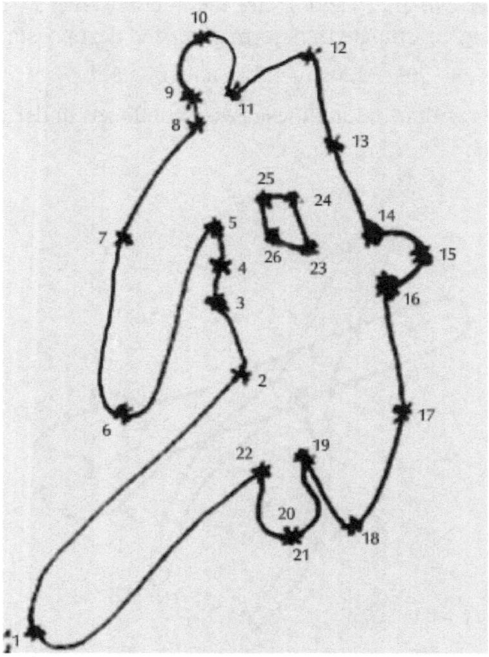

FIGURE 10 Spline approximation of scene.

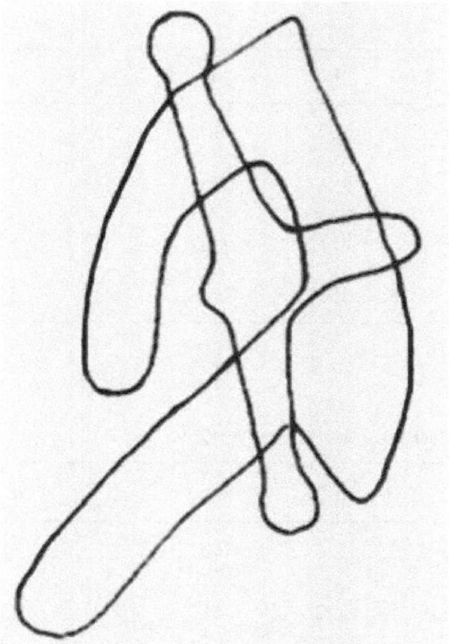

FIGURE 11 Location of model 1 and model 2 found in the scene.

Table 5 and Table 6 show matches that make up compatible structures for the model object 1 and model object 2. Ultimately, Figure 11 represents the location of two occluded objects.

To test the hypothesis, we can go back to the polygonal (smoothened polygons) representation of the original scene and the scene formed by the hypothesized model objects. These two polygons should contain the same area if the hypothesized objects appear in the scene. The difference between the two polygonal areas represents the measure of hypothesis consistency. If the difference in area consistency is too high the particular hypothesis is rejected. The computation of area consistency involves the use of polygonal operations such as union, area, and polygon transformation [15, 73, 153]. An alternative approach for hypothesis verification is stated in [12].

TABLE 5 Cluster (using minimal spanning tree) for determining the model object 1 in the scene represented by Figure 7.

Scene	Model	J	θ	t_x	t_y	θ'	t'_x	t'_y
1	1	0.01	350	–150	250	353	–149	251
18	3	0.005	354	–145	252	353	–148	250
17	4	0.03	351	–141	254	353	–149	251
13	5	0.012	352	–149	251	353	–147	252

TABLE 5 *(Continued)*

Scene	Model	J	θ	t_x	t_y	θ'	t'_x	t'_y
12	6	0.004	357	–145	253	353	–145	250
7	7	0.01	356	–146	255	353	–148	253
6	8	0.005	353	–147	252	353	–149	251
24	10	0.004	351	–148	250	353	–148	250

$\theta'==353, T_x==-148, T_y==251$

TABLE 6 Cluster of compatible matches form odelo bject 2.

Scene	Model	J	θ	t_x	t_y	θ'	t'_x	t'_y
21	1	0.03	320	40	80	320	65	75
3	4	0.0125	322	41	78	320	64	78
4	5	0.023	318	60	65	320	64	76
9	6	0.002	319	90	68	320	64	77
10	7	0.001	320	75	60	320	65	77
26	9	0.015	321	80	72	320	63	77
15	10	0.004	319	68	51	320	61	75
20	12	0.017	318	90	65	320	63	78

$\theta' = = 320, T_x = = 64, T_y = = 77$

19.9 CONCLUSION

We have presented a model based vision scheme to locate and identify occluded two-dimensional rigid objects. The hypothesis generation and verification algorithm depend on both the smoothed polygonal approximation and spline representation of the boundaries of the objects and scene. One of the major novelties of the present vision scheme is to determine the local features (corner points) of the scene and the model objects using the concept of differential geometry. This method of feature extraction is so generalized that just by considering two more parameters, example torsion and binormal, (see equation (1)) we can extend the results of two-dimensional world to three-dimensional world. The validity of the present vision scheme is tested on two examples and very promising results are obtained.

KEYWORDS

- **Hypothesis generation**
- **Hough transform Occluded objects**
- **Polygonal approximation**
- **Vision system**

20 Object Recognition with Belief Revision: Hypothesis Generation and Belief Revision Paradigm

CONTENTS

20.1 INTRODUCTION

In this chapter we present a new approach to recognize and locate partially occluded rigid objects from a given scene based on belief revision. We generate a belief about the scene using assumption-based truth maintenance (ATM) system. The ATM system is basically a tool for belief revision. It explores multiple potential solutions and can work out efficiently with inconsistent information. In practice, sometimes occlusion of objects in a 2D scene may occur due to the presence of objects which are not described in our primary knowledge base and which may appear to be an object, in addition to the model objects of our primary knowledge base. Hence, after detection of such an event, question of revising belief about the scene may arise to establish a new belief. The present approach [66] to recognize and locate an occluded scene is completely different from the existing paradigm based on the concept of hypothesis generation and verification which is already discussed.

To increase the efficiency in industrial inspection by an automated computer vision (CV) (robot vision) system it is a common task to recognize and locate an object in a given scene properly. It is very much simple to recognize an object with complete view. However, the task becomes a difficult one when the object is seen partially in a scene due to occlusion.

The recognition of individual object with partial occlusion has been studied for a long time. Briefly, the objects of a scene are recognized by "generate-test paradigm" [21, 74]. In a model-based recognition scheme [11, 12, 74, 131] the objects of a scene are compared with some model objects. This set of model objects are known prior to the vision system. If a certain portion of a model object matches to a particular object of a scene then the vision system generates a hypothesis that the model object is present in the scene and then locates the position of the model object in the scene. During the positioning of the model object into the scene, if a wide difference in coordinates of the matched point between model object and scene object does not occur then the vision system concludes that the hypothesized scene consists of the model object is a correct representation of real scene to be recognized [21, 74].

The task to match the model objects to the scene objects is performed by extracting features from the boundary of the objects of both model and scene. The features are calculated from the part of the image boundary, because the global features like area, perimeter and so on could not be calculated correctly due to occlusion. That is why local features [74, 131] are calculated which do not depend upon the whole part of the objects.

In the present approach [66], which is different from the existing paradigm of hypothesis generation and verification [21, 74], the vision system initially, generates a hypothesis for a given scene and gradually modifies it towards the correctness. Finally it draws a belief that the scene consists of some model objects. There may be multiple beliefs about a scene. A particular belief may be revised non-monotonically if contradictory information arises. We discuss the belief revision [32, 35, 134] concept later.

It is quite impossible to consider all the objects in the universe as primary models of the knowledge base of our vision system. That is why we consider a set of a few model objects. That means, the primary knowledge of the vision system is closed. Here the vision system draws the belief (Conclusion) about the scene with the help of only those model objects which are currently present in the knowledge base. If the vision system finds a new object then the knowledge base will be modified or expanded and the belief about the scene will be revised.

20.2 STATE OF ART

The present state of art [21, 74] for recognition of occluded scene is divided into two parts. The first part is hypothesis generation. This is based on matching of features between model objects and scene. If a reasonable portion of the model objects matches to the scene objects we create a hypothesized scene by positioning the model objects into the scene. Then in the second part that hypothesis is verified. The verification is made by calculating area of the object. The area of the hypothesized scene is compared with the area of the original scene. If these two areas are almost equal the vision system concludes that the recognition is successful [21, 74].

Unfortunately, in our real world there are various objects of different shapes and some of them are alike. Due to this cruelty of our natural world sometime the conclusion taken by the vision system may not be correct.

20.2.1 Dilemma in the Present State of Art

Let us consider the following example. Here two model objects (Figure 1(a), (b)) and one scene (Figure 1(c)) are shown as:

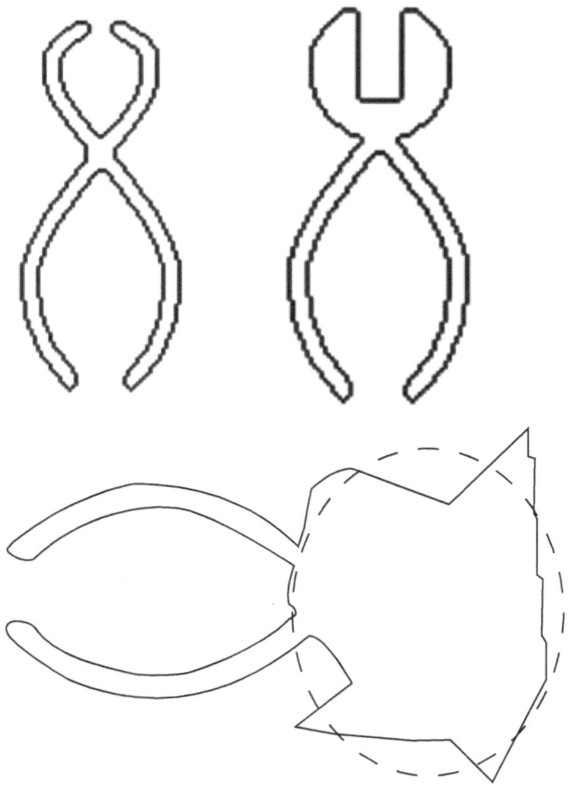

FIGURE 1 Model objects and scene.

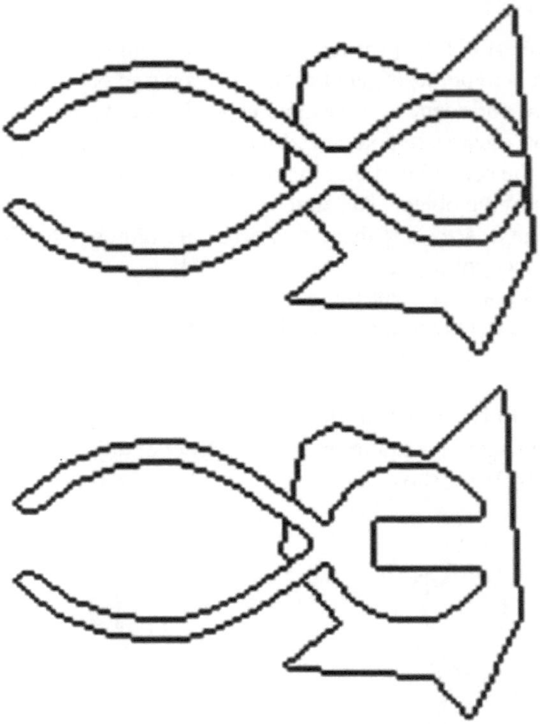

FIGURE 2 Two interpretations of Figure 1(c).

The model objects are alike in shape. In the scene the uniquely identifiable distinct features of these two model objects are occluded (Marked by dotted line in Figure 1(c)) by an unknown object. In this case, the hypothesis generation by matching features between these two model objects and scene object will be successful. There will be two generated hypotheses. After positioning the model objects into the scene the hypothesized scenes are generated (Figure 2(d), (e)). In both the cases the area between hypothesized scene and original scene do not differ widely and the vision system does not come to a single conclusion about the scene. That is the vision system may decide that the scene consists of both the model objects. Or it can be said that due to occlusion the vision system cannot differentiate the essential difference between the two model objects and recognizes two different model objects in the scene to be the same object which is an inconsistent recognition. Thus, the hypothesis generation and verification (test) paradigm [21, 74] fails. Actually, the verification component of the said paradigm is a second hypothesis which essentially tries to support the first one to enhance the vote of thanks in favor of any decision (right or wrong).

20.3 PROPOSED CONCEPT

To resolve the problem get a consistent interpretation about the scene, we consider (ATM) system [32] which is basically a tool for belief revision and which can handle inconsistent information. The ATM system can handle multiple potential information and can generate multiple beliefs about the scene rather than concluding a single decision. It can revise an existing belief about the scene if any contrary information arises. That means belief can be changed non-monotonically [32, 35, 134]. With the help of non-monotonic reasoning technique, we can modify or expand the knowledge base of the vision system.

20.3.1 Generation of Environmental Lattice [32]

The essence of the ATM system is pivoted on the concept of environmental lattice where all possible solutions of a given situation can be assumed and governed by a set of justifications stated in terms of rules. We may obtain environments by a process which starts from an empty environment. Then we add one assumption (in all possible ways) and then two assumptions (in all possible ways) and so on. This increment process may be represented by an environmental lattice which represents the inclusion relation among sets of assumptions.

Every consistent environment characterizes a context. If there are *n* assumptions, then there are potentially 2^n contexts. There are (n_k) environments having k assumptions. Figure 3 illustrates the environment lattice for assumptions {A, B, C}. The nodes represent environments. The upward edges from an environment represent subset relationships with environments one larger in size. Conversely, the downward edges from an environment represent superset relationships with environments one smaller. All the supersets of an environment can be found by tracing upward through the lattice, and all the subsets of an environment can be found by tracing downwards through the lattice. Thus all environments are subsets of the top-most node {A, B, C}, and all environments are supersets of the bottom-most node *{}*.

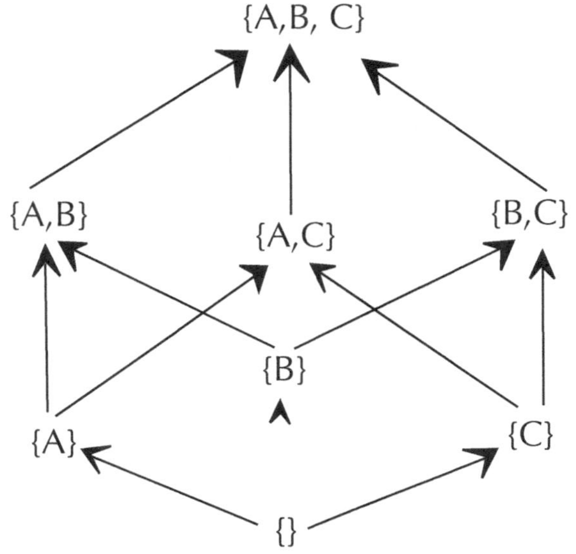

FIGURE 3 Environmental lattice.

In an environmental lattice all sets of assumptions cannot be considered. For example A may correspond to an assertion and C to its negation. Thus, they contradict each other and cannot be considered together. On the other hand, it may happen that A corresponds to a fact in the domain which cannot be combined with the fact represented by B. For example, A = "X is very hot" and C = "The temperature of X is very low". In such case, the environments which contain both A and C are no good (marked with a bold frame in the example). If an environment is no good, then all of its superset environments are no good as well. The no good node {A, B, C} of Figure 4 is the result of the no good node {A, B}.

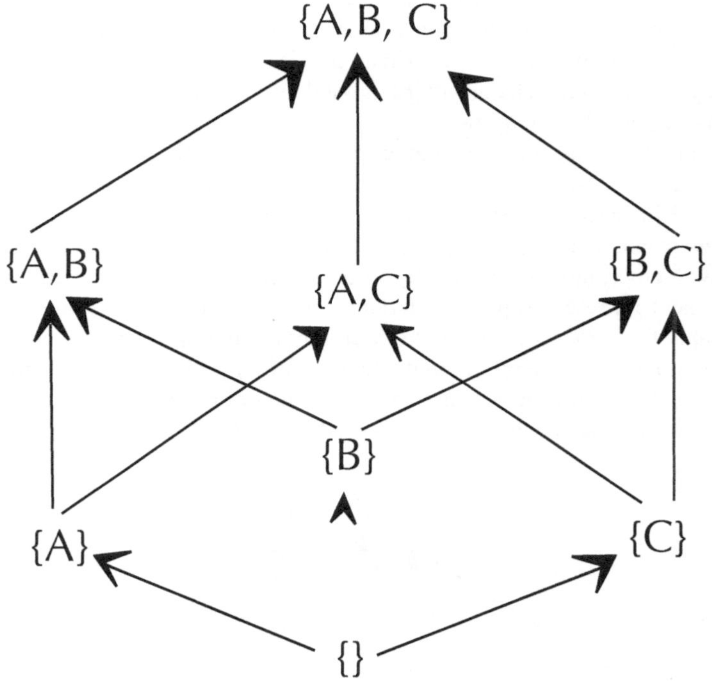

FIGURE 4 Environmental lattice with no good nodes.

Before recognizing an object in a scene, we generate a belief about the objects in the scene with the help of model objects which are primary knowledge base. This belief is nothing but a basic knowledge about the scene. The knowledge may be revised non-monotonically, if any contrary information due to the presence of any unknown object arises. As for example this situation can be resolved two ways:

1. If the unknown object is removed from the scene then the vision system should be able to come to a single conclusion.
2. Or if any additional information is provided to the vision system, like "This scene does not contained model one (or model object two)" then also the vision system should be able to conclude a single decision.

However to generate a belief, first of all we have to generate an environmental lattice controlled by a set of rules. This environmental lattice essentially represents all possible combinations of model objects which may generate the scene. As in the present context of object recognition, the rules are essentially concerned with the matching of features and area between model objects and scene we proceed through the following sections before we arrive at section 20.2.5 where specific rules for object recognition are considered.

20.3.2 Features Matching

A model m from the model set M has *p* features, M_k (M_{xk}, M_{yk}, Curvature) (M_{xk} ≡ x–coordinate of the kth feature and M_{yk} ≡ y–coordinate of the kth feature) for $k = 1$, p and a scene *S* has q features S_l(S_{xl}, S_{yl}, Curvature) (S_{xl} ≡ x-coordinate of the lth feature and S_{yl} ≡ y–coordinate of the lth feat) for $l = 1$, q. These features refer to as curvature of distinct corner points of an object of both model and scene. To get a match between model and scene object we compare all the features of the model corners to those of entire scene corners. Model and scene corners with dissimilarity less than T_j form a match. Here T_j is a threshold value of matching between the model and scene features. In this study, we set the value of T_j as 0.001. The system first makes a list of matches between the features of the scene corners and those of possible model corners that appear in the scene. This is a primary match list. This list is revised when a final match list is generated by checking some compatibility check which is discussed.

20.3.3 Shape Matching

To get the exact matches between the model object and the scene object it is required to match the shape of both the objects. For shape matching mutual compatibility constrains extract a set of consistent matches from a model's match list. The following four rules test the three matches K_1, K_2, and K_3 for mutual compatibility [74]:

1. They do not contain the same scene corner ($S_1 \neq S_2 \neq S_3$).
2. They do not contain the same model corner ($M_1 \neq M_2 \neq M_3$).
3. These scene corners refer to the model corners belonging to the same model (i.e. M_1, M_2, and M_3 belong to the same model).
4. The structure between the three scene features equals the structure between the three model features.

Any three matches which satisfy the all four compatibility constrains (mentioned above) are considered for shape matching.

The first rule takes three unique scene features occupying different location in the scene. Rule two deals with three model features for the same. Rule three prevents the combination of three model features from different models because no prior knowledge about the relationship between models exists. To illustrate constraint four, suppose that there exist three scene features S_1, S_2, and S_3. The three model features M_1, M_2, and M_3 should have the same structure as the scene features. To do so we apply simple geometry of equality property/congruent property of a triangle.

Consider the said three points of both model and scene objects and construct two triangles, one on the model object and one on the scene. If any two distances between any two points of both the triangles and any angle constructed by these three points of both model and scene objects match exactly or by a threshold value then it is said that the three matches are in the same structure in both the model and scene objects. And these three matches will be listed as final match list. Here we fix the threshold value as 0.001 for angle and 1.414 for distance between two points. This check may be applied for more than three matches. For structure matching more matches result more accurate matching between model and scene objects.

To illustrate the concept of shape matching we consider the following Example 1.

Example 20.1

The following figure represents one model object (Wrench) and a scene (Scene 7) with the feature points marked by dot and index number.

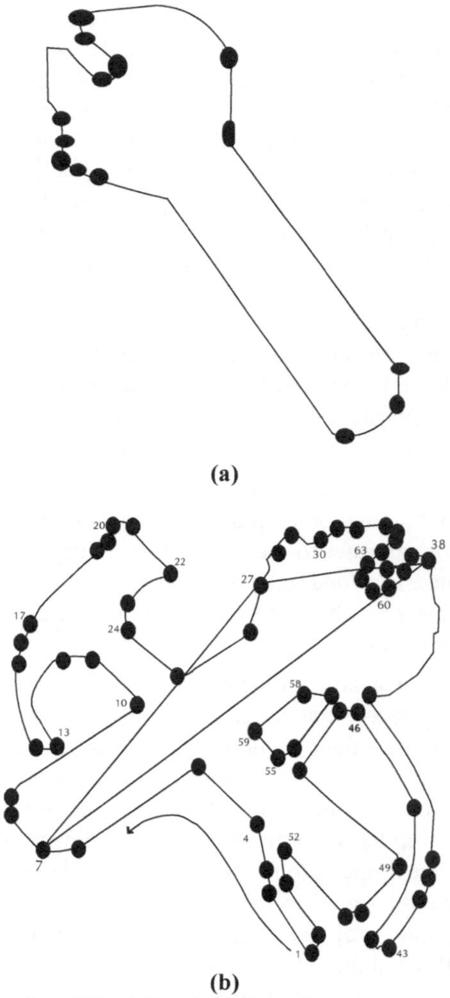

FIGURE 5 (a) Model object (Wrench) and (b) Scene image.

To get the set of points of final match (actual match) from the set of primary match points, the system verifies the matches by shape. This is done by properties of geometry. First the system finds a set of points of both scene and model objects as primary match. Like $K\{(M_i, S_j)\}$, here M_i and S_j (where $1 \leq I \leq n, 1 \leq i \leq m$ and n, m are the number of boundary points of both model and scene respectively) are the model and scene points respectively. From this set of matches the system verifies the shape creating triangles choosing any three points from both model and scene. If the triangles $T_m(M_a, M_b, M_c)$ (where $1 \leq a \leq n$, $1 \leq b \leq n, 1 \leq c \leq n$ and $a \neq b \neq c$) from model object are equal by threshold value (as mentioned earlier) with the corresponding triangles $T_s(S_x, S_y, S_z)$ (where $1 \leq x \leq m$, $1 \leq y \leq m$, $1 \leq z \leq m$ and $x \neq y \neq z$) from scene object, then the system picks those points as final match. So the set of final match should be K_f $\{(M_a, S_x), (M_b, S_y), (M_c, S_z)\}$.

Let us consider the Figure 5. Based on the discussion we obtain the following results:

1. Primary match set $K = \{(1, 23), (2, 7), (18, 41), (18, 53)\}$. Number of matches is 66.
2. The triangles, which are equal for both model and scene object areas, T_m (2, 6, 18) and T_s (7, 38, 27) (as displayed in the figure), T_m (6, 7, 15), and T_s (38, 37, 30), T_m (15, 16, 17) and T_s (30, 29, 28) and so on. These points from both model and scene object belong to final match (actual match) set.
3. So, the final match set K_f $\{(2, 7), (18, 27), (6, 38)\}$. Number of matches are 7. Here all the numbers are indicating the index of feature points for both model and scene object (see Table 7 of Section 20.3.4).

20.3.4 Area Matching

For a successful recognition of objects it is not suffice the matching of features only. We also match area between model objects and scene objects. In our approach the area of an object is calculated by a modified computer graphics filling algorithm.

Algorithm (Area Calculation)

Variable

1. A global variable "count" is set to zero which count the number of pixels.
2. Choose a color "col" expect the boundary pixels' color by which the inner pixels of an image boundary will be colored.

Assumption/Pre-requisite:

1. Image boundary should be closed. That means an object may have more than one curve but each curve should be continuous and close.
2. Normally the color of the boundary pixels should be white. That means it will be different from the boundary pixels' color and "col".

Steps

Step 1: Find a white pixel inside the image boundary. (It can be done by scanning the image by "left to right manner" from bottom of the image)
Step 2: Color the pixel by "col" and increment the "count" by one.

Step 3: Recursively travel (cover) the inner pixels of the image boundary as following manner. (Step 4–6)
Step 4: Travel upwards by one pixel and color the pixel by "col" and increment the "count" by one.
Step 5: Travel leftwards by one pixel and color the pixel by "col" and increment the "count" by one.
Step 6: Travel down-rightwards by one pixel and color the pixel by "col" and increment the "count" by one.
Step 7: Repeat all the steps (Step 1–6) until no white pixels closed by the boundary of the image are found.
Step 8: Add the number of boundary pixels to "count".

The area of an object is nothing but the sum of the number of pixels within the boundary of an object represented by a closed curve and the number of pixels on the boundary of the said closed curve.

20.3.5 Initial Hypothesis

Initial hypothesis about a given scene is nothing, but the raw belief about that scene. This belief is generated under the control of environmental lattice and its rules as mentioned in section 20.2.1.

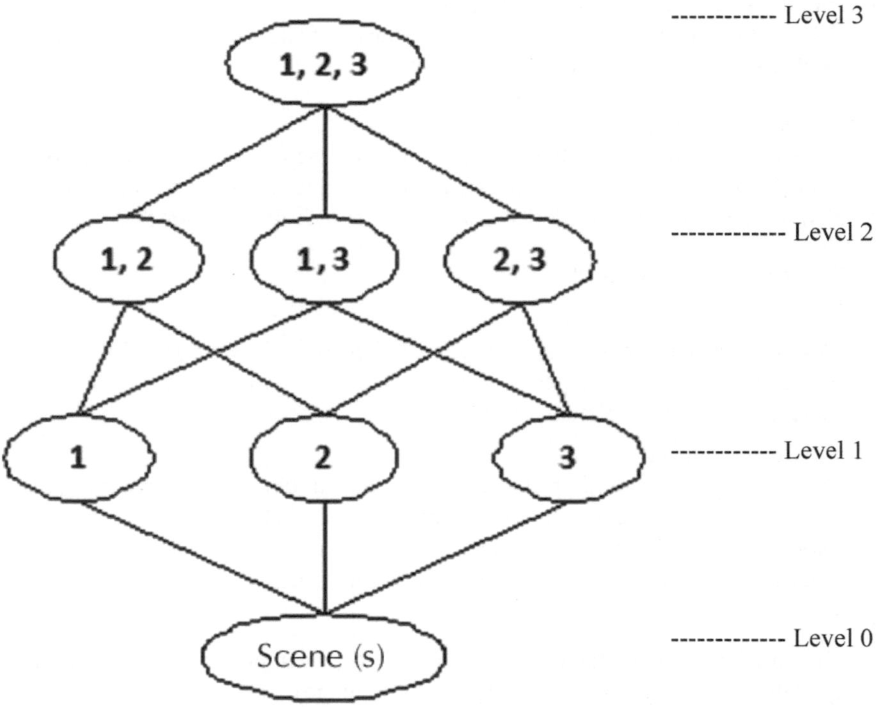

FIGURE 6 Environmental lattice with tree model objects.

This assumption of presence of models in the scene will take place step by step and the steps are controlled by a graph structure constructed by the model objects. The graph structure is nothing, but the environmental lattice [32]. In Figure 6 we explain this process using our closed world model database. For simplicity of demonstration of environmental lattice we assume that we have three model objects (i.e. 1, 2, and 3) as our primary knowledge base.

In the Figure 6 the lowest node (at level 0) represents an empty environment. In the present situation it represents the scene itself. Other nodes represent model or combination of models. The number in each node represents different models. At the level 1 the nodes of the environmental lattice will be considered by a set of rule. In this level each model object from the knowledge base is considered separately.

These rules are as follows:

(i) 15% features of the model object match the features of the scene objects.

(ii) The area of the model object is either equal or less than the area of the scene. For the nodes of any level of the environmental lattice higher than the level 1 are considered by a different set of rules if and only if the corresponding lower level nodes satisfy their corresponding rules. As for example the node (1, 2) at the level 2 will be considered if and only if the node (1) and (2) at level 1 are satisfied. The rules of the nodes of the higher level than the level 1 are as follows:

(iii) 70% area of the sum of the areas of all the corresponding model objects (mentioned in the node) is less than or equal to the scene area.

(iv) Sum of all the areas of all the corresponding model objects (mentioned in the node) is less than or equal to the scene area.

By these two sets of rules the initial assumption about the existence of model objects in the scene is established. Thus, the initial hypothesis will be generated.

So in brief at each node at level 1 of the environmental lattice, different model object from the knowledge base is considered separately by a set of rules (rule I and II). On the other nodes at any level higher than the level 1, a combination of model objects is considered by another set of rules (rule III and IV).

In the next section we discuss about the generation of final hypothesis or final belief based on the initial belief (hypothesis).

20.3.6 Final Hypothesis

Each time we get an initial hypothesis (as discussed in the section 20.4.5) we check it whether the belief is true and final. This verification is done by plotting the model (s) of the nodes of the environmental lattice at which the belief has been generated onto the scene. As for example if the initial hypothesis (raw belief) is generated at the node (2, 3) at the level 2 then the model 2 and the model 3 will be plotted onto the scene for verification of truth or for taking it as a final hypothesis (belief).

Coordinate Transformation of Model Objects for Final Hypothesis (belief)

To plot a model object onto the scene it is required to transform the coordinates of the model object on the coordinate of the matched points of the scene object. The transformation scheme works as follows:

Let the variable set θ, T_x, T_y represents the coordinate transform factor. The variable θ describes the rotation around the model coordinate system to map the model corner onto the scene corner. The variable T_x and T_y give the translations, along the indicated axis, that follow the rotation. To create the hypothesized image, that is to plot the model points onto the scene image we consider a pair of valid matches between model image and scene image.

Let K_1 and K_2 are those two matches with following details:

K_1 and K_2 correspond (M_i, S_a) and (M_j, S_b) respectively. Where $M_k|k = i, j$ and $S_l|l = a, b$ are model image pairs and scene image pairs respectively.

Now, M_k and S_l will be defined by natural coordinate parameter as follows:

$$M_i = (M_{xi}, M_{yi}) \qquad S_a = (S_{xa}, S_{ya})$$

$$M_j = (M_{xj}, M_{yj}) \qquad S_b = (S_{xb}, S_{yb})$$

After using the following algorithm we get the new coordinate M_{xm} and M_{ym} for the model points onto the scene.

Algorithm (Coordinate Transform)

Step 1: Calculate T_x and T_y , where $T_x = S_{xa} - M_{xi}$ and $T_x = S_{ya} - M_{yi}$.

Step 2: Calculate θ, where θ = Gradient (Gradient is nothing but the angle between the positive side of *X* axis of the Cartesian coordinate system and the straight line constructed by these two points) of (M_i, M_j) – Gradient of (S_a, S_b).

Step 3: Calculate the distance *(dst)* between M_i and any one pixel of M_m, where *m* = *1, 2, N* and *N* is number of pixel in the model image. Here M_m will be transformed to new coordinate and M_i is the center for rotation and from this point the translation factor is calculated.

Step 4: Calculate Gradient *(Gr)* of (M_i, M_m).

Step 5: Calculate new coordinate (M_{xm}, M_{ym}) of pixel.

where $M_{xm} = M_{xi} + T_x + dst * \cos(Gr - \theta)$

and $M_{ym} = M_{yi} + T_y + dst * \sin(Gr - \theta)$

(M_{xm} and M_{ym} are integers)

Step 5: Repeat Step 3 and 4 for all non-zero pixels M_m in the model image except M_i.

Step 6: For the pixel M_i, $M_{xm} = M_{xi} + T_x$ and $M_{ym} = M_{yi} + T_y$.

Step 7: Delete single pixels and corner pixels.

The time complexity of this algorithm is *O*(n), where n is the number of points in the model image.

The transformation of the coordinates of the model object(s) on to the objects of the scene image become successful if we see that the number of pixels, say less than the threshold value 5, of the model object(s) with white color is out of the scene image (see section 20.3.4). Thus, the initial (raw) belief is true.

The scene obtained after the successful plotting of the model object(s) onto the actual scene, by the said process of coordinate transformation, is called a hypothesized scene with final belief.

According to the node of the environmental lattice usually the last true initial (raw) belief will be considered as a final belief about the given scene formed by the model object(s). After getting the hypothesized scene with final belief we may attach interpretation to our final belief about the scene.

In the next section we consider our design study based on the concept of hypothesized scene.

20.4 DESIGN STUDY

Basic design philosophy of the present study is as follows:

We can recognize an object which we have seen earlier.

Basic assumptions considered for the design study are as follows:

(i) Closed World Assumption.
(ii) Rigid Body Assumption.
(iii) Certain percentage of occlusion of the model object in the scene is allowed.

Remark 20.1

If any contrary information arises outside the closed world assumption: the question of belief revision comes into the picture.

20.4.1 Models Considered

In the present design study a set of model objects (boundary image of 2D view of objects) is known as primary knowledge base to the vision system. Our algorithm tries to recognize and locate each object in a given scene and generate a belief about the scene.

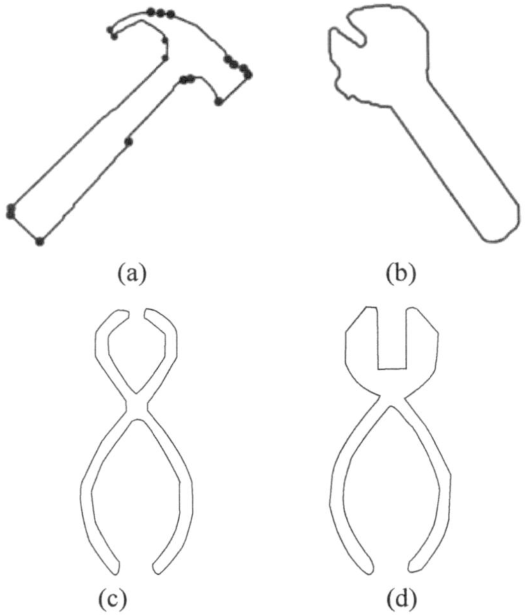

FIGURE 7 (a) Model object Hammer, (b) Model object Wrench, (c) Model object Clip, and (d) Model object Cutter.

We consider a set of four model objects. The model set M = {Hammer, Wrench, Clip, and Cutter} is shown in the Figure 7. Here the models Clip and Cutter are alike and they can generate multiple believes for a scene. In our approach the local features at distinct corner points on the boundary and the coordinates of all the points of image boundary are priory known to the vision system.

20.4.2 Image Pre-operation

As all the points on a curve (boundary) of an image boundary are not so important as a features, we extract some distinct points as features of 2D images. In our approach we mark some distinct points on a curve, at which the local features will be calculated. We apply a simple threshold algorithm by calculating slope or gradient for this purpose. By this algorithm only a few distinct points on the boundary of a digital image where the curve changes its gradient abruptly will be preserved by measuring the slope of the curve.

Algorithm (Finding Distinct Points of the Edge)

Variable

1. Start point is S_p. Initially $S_p \equiv P_0$. Here P_i (Where $i = 0$ N, N is the number of points on a curve) is the all points on an image boundary. In or approach P_0 is the first left point form bottom of the image. It can be obtained by scanning the image by "left to right" manner from bottom of the image.

Assumption/Prerequisite

1. In our approach all the curves are continuous and closed.

Steps

Step 1: Preserve S_p as a distinct point.
Step 2: Calculate G_m = Gradient of (P_{i+1}, S_p).
Step 3: Calculate G = Gradient of (P_{i+2}, S_p).
Step 4: Calculate G_l = Gradient of (P_{i+1}, P_i).
Step 5: Preserved the point P_i as distinct point and assign it to S_p if $|G - G_m| > 1^0$ and $|G_l - G| > 50^0$.
Step 6: Repeat Step 3 – 5 until a new S_p is found.
Step 7: If a new S_p is found then repeat step 2 – 6 for all on P_i on the curve. (Here P_i is chosen from P_0 to P_N by considering the nest neighbor point of previous point.)
Step 8: If two consecutive points are preserved then consider any one.
Step 9: If three consecutive points are preserved then consider the middle one.
Step 10: If more than three consecutive points are preserved then consider only the first and last point.
Step 11: If more than three consecutive preserved points exit on same straight line then consider only the first and last point.
Step 12: Record only preserved points as distinct points on the curve.

The time complexity of this algorithm is O(n), where n is the number of black pixels of the boundary curve of the objects image.

By this algorithm we get a few distinct points of the model objects (Figure 8(a), (b), (c), (d)) at which the features are calculated.

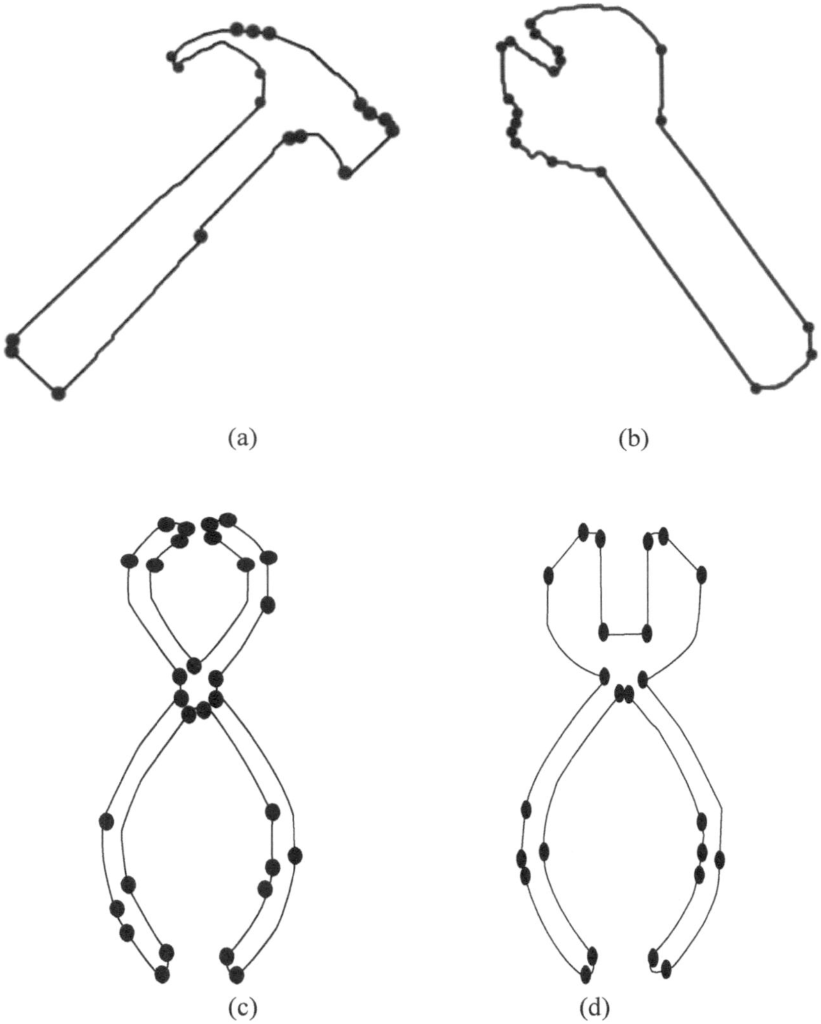

FIGURE 8 Model objects with distinct feature points (marked by dots).

In the Figure 8 the black doted pixels are considered as distinct points of each model object. These points can be considered as corners of a curve.

20.4.3 Extraction Features

After getting some corner points on the curve of an image boundary, we calculate the features on those points only. We calculate the features by means of curvature which is the sharpness of a corner. This feature is extracted locally. Hence, it is considered as local features of an object.

The local features are calculated in two steps:

Internal Angle

In a digital image the curves is defined as a set (P) of pixels and the curves are one pixel wide with two neighbors (except cross point of two or more curves). The set contains integer coordinate points P_i. So, $P = \{P_i | P_i = (x_i, y_i) and 1 \leq i \leq n\}$ where n is the number of distinct pixels of the curve. Hence P_{i+1} and P_{i-1} are two neighbors of P_i which are also distinct points. That is in this process only the distinct points are considered.

Internal angle is an interior angle at one point by its two neighbors. Internal angle (c_i) at point P_i by its two neighbors is defined as:

$$c_i = \cos^{-1}(a_i \bullet b_i)/|a_i||b_i| .$$

Here $a_i = (\delta x_{i+}, \delta y_{i+})$, where $\delta x_{i+} = x_i - x_{i+1}$ and $\delta y_{i+} = y_i - y_{i+1}$,
And $b_i = (\delta x_{i-}, \delta y_{i-})$, where $\delta x_{i-} = x_i - x_{i-1}$ and $\delta y_{i-} = y_i - y_{i-1}$ [139].

Curvature

If the internal angle (c_i) at point P_i tends to $+\Pi$, then the points P_{i-1}, P_i and P_{i+1} tends to construct a straight line. That means curvature (ρ_i) at point P_i should be smaller. So, curvature at any point relates to its internal angle as:

Bigger internal angle means smaller curvature and *vice-versa.*

Now, the expression of curvature is defined as follows:

Curvature $(\rho_i) = (\Pi - c_i)/(D_p + D_n)$

Where, $D_n = |a|$, $D_p = |b|$ and D_p, D_n are the distance from next neighbor towards both the directions from point p_i and if D_p, D_n exceed by $2\sqrt{2}$ unit than the values of these two distance are set to $2\sqrt{2}$.

After calculating the features at the distinct points of a curve we stored only those points with its curvature whose curvature value is a threshold value of 0.035.

The features of model objects are known as primary knowledge of the vision system. It provides the detail information about the model objects.

Design Parameters

For the present design study we consider four model objects, *viz.* hammer, wrench, clip, and cutter. Details of the model objects are given in Table 1. The features of the model objects are considered either in clockwise or anti-clockwise direction from the extreme bottom-left corner of the boundary of each model object. Figure 9 shows the direction of traversal through the contour of the image of each model object. The feature points of the model objects are shown on the figures of Figure 9.

TABLE 1 Details of the model objects.

Model	No. of Boundary Points	No. of Features	Area (Total Pixels)
 **Hammer**	311	18	2882
 **Wrench**	314	19	4129
 **Clip**	465	30	1565
 **Cutter**	438	24	1892

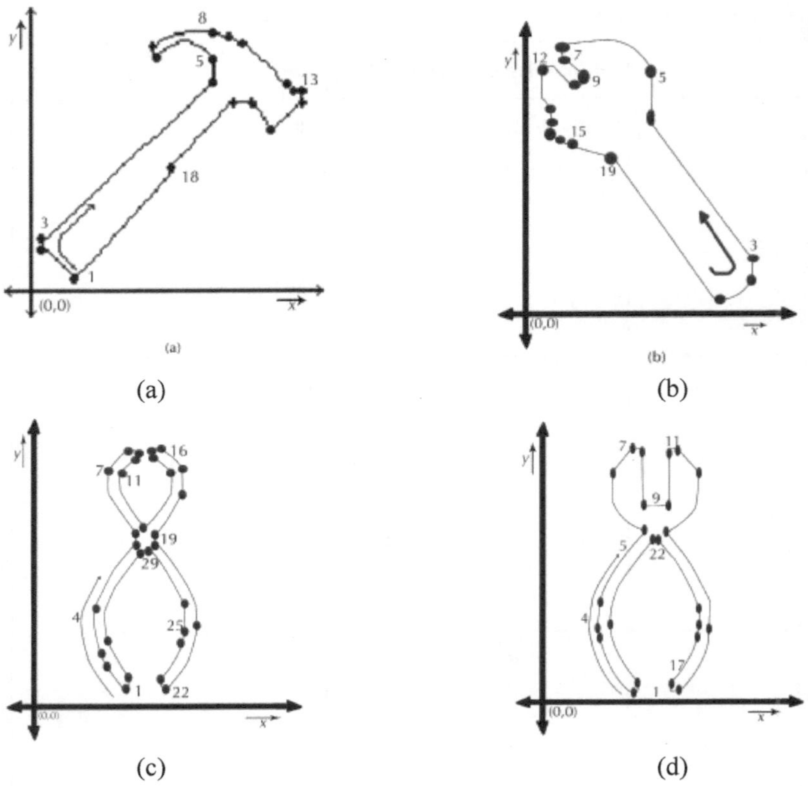

FIGURE 9 Polygonal approximation of the model objects.

Figure 10 shows the graph of the features of the model objects.

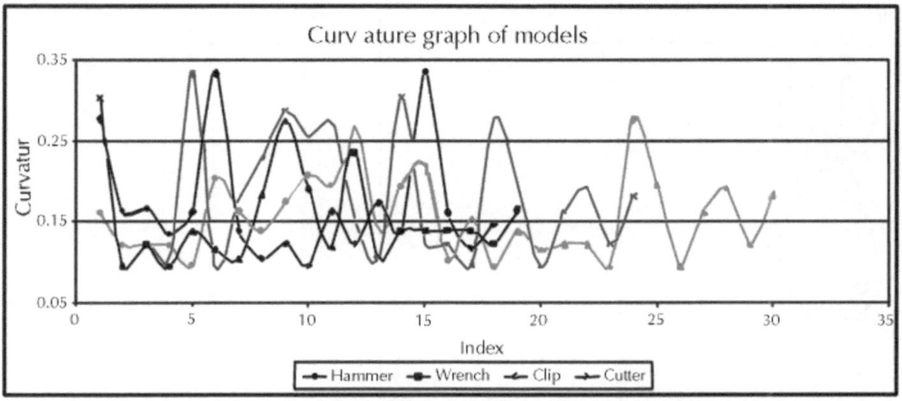

FIGURE 10 Graph of the features of the model objects.

Figure 11 shows the features (curvature) of the model objects (Clip and Cutter) which look alike. And the alike portion are representing by the first four and the last ten points in each graph.

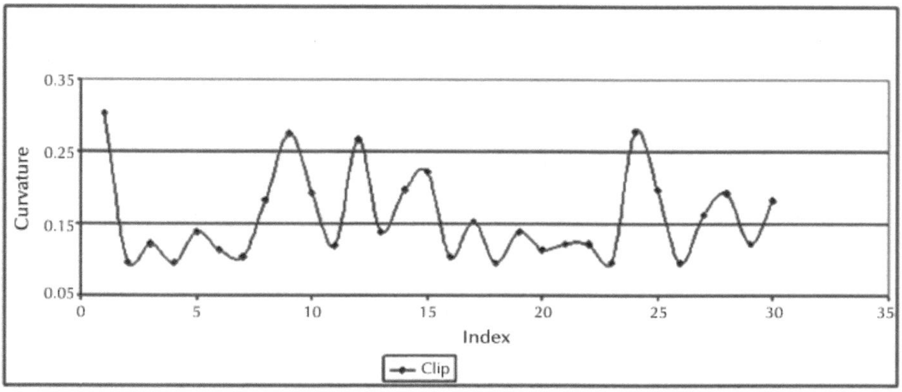

FIGURE 11 (a) Curvature graph of Clip.

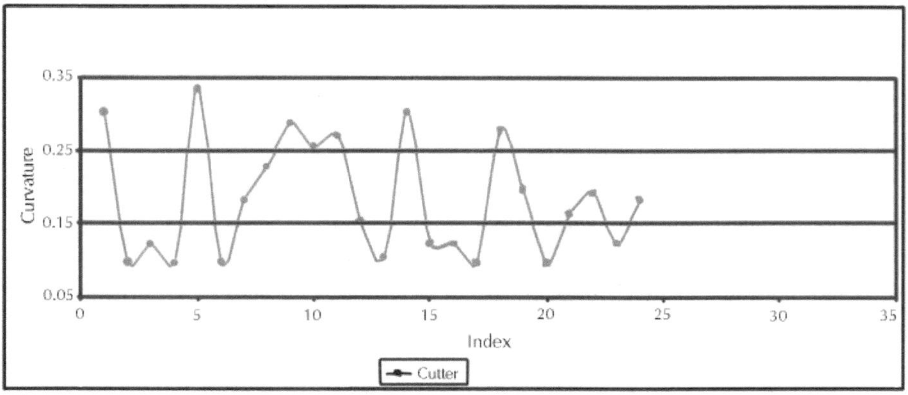

FIGURE 11 (b) Curvature graph of Cutter.

20.4.4 Belief/Multiple Beliefs Based on Hypothesized Scene

After getting the hypothesized scene (see section 20.3.6) the vision system may arrive at single belief or multiple beliefs. We discuss such situations through different design studies.

Case Study 1

Consider the scene in the figure (Figure 12(a)). Here the uniquely identifiable distinct part of the two model objects Clip and Cutter are occluded by the wrench. After applying all the algorithms and schemes described in the sections the vision system obtains a hypothesized scene as shown in Figure 12(b). From the hypothesized scene we may conclude either, "The scene consists of wrench and clip" or "the scene consists of wrench and cutter".

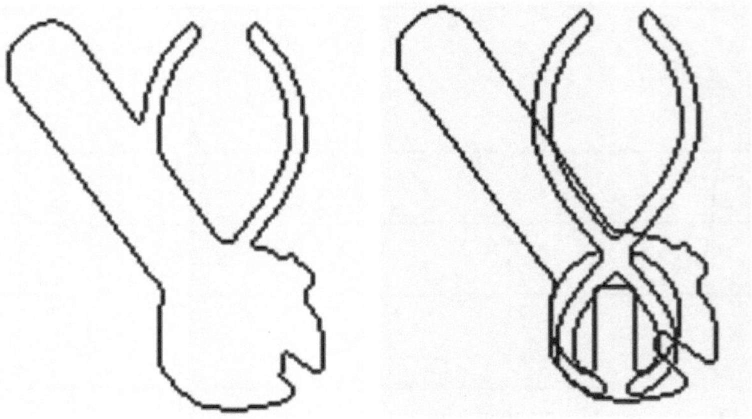

(a) (b)

FIGURE 12 (a) Scene and (b) Hypothesized scene.

This problem can be resolved (i.e. the vision system will arrive at a single belief) if any additional information is provided to the vision system or the objects are viewed separately by moving it from the scene. Hence, the concept of belief revision (by non-monotonic reasoning approach) becomes essential.

Case Study 2

Consider the input scene images as shown in the Figure 13(a) and Figure 14(a). For these scenes the distinct corner points are marked by black dots as shown in the Figure 10(b) and 11(b) and the hypothesized scenes are shown in the Figure 13(c) and Figure 14(c).

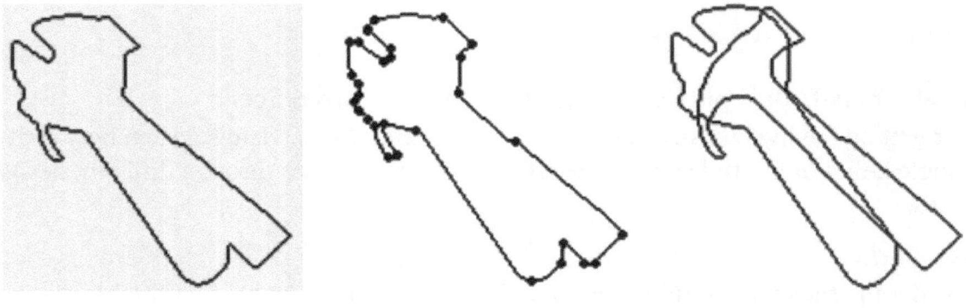

(a) (b) (c)

FIGURE 13 (a) Scene, (b) Scene with marked feature points, and (c) Hypothesized scene.

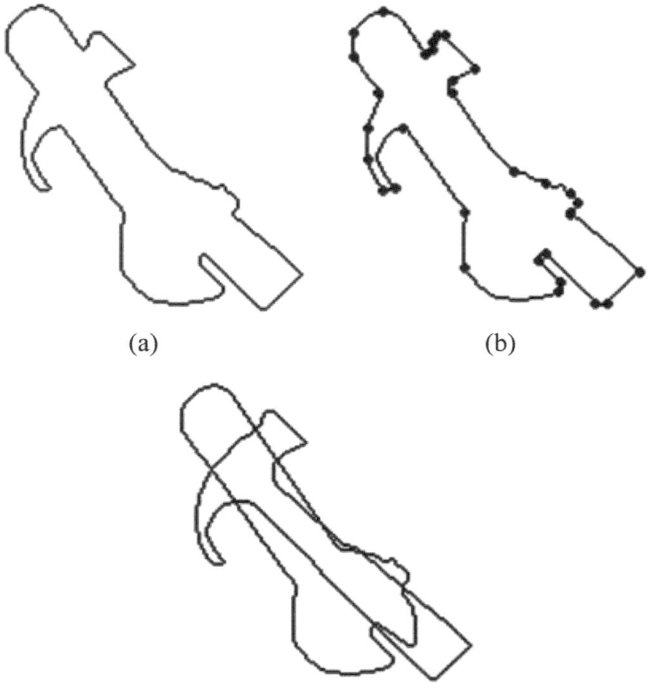

(a) (b) (c)

FIGURE 14 (a) Scene, (b) Scene with marked feature points, and (c) Hypothesized scene.

The detail experiment results on scene image 13(a) are;

(a) Model object (Hammer) and scene image 13(a) The following figure (Figure 12) represents the feature points of model object (Hammer) and scene image 13(a).

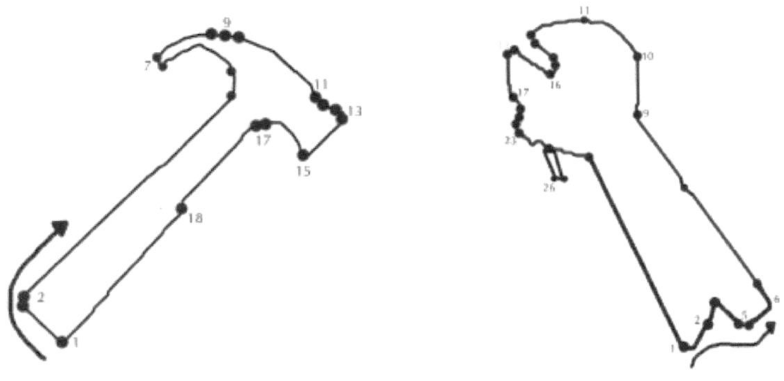

(a) Hammer (b) Scene Image

FIGURE 15 Feature points of (a) model object (Hammer) and (b) scene image 10 (a).

We first consider the primary match and final match between the model object (Hammer) and the corresponding scene object of the scene image of Figure 15(b). Table 1 represents the primary and final match details. Here the bottom and left side indices are the feature point index of the model object and scene object respectively as shown in Figure 15.

TABLE 2 Match matrices between Hammer and the corresponding object of the scene image.

1	.	.	.	.	.	.	.	.	.	.	.	.	.	.	.	@	.	.
2	.	.	.	.	.	.	.	.	.	.	.	.	.	.	.	.	.	.
3	.	.	.	.	.	.	.	.	.	.	.	.	.	.	.	.	.	.
4	.	@	.	.	@	.	.	.	.	.	@	.	.	.	.	.	.	.
5	.	**(@)**	.	.	@	.	.	.	.	.	@	.	.	.	.	.	.	.
6	.	.	.	.	.	.	.	.	.	.	.	.	.	.	.	.	.	.
7	.	.	.	.	.	.	.	.	.	.	.	.	.	.	.	.	.	.
8	.	.	.	.	.	.	.	.	.	.	.	.	.	.	.	.	.	.
9	.	.	.	.	.	.	@	.	.	.	.	.	.	@	.	.	.	.
10	.	.	.	.	.	@	.	.	.	.	.	.	.	.	**(@)**	.	.	.
11	.	.	.	.	.	.	@	.	.	.	.	.	.	**(@)**	.	.	.	.
12	.	.	.	.	.	.	.	.	.	.	.	.	.	.	.	.	.	.
13	.	@	.	.	@	.	.	.	.	.	@	.	.	.	.	.	.	.
14	.	.	.	.	.	.	@	.	.	.	.	.	.	@	.	.	.	.
15	.	.	.	.	.	.	.	.	.	.	.	.	@	.	.	.	.	.
16	.	.	.	.	.	.	.	.	.	.	.	.	.	.	.	.	.	.
17	.	.	.	.	.	.	.	.	.	.	.	.	.	.	.	.	.	.
18	.	.	.	.	.	.	.	.	.	.	.	.	.	.	.	.	.	.
19	.	.	.	.	.	.	.	@	.	.	.	.	.	.	.	.	.	.
20	.	.	.	.	.	.	@	.	.	.	.	.	.	@	.	.	.	.
21	.	.	.	.	.	.	@	.	.	.	.	.	.	@	.	.	.	.
22	.	.	.	.	.	.	@	.	.	.	.	.	.	@	.	.	.	.
23	.	.	.	.	.	.	@	.	.	.	.	.	.	@	.	.	.	.
24	.	.	.	.	.	.	.	.	.	.	.	.	.	.	.	.	.	.
25	.	.	.	.	.	.	.	.	.	.	.	.	.	.	.	.	.	.
26	.	.	.	.	.	.	.	.	.	.	.	.	.	.	.	.	.	.
27	.	.	.	.	.	.	.	.	.	.	.	.	.	.	.	.	.	.
28	.	.	.	.	.	.	.	.	.	.	.	.	.	.	.	.	.	.
29	.	.	.	.	.	.	.	.	.	.	.	.	.	.	.	.	.	.
	1	2	3	4	5	6	7	8	9	10	11	12	13	14	15	16	17	18

Key: **@** ≡ primary match
(@) ≡ Final match

Here the percentage of match of the model object with the corresponding object of the scene is as follows:

* No. of Model (Hammer) features: 18
** Percentage of match:(3/18) = 16.67%

Table 3 represents the number of the final matches and their corresponding coordinates.

TABLE 3 Coordinates of final match points.

No. of Match	Scene # (x, y)	Model (Hammer) # (x, y)
1	5 (109, 10)	2 (5, 20)
*** 2**	**10 (57, 109)**	**15 (104, 72)**
3	11 (44, 122)	14 (117, 85)

Match no. 2 of Table 2 is considered as the centre for rotational transformation.

(b) Model object (Wrench) and scene image 13(a). The following figure (Figure 16) is represents the feature points of model object (Wrench) and scene Image 13(a).

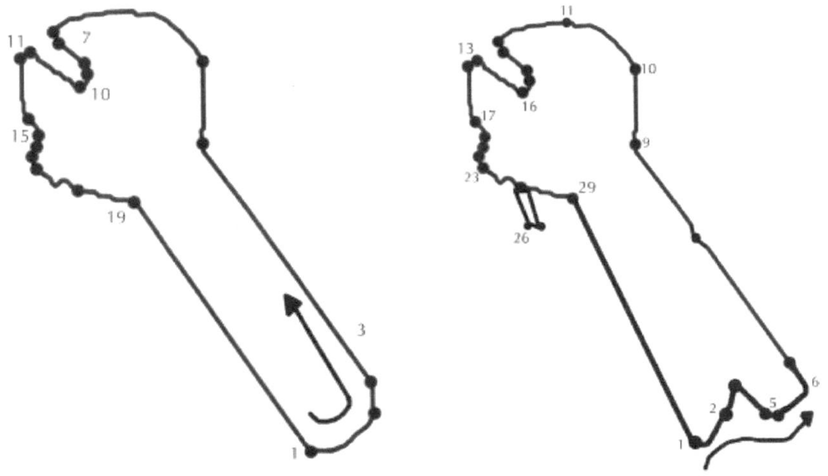

(a) Wrench (b) Scene Image

FIGURE 16 Feature points of (a) model object (Wrench) and (b) scene image 13(a).

Table 4 represents the primary and final match between the model object (Wrench) and the corresponding object of the scene image.

TABLE 4 Match matrices between Wrench and the corresponding object of the scene image.

	1	2	3	4	5	6	7	8	9	10	11	12	13	14	15	16	17	18	19
1	(@)	.	.	.	.	.	.	.	.	.	.	.	.	.	.	.	.	.	.
2	.	.	.	.	.	.	.	.	.	.	.	.	.	.	.	.	.	.	.
3	.	.	.	.	.	.	.	.	.	.	.	.	.	.	.	.	.	.	.
4	.	.	.	.	.	.	@	.	.	.	.	.	.	.	.	.	.	.	.
5	.	.	.	.	.	.	@	.	.	.	.	.	.	.	.	.	.	.	.
6	.	.	.	.	.	.	.	.	.	.	.	.	.	.	.	.	.	.	.
7	.	.	.	.	.	.	.	.	.	.	.	.	.	.	.	.	.	.	.
8	.	.	.	.	.	.	.	.	.	.	.	.	.	.	.	.	.	.	.
9	.	.	.	.	.	.	.	@	.	.	.	.	.	@	@	@	@	.	.
10	.	.	.	.	.	.	.	.	.	.	.	.	.	.	.	.	.	.	.
11	.	.	.	.	.	.	.	@	.	.	.	.	.	@	@	@	@	.	.
12	.	.	.	.	.	(@)	.	.	.	.	.	.	.	.	.	.	.	.	.
13	.	.	.	.	.	.	(@)	.	.	.	.	.	.	.	.	.	.	.	.
14	.	.	.	.	.	.	.	(@)	.	.	.	.	.	@	@	@	@	.	.
15	.	.	.	.	.	.	.	.	(@)	.	.	.	.	.	.	.	.	.	.
16	.	.	.	.	.	.	.	.	.	(@)	.	.	.	.	.	.	.	.	.
17	.	.	.	.	.	.	.	.	.	.	(@)	.	.	.	.	.	.	.	.
18	.	.	.	.	.	.	.	.	.	.	.	(@)	.	.	.	.	.	.	.
19	.	.	.	.	.	.	.	.	.	.	.	.	(@)	.	.	.	.	.	.
20	.	.	.	.	.	.	.	@	.	.	.	.	.	(@)	@	@	@	.	.
21	.	.	.	.	.	.	.	@	.	.	.	.	.	@	(@)	@	@	.	.
22	.	.	.	.	.	.	.	@	.	.	.	.	.	@	@	(@)	@	.	.
23	.	.	.	.	.	.	.	@	.	.	.	.	.	@	@	@	(@)	.	.
24	.	.	.	.	.	.	.	.	.	.	.	.	.	.	.	.	.	.	.
25	.	.	.	.	.	.	.	.	.	.	.	.	.	.	.	.	.	.	.
26	.	.	.	.	.	.	.	.	.	.	.	.	.	.	.	.	.	.	.
27	.	.	.	.	.	.	.	.	.	.	.	.	.	.	.	.	.	.	.
28	.	.	.	.	.	.	.	.	.	.	.	.	.	.	.	.	.	.	.
29	.	.	.	.	.	.	.	.	.	.	.	.	.	.	.	.	.	.	(@)

Key: @ ≡ primary match
(@) ≡ final match

The left index of Table 4 indicates the scene features and bottom index of Table 4 indicates the model features. For the model object wrench the percentage of match with the corresponding object of the scene image is as follows:

* No. of Model (Hammer) features: 19
** Percentage of match: (14/19)%
= 73.68%

Table 5 represents the number of the final matches and their corresponding coordinates.

TABLE 5 Coordinates of final match points.

No. of Match	Scene # (x, y)	Model (Wrench) # (x, y)
1	1 (82, 2)	1 (94, 3)
*** 2**	**12 (11, 118)**	**6 (23, 119)**
3	13 (11, 116)	7 (23, 117)
4	14 (20, 107)	8 (32, 108)
5	15 (20, 104)	9 (32, 105)
6	16 (17, 102)	10 (29, 103)
7	17 (5, 111)	11 (17, 112)
8	18 (2, 110)	12 (14, 111)
9	19 (3, 95)	13 (15, 96)
10	20 (7, 90)	14 (19, 91)
11	21 (7, 86)	15 (19, 87)
12	22 (5, 84)	16 (17, 85)
13	23 (5, 81)	17 (17, 82)
14	29 (32, 71)	19 (44, 72)

Match no. 2 of Table 5 is considered as the centre for rotational transformation.

The final belief about the scene 13(a) is "The scene consists of hammer and wrench".

Figure 17(a) represents one scene and 17(b) represents its corresponding hypothesized version.

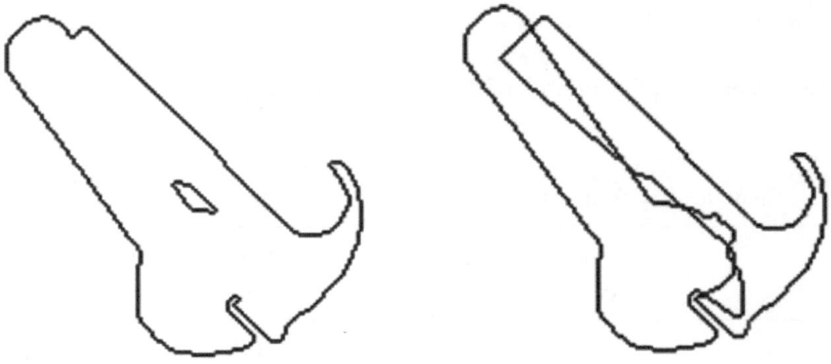

(a) (b)

FIGURE 17 (a) Scene and (b) Hypothesized scene.

The belief about the scene is "The scene consists of wrench and hammer".

Another scene and its corresponding hypothesized version are shown in the Figure 18(a, b)

(a) (b)

FIGURE 18 (a) Scene and (b) Hypothesized scene.

Here, the belief about the scene is "The scene consists of wrench and clip".

Case Study 3

In Figure 19 and Figure 20 we consider more complex scenes and its hypothesized version.

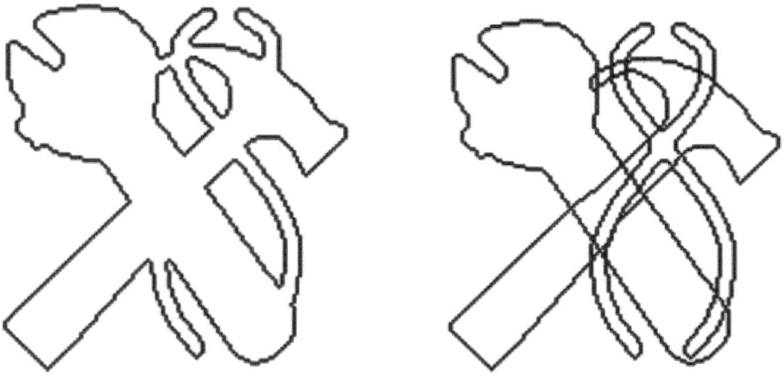

(a) (b)

FIGURE 19 (a) Scene and (b) Hypothesized scene.

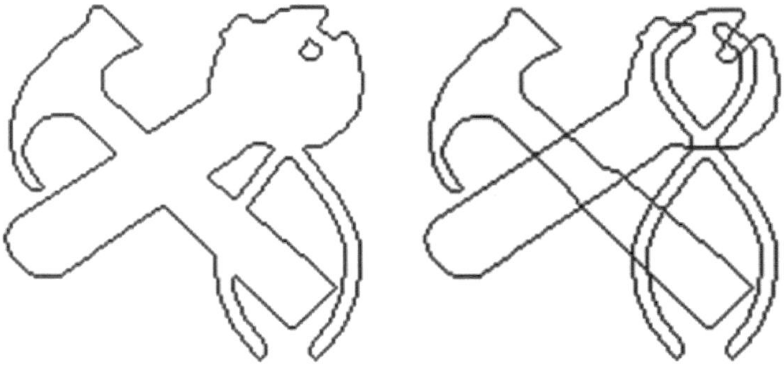

(a) (b)

FIGURE 20 (a) Scene and (b) Hypothesized scene.

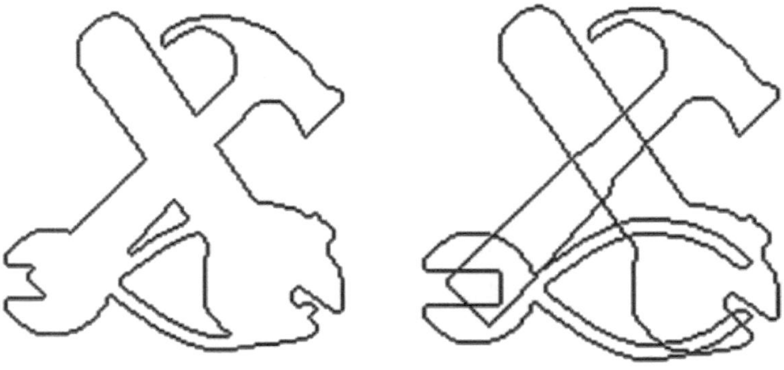

(a) (b)

FIGURE 21 (a) Scene and (b) Hypothesized scene.

Detail experimental results on scene image 20(a) are:

(a) Model object (Hammer) and scene image 20(a). The figure (Figure 22) represents the feature points of model object (Hammer) and scene image 20(a).

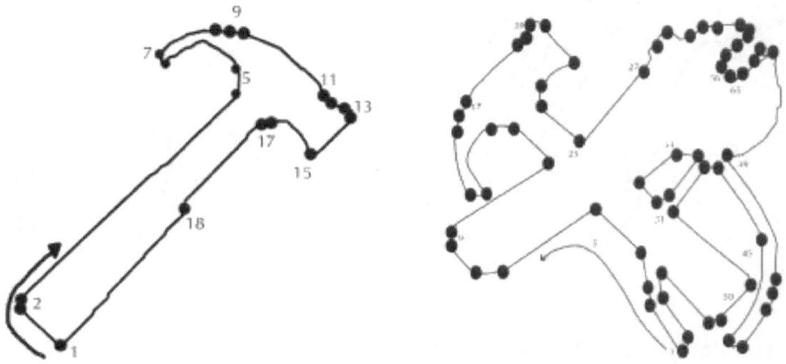

(a) Hammer (b) Scene Image

FIGURE 22 Feature points of (a) model object (Hammer) and (b) scene image 17 (a).

Table 6 represents the primary and final match between the model object (Hammer) and the corresponding object of the scene image.

TABLE 6 Match matrices between Hammer and the corresponding object of the scene image.

1	. . . .
2	. . . . @ . . . .
3	. . . . @ . . @ . . . . . .
4	. . . . . . @ . . . . . . @ . . . .
5	. . . .
6	. . . .
7	. . . .
8	. . . . @ . . . .
9	. . . . @ . . . .
10	. . . .
11	. . . .
12	. (@) . . (@) . . . . . @ . . . .
13	. . . . . (@) . . . . @ . . .
14	. . . . . . (@) . . . . . . @ . . . .
15	. . . . (@) . . . .
16	. . . . @ . . @ . . . . . .
17	. . . . @ . . . .
18	. @ . . @ . . . . . (@) . . . .

```
19 . . . . . . . . @ . . (@) . . . . . .
20 . . . . . . . . . . . . (@) . . . . .
21 . . . . . . @ . . . . . . (@) . . . .
22 . . . . . @ . . . . . . . . (@) . . .
23 . . . . . . . . . . . . . . . (@) . .
24 . . . . . . . . . . . . . . . . (@) .
25 . . . . . . . . . . . . . . . . . .
26 . . . . . . . . . . . . . . . . . .
27 . . . . . . . . @ . . @ . . . . . .
28 . . . . . . @ . . . . . . @ . . . .
29 . . . . . . @ . . . . . . @ . . . .
30 . . . . . . @ . . . . . . @ . . . .
31 . . . . . . . @ . . . . . . . . . .
32 . . . . . . . . . @ . . . . . . . .
33 . . . . . . . . . . . . . . . . . .
34 . . . . . . . . @ . . @ . . . . . .
35 . . . . . . . . . . . . . . . . . .
36 . . . . . . . . . . . . . . . . . .
37 . @ . . @ . . . . . @ . . . . . . .
38 . . . . . . . . . . . . . . . . . .
39 . . . . . . . . . . . . . . . . . .
40 . . . . . . . . @ . . @ . . . . . .
41 . . . . . . . . (@) . . @ . . . . . .
42 . . . . . . . . . (@) . . . . . . . .
43 @ . . . . . . . . . . . . . . . . .
44 . . . . . . . . . . . . . . . . . .
45 . . . . . . . . . @ . . . . . . . .
46 . @ . . @ . . . . . @ . . . . . . .
47 . . . . . . . . . . . . . . . . . .
48 . . . . . . . . . . . . . . . . . .
49 (@) . . . . . . . . . . . . . . . . .
50 . @ . . @ . . . . . @ . . . . . . .
51 . . . . . . . . . . . . . . . . . .
52 . . . . . . . . . . . . . . . . . .
53 . . . . . . . . @ . . @ . . . . . .
54 . . . . . . . . . . . . . . . . . .
55 . . . . . . . . . . . . . . . . . .
56 . . . . . . . . . . . . . . . . . .
57 . . . . . . @ . . . . . . @ . . . .
```

58 . . . .
59 . . . .
60 . . . .
61 . @ . . @ . . . . . @ . . . .
62 . . . .
63 . . . .
64 . . . .
65 . . . .
2 3 4 5 6 7 8 9 10 11 12 13 14 15 16 17 18

Key: @ ≡ primary match
(@) ≡ final match

The left index of Table 6 indicates the scene features and bottom index of Table 5 indicates the model features. Here the percentage of match of model object (Hammer) with the corresponding object of the scene image is as follows;

* No. of Model (Hammer) features: 18
** Percentage of match: (15/18)%
=83.33%

Table 7 represents the number of the final matches and their corresponding coordinates.

TABLE 7 Coordinates of the final match points.

No. of Match	Scene # (x, y)	Model (Hammer) # (x, y)
1	12 (15, 86)	2 (5, 20)
2	41 (118, 18)	9 (85, 113)
3	42 (118, 16)	10 (87, 113)
4	12 (15, 86)	5 (78, 105)
*** 5**	**13 (14, 61)**	**6 (53, 106)**
6	14 (11, 61)	7 (53, 109)
7	15 (5, 90)	8 (82, 115)
8	18 (30, 120)	11 (112, 90)
9	19 (30, 122)	12 (114, 90)
10	20 (32, 125)	13 (117, 88)

TABLE 7 *(Continued)*

No. of Match	Scene # (x, y)	Model (Hammer) # (x, y)
11	21 (35, 125)	14 (117, 85)
12	22 (48, 112)	15 (104, 72)
13	23 (35, 103)	16 (95, 85)
14	24 (36, 95)	17 (87, 84)
15	49 (114, 27)	1 (19, 6)

Match no. 5 of Table 7 is considered as the centre for rotational transformation.

(a) The figure (Figure 20) represents the features points of model object (Wrench) and scene Image 17(a).

(b) Model object (Wrench) and scene image 17

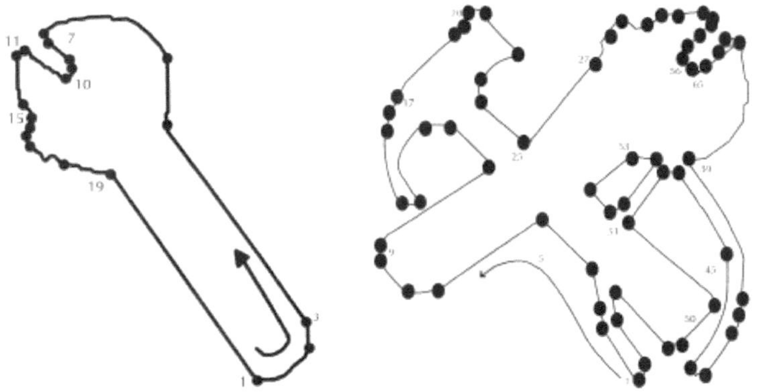

(a) Wrench (b) Scene Image

FIGURE 23 Feature points of (a) model object (Wrench) and (b) scene image 17(a).

Table 8 represents the primary and final match between the model object (Wrench) and the corresponding object of the scene image.

TABLE 8 Match matrices between Wrench and the corresponding object of the scene image.

1	. . . .
2	. . . .
3	. . @ . . . . @ .
4	. . . . @ . . . . . @ @ @ @ . .

```
5  . . . . . . . . . . @ . . . . . . . .
6  . . . . . . . . . . . . . . . . . . .
7  . (@) . . . . . . . . . . . . . . . . .
8  . . . . . . . . . . . . . . . . . . .
9  . . . . . . . . . . . . . . . . . . .
10 . . . . . . . . . . . . . . . . . . .
11 . . . . . . . . . . . . . . . . . . .
12 . . . . . . @ . . . . . . . . . . . .
13 . . . . . . . . . . . . . . . . . . .
14 . . . . . . . @ . . . . . @ @ @ @ . .
15 . . . . . . . . . . . . @ . . . . . .
16 . . @ . . . . . . . . . . . . . . @ .
17 . . . . . . . . . . . . . . . . . . .
18 . . . . . . @ . . . . . . . . . . . .
19 . . @ . . . . . . . . . . . . . . @ .
20 . . . . . . . . @ . . . . . . . . . .
21 . . . . . . . @ . . . . . @ @ @ @ . .
22 . . . . . . . . . . . . . . . . . . .
23 @ . . . . . . . . . . . . . . . . . .
24 . . . . . . . . . . . . . . . . . . .
25 . . . . . . . . . . @ . . . . . . . .
26 . . . . . . . . . . . . . . . . . . .
27 . . @ . . . . . . . . . . . . . . (@) .
28 . . . . . . . @ . . . . . @ @ @ (@) . .
29 . . . . . . . @ . . . . . @ @ (@) @ . .
30 . . . . . . . @ . . . . . @ (@) @ @ . .
31 . . . . . . . . . . . . @ . . . . . .
32 . . . . . . . . . . . . . . . . . . .
33 . . . . . . . . . . . . . . . . . . .
34 . . @ . . . . . . . . . . . . . . @ .
35 . . . . . . . . . @ . . . . . . . . .
36 . . . . . . . . . . . . . . . . . . .
37 . . . . . . (@) . . . . . . . . . . . .
38 . . . . . (@) . . . . . . . . . . . . .
39 . . . . . . . . . . . . . . . . . . .
40 . . @ . . . . . . . . . . . . . . @ .
41 . . @ . . . . . . . . . . . . . . @ .
42 . . . . . . . . . . . . . . . . . . .
43 . . . . . . . . . . . . . . . . . . .
```

44 . . . . @ . . . .

45 . . . .

46 . . . . . . @ . . . .

47 . . . .

48 . . . .

49 . . . .

50 . . . . . . @ . . . .

51 . . . .

52 . . . .

53 . . @ . . . . @ .

54 . . . .

55 . . . .

56 . . . .

57 . . . . @ . . . . . @ @ @ @ . .

58 . . . .

59 . . . .

60 . . . .

61 . . . . . . @ . . . .

62 . . . . @ . . . .

63 . . . .

64 . . . .

65 . . . .

2 3 4 5 6 7 8 9 10 11 12 13 14 15 16 17 18 19

Key: @ ≡ primary match
(@) ≡ final match

The left index of Table 8 indicates the scene features and bottom index of Table 7 indicates the model features. Here the percentage of match of model object (Wrench) with the corresponding object of the scene image is as follows:

* No. of Model (Wrench) features:	19
** Percentage of match:	(7/19)%
	= 36.84%

Table 9 represents the number of the final matches and their corresponding coordinates.

TABLE 9 Coordinates of the final match points.

No. of Match	Scene # (x, y)	Model (Wrench) # (x, y)
1	7 (11, 32)	2 (107, 12)
*** 2**	**27 (76, 112)**	**18 (27, 77)**
3	28 (81, 122)	17 (17, 82)
4	29 (84, 122)	16 (17, 85)
5	30 (86, 120)	15 (19, 87)
6	37 (116, 116)	7 (23, 117)
7	38 (118, 116)	6 (23, 119)

Match no. 2 of Table 9 is considered as the centre for rotational transformation.

(c) Model object (Clip) and scene image 20(a). The figure (Figure 24) represents the feature points of model object (Clip) and scene image 20(a).

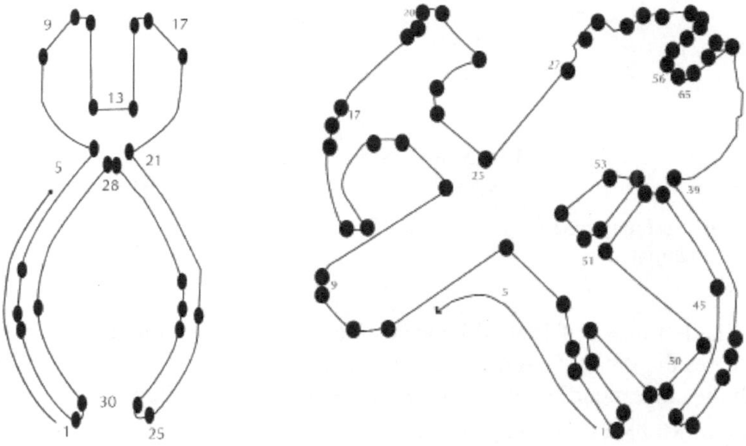

(a) Clip (b) Scene Image

FIGURE 24 Feature points of (a) model object (Clip) and (b) scene image 20(a).

Table 10 represents the primary and final match between the model object (Clip) and the corresponding object of the scene image.

TABLE 10 Match matrices between Clip and the corresponding object of the scene image.

1 . . . .

2 . **(@)** . @ . . . . @ . . . . @ . . @ . . . .

3 . . **(@)** . . . . @ @ . . . . . . @ .

4 . . . . @ . . . . @ . . . . . @ . . . .

5 . . . . @ . . . . @ . . . . .

6 . . . .

7 . . . .

8 . @ . @ . . . . @ . . . . @ . . @ . . . .

9 . @ . @ . . . . @ . . . . @ . . @ . . . .

10 . . . .

11 . . . .

12 . . . . @ . . .

13 . . . .

14 . . . . @ . . . . @ . . . . . @ . . . .

15 . . . . . . @ . . . . @ . . . .

16 . . @ . . . . @ @ . . . . . . @ .

17 . @ . @ . . . . @ . . . . @ . . @ . . . .

18 . . . . @ . . .

19 . . @ . . . . @ @ . . . . . . @ .

20 . . . .

21 . . . . @ . . . . @ . . . . . @ . . . .

22 . . . .

23 . . . .

24 . . . .

25 . . . . @ . . . . @ . . . . .

26 . . . .

27 . . @ . . . . @ @ . . . . . . @ .

28 . . . . @ . . . . @ . . . . . @ . . . .

29 . . . . @ . . . . @ . . . . . @ . . . .

30 . . . . @ . . . . @ . . . . . @ . . . .

31 . . . . . . @ . . . . **(@)** . . . .

32 . @ . @ . . . . @ . . . . @ . . @ . . . .

33 . . . .

34 . . @ . . . . @ @ . . . . . . @ .

35 . . . .

36 . . . .

37 . . . . @ . . .

38 . . . .

```
39 . . . . . . . . . . . . . . . . . . . . . . . . . . . .
40 . . @ . . . . . . . . . . . . . . . . (@) @ . . . . . . @ .
41 . . @ . . . . . . . . . . . . . . . . @ (@) . . . . . . @ .
42 . @ . @ . . . . . . . . . . . . @ . . . . (@) . . @ . . . .
43 . . . . . . . . . . . . . . . . . . . . . (@) . . . . . .
44 . . . . . . . . . . . . @ . . . . . . . . . (@) . . . . .
45 . @ . @ . . . . . . . . . . . . @ . . . . @ . . (@) . . . .
46 . . . . . . . . . . . . . . . . . . . . . . . . (@) . . .
47 . . . . . . . . @ . . . . . . . . . . . . . . . (@) . .
48 . . . . . . . . . . . . . . . . . . . . . . . . . . . .
49 . . . . . . . . . . . . . . . . . . . . . @ . . . . . .
50 . . . . . . . . . . . . . . . . . . . . . . . . @ . . .
51 . . . . . . . . . . . . . . . . . . . . . . . . . . . .
52 . . . . . . . . . . . . . . . . . . . . . . . . . . . .
53 . . @ . . . . . . . . . . . . . . . . @ @ . . . . . . (@) .
54 . . . . . . . . . . . . . . . . . . . . . . . . . . . .
55 . . . . . . . . . . . . . . . . . . . . . . . . . . . .
56 . . . . . . . . . . . . . . . . . . . . . . . . . . . .
57 . . . . @ . . . . . . . @ . . . . . @ . . . . . . . . . .
58 . . . . . . . . . . . . . . . . . . . . . . . . . . . .
59 . . . . . . . . . . . . . . . . . . . . . . . . . . . .
60 . . . . . . . . . . . . . . . . . . . . . . . . . . . .
61 . . . . . . . . . . . . . . . . . . . . . . . . @ . . .
62 . . . . . . . . . . . . @ . . . . . . . . . @ . . . . .
63 . . . . . . . . . . . . . . . . . . . . . . . . . . . .
64 . . . . . . . . . . . . . . . . . . . . . . . . . . . .
65 . . . . . . . . . . . . . . . . . . . . . . . . . . . .
2 3 4 5 6 7 8 9 10 11 12 13 14 15 16 17 18 19 20 21 22 23 24 25 26 27 28 29 30
```

Key: @ ≡ primary match

(@) ≡ final match

The left index of Table 10 indicates the scene features and bottom index of Table 9 indicates the model features. Here the percentage of match of model object (Clip) with the corresponding object of the scene image is as follows:

* No. of Model (Clip) features: 30

** Percentage of match: (12/30)%

= 40.00%

Table 11 represents the number of the final matches and their corresponding co-ordinates.

TABLE 11 Coordinates of the final match points.

No. of Match	Scene # (x, y)	Model (Clip) # (x, y)
1	2 (77, 17)	2 (43, 19)
*** 2**	**3 (77, 19)**	**3 (43, 21)**
3	40 (120, 21)	21 (86, 23)
4	41 (118, 18)	22 (84, 20)
5	42 (118, 16)	23 (84, 18)
6	43 (107, 2)	24 (73, 4)
7	44 (105, 4)	25 (71, 6)
8	45 (115, 48)	26 (81, 50)
9	46 (99, 72)	27 (65, 74)
10	47 (97, 72)	28 (63, 74)
11	53 (80, 23)	29 (46, 25)
12	31 (90, 120)	16 (71, 122)

Match no. 2 of Table 10 is considered as the centre for rotational transformation.

(d) Model object (Cutter) and scene image 20(a): The figure (Figure 25) represents the feature points of model object (Cutter) and scene image 20(a).

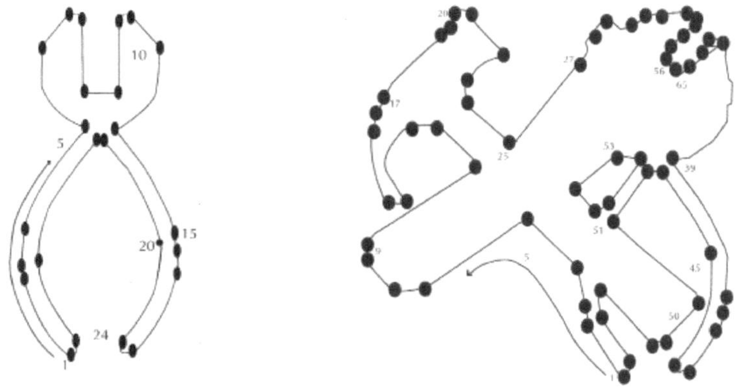

(a) Cutter (b) Scene Image

FIGURE 25 Feature points of (a) model object (Cutter) and (b) scene image 17 (a).

Table 12 represents the primary and final matches between the model object (cutter) and the corresponding object of the scene image.

TABLE 12 Match matrices between Cutter and the corresponding object of the scene image.

1 . . . .
2 . (@) . @ . . . . @ . . @ . . . .
3 . . (@) . . . . @ @ . . . . . . @ .
4 . . . .
5 . . . . @ . . . . .
6 . . . .
7 . . . .
8 . @ . @ . . . . @ . . @ . . . .
9 . @ . @ . . . . @ . . @ . . . .
10 . . . .
11 . . . .
12 . . . . @ . . .
13 . . . . @ . . . .
14 . . . .
15 . . . . @ . . . .
16 . . @ . . . . @ @ . . . . . . @ .
17 . @ . @ . . . . @ . . @ . . . .
18 . . . . @ . . .
19 . . @ . . . . @ @ . . . . . . @ .
20 . . . .
21 . . . .
22 . . . . @ . . . .
23 . . . .
24 . . . .
25 . . . . @ . . . . .
26 . . . .
27 . . @ . . . . @ @ . . . . . . @ .
28 . . . .
29 . . . .
30 . . . .
31 . . . . @ . . . .
32 . @ . @ . . . . @ . . @ . . . .
33 . . . .
34 . . @ . . . . @ @ . . . . . . @ .
35 . . . .

```
36 . . . . . . . . . . . . . . . . . . . . . . . .
37 . . . . . . . . . . . . . . . . . . . . @ . . .
38 . . . . . . . . . . . . . . . . . . . . . . . .
39 . . . . . . . . . . . . . . . . . . . . . . . .
40 . . @ . . . . . . . . . . . (@) @ . . . . . . @ .
41 . . @ . . . . . . . . . . . @ (@) . . . . . . @ .
42 . @ . @ . . . . . . . . . . . . . (@) . . @ . . . .
43 . . . . . . . . . . . . . . . . . (@) . . . . . .
44 . . . . . . . . . . . . . . . . . . (@) . . . . .
45 . @ . @ . . . . . . . . . . . . . @ . . (@) . . . .
46 . . . . . . . . . . . . . . . . . . . . (@) . . .
47 . . . . . . . . . . . . . . . . . . . . . (@) . .
48 . . . . . . . . . . . . . . . . . . . . . . . .
49 . . . . . . . . . . . . . . . . . @ . . . . . .
50 . . . . . . . . . . . . . . . . . . . . @ . . .
51 . . . . . . . . . . . . . . . . . . . . . . . .
52 . . . . . . . . . . . . . . . . . . . . . . . .
53 . . @ . . . . . . . . . . . @ @ . . . . . . (@) .
54 . . . . . . . . . . . . . . . . . . . . . . . .
55 . . . . . . . . . . . . . . . . . . . . . . . .
56 . . . . . . . . . . . . . . . . . . . . . . . .
57 . . . . . . . . . . . . . . . . . . . . . . . .
58 . . . . . . . . . . . . . . . . . . . . . . . .
59 . . . . . . . . . . . . . . . . . . . . . . . .
60 . . . . . . . . . . . . . . . . . . . . . . . .
61 . . . . . . . . . . . . . . . . . . . . @ . . .
62 . . . . . . . . . . . . . . . . . . @ . . . . .
63 . . . . . . . . . . . . . . . . . . . . . . . .
64 . . . . . . . . . . . . . . . . . . . . . . . .
65 . . . . . . . . . . . . . . . . . . . . . . . .
2 3 4 5 6 7 8 9 10 11 12 13 14 15 16 17 18 19 20 21 22 23 24
```

Key: @ ≡ primary match
(@) ≡ final match

The left index of Table 12 indicates the scene features and bottom index of Table 11 indicates the model features. Here the percentage of match of model object (Cutter) with the corresponding object of the scene image is as follows:

* No. of Model (Cutter) features:	24
** Percentage of match:	(11/24)%
	= 45.83%

Table 13 represents the number of final matches and their corresponding coordinates.

TABLE 13 Coordinates of the final match points.

No. of Match	Scene # (x, y)	Model (Cutter) # (x, y)
1	2 (77, 17)	2 (43, 19)
2	3 (77, 19)	3 (43, 21)
3	40 (120, 21)	15 (86, 23)
4	41 (118, 18)	16 (84, 20)
5	42 (118, 16)	17 (84, 18)
6	43 (107, 2)	18 (73, 4)
7	44 (105, 4)	19 (71, 6)
8	45 (115, 48)	20 (81, 50)
9	46 (99, 72)	21 (65, 74)
10	47 (97, 72)	22 (63, 74)
11	53 (80, 23)	23 (46, 25)

For this model (cutter) and scene image the system finds a set of final matches, but during plotting the model (cutter) onto the scene it rejects this model because of the following reasons:

As per our algorithm to a plot model object onto the scene first the system separates the pixels surrounded by the border of the scene by any color (Figure 26) and then plots the model onto the scene. If the system finds some pixels of the model object are outside the boundary pixels of the scene then the system reject the model as it does not consist of that particular model. For the present design study the threshold value of the number of pixels of the model object outside the scene is ≥ 5.

Here, for the particular scene (Figure 26(b) and Figure 26(c)) and model (cutter), the number of pixels outside the scene area is 5. So, the system rejects this model though 45.83% of its features are matched with the scene.

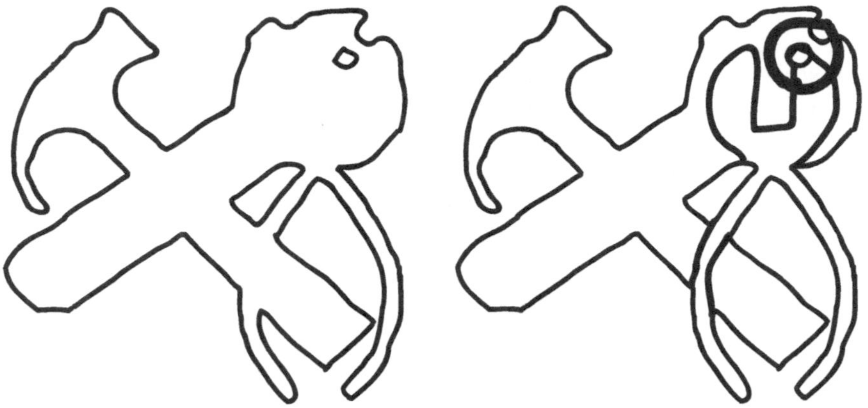

(a) Scene image

(b) Hypothesized scene image with ictitious hypothesis that model object cutter is a part of the scene.

(c)

FIGURE 26 (a) Scene image 17 (a), (b) Hypothesized scene image, and (c) Number of pixels (≥5) of the model object (cutter) is outside the scene boundary.

In other way it can be said that the system verifies the initial hypothesis to get a final hypothesis by plotting each of the model object onto the scene, which are matched with the scene objects by features. So, inspite of satisfactory matching of features finally the system comes to a conclusion that the scene does not consist of cutter, which is a correct conclusion.

Here, fortunately, the belief about the scene is, "The scene consists of hammer, wrench and clip". But if somehow, the count of the number of pixels of the model object outside the scene is less than the threshold value, say 4, the system immediately goes for the multiple beliefs. For instance, in this particular case, the system may belief that either (i) the scene consists of the model objects hammer, wrench and clip or (ii) the scene consists of the model objects hammer, wrench and cutter.

Now, fortunately/unfortunately, if the system picks up the second choice for a final belief about the scene and if ultimately it appears that the first choice that is belief (i) is correct, then the system must go for revising the belief for any final decision making. Thus, the notion of belief revision, is an added flexibility for the vision system to try to mimic the human decision making approach in reality.

Case Study 4

Figure 27 represents another very complex scene.

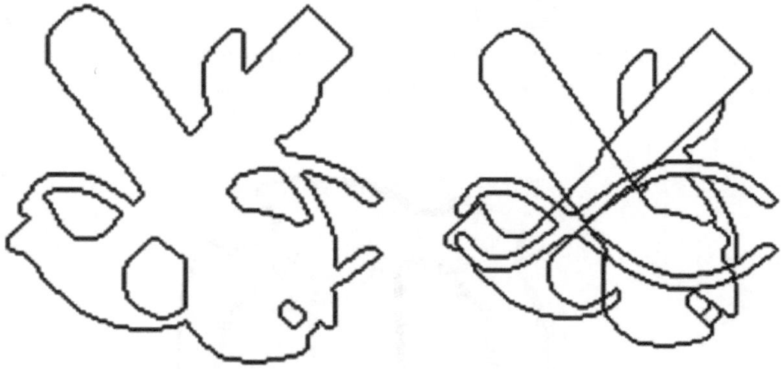

(a) (b)

FIGURE 27 (a) Scene image and (b) Hypothesized scene image.

In the scene shown (Figure 27(a)) the system is unable to recognize the model object cutter. To explain this inability we consider the following results:

Table 14 represents the number of final matches and their corresponding coordinates for the model object clip and the corresponding object of the scene image 24(a).

TABLE 14 Coordinates of the final match points.

No. of Match	Scene # (x, y)	Model (Clip) # (x, y)
1*	23 (123, 40)	30 (56, 6)
# 2*	**28 (123, 55)**	**25 (71, 6)**
3*	30 (111, 68)	23 (84, 18)
4*	31 (109, 68)	22 (84, 20)

TABLE 14 *(Continued)*

No. of Match	Scene # (x, y)	Model (Clip) # (x, y)
5	71 (18, 60)	13 (76, 111)
6*	27 (105, 65)	29 (46, 25)
7*	30 (111, 68)	2 (43, 19)
8*	31 (109, 68)	3 (43, 21)

Here, match no. 2 (Model point no. 25) is considered as the center for rotational transformation.

Now consider the Table 14 which represents the number of final matches and their corresponding coordinates for the model object cutter and the corresponding object of the scene image 27(a).

TABLE 15 Coordinates of the final match points.

No. of Match	Scene # (x, y)	Model (Cutter) # (x, y)
1	46 (16, 103)	3 (43, 21)
2*	23 (123, 40)	24 (56, 6)
#3*	28 (123, 55)	19 (71, 6)
4*	30 (111, 68)	17 (84, 18)
5*	31 (109, 68)	16 (84, 20)
6*	27 (105, 65)	23 (46, 25)
7*	30 (111, 68)	2 (43, 19)

In both the tables the entries marked by "star" (*) are matches found after final match. Actually these matches belong to the model object Clip. For Cutter the system finds matches (final) from the alike portion of both the model object clip and cutter.

For this particular scene and model object (cutter), unfortunately, the system finds only one match from distinguishable portion and select randomly third match (marked by "#" in Table 14) for translation. If the system would find at least two matches from distinguishable portion and would select those matches for translation then it would be possible to recognize the model cutter.

Under such circumstances, the vision system may either go by degraded belief (degradation from the highest level to the next lower level in the environmental lattice) or may relax (revise) the threshold values of different parameters which were initially chosen for primary match and final match to include some additional match points (final match) for better result. In case of degraded belief we get the following result for scenes 27 (a):

"The scene consists of hammer, wrench and clip".

20.5 VISION ALGORITHM AND COMPREHENSIVE RESULTS OF THE DESIGN STUDY

We state the vision algorithm as follows:

Step 1: Match features and shape of model objects with the scene objects. Model objects are known prior to the system.

Step 2: If the system finds certain number (equal to or above 15% of the number of model features) of model features matches with the scene features (section 20.3.2 and section 20.3.3), then the system initially decide that the scene may consists of the model whose features are matched.

Step 3: Performing the system finds a set of model objects (except those scenes which are containing single object) and tries to conclude a belief about the scene comparing the area of model objects with the area of scene objects (Section 20.3.4). This step is controlled by environmental lattice (Section 20.3.5). After this step the system comes to a draft of conclusion about the scene (e.g. The scene consists of $model_0$, $model_1$, model) which is called here initial hypothesis (Section 20.3.5).

Step 4: Now, to verify the initial hypothesis, the system plots each model onto the scene. To do this the system picks any two matches to calculate the transformation factors (Translation and rotational factors) (Section 20.3.6). During this process if the system finds certain number (say ≥ 5) of pixels of model objects is outside of the scene area then the system decides that the particular model objects is not present in the scene and roll back the plotting process, because it may so happen that some other model object alike to that particular model object is present in the scene and that is why the system finds some primary matches in terms of features.

Step 5: The vision system gives some interpretation based on the hypothesized scene.

Table 16 provides the comprehensive results of the design study.

20.6 COMPARATIVE STUDY

The idea of belief revision is basically an area of Artificial Intelligence (AI). Object recognition is a component of computer vision (CV). Both these areas were independently developed in eighties. The existing state of art of recognition of partially occluded object is essentially based on the concept of hypothesis generation and verification as stated at the beginning of the text. The present text differs from the past paradigm of vision research and essentially based on the fusion methodology, fusion between the belief revision concept of AI and model based object recognition of CV. This state of art adds some additional features which are discussed during the course

TABLE 16 Comprehensive results of the design study.

Models (Matches in %)	mod0.bmp (Hammer)			Mod1.bmp (Wrench)			Mod2.bmp (Clip)			Mod3.bmp (Cutter)		
	Translation		Rotation (Degree)	Translation		Rotation (Degree)	Translation		Rotation (Degree)	Translation		Rotation (Degree)
Scene (Area)	T_x	T_y	θ	T_x	T_y	θ	T_x	T_y	θ	T_x	T_y	θ
 Scene 1 (4194)	1	7	0.00	Not present			–64	51	–89.96	Not present		
	72.22				96.67			50.00				
 Scene 2 (5496)	–47	37	89.96	–12	–1	0.00	Not present			Not present		
	16.67			73.68								

TABLE 16 *(Continued)*

Models (Matches in %)	mod0.bmp (Hammer)			Mod1.bmp (Wrench)			Mod2.bmp (Clip)			Mod3.bmp (Cutter)		
	Translation		Rotation (Degree)	Translation		Rotation (Degree)	Translation		Rotation (Degree)	Translation		Rotation (Degree)
Scene (Area)	T_x	T_y	θ	T_x	T_y	θ	T_x	T_y	θ	T_x	T_y	θ
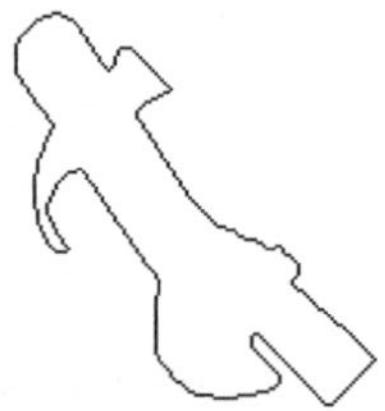 Scene 3 (5177)	–49 44.44	29	89.96	79 47.37	–36	179.92	Not present			Not present		
 Scene 4 (6301)	–20 33.33	–81	–89.96	66 42.11	–108	179.92	Not present			Not present		

TABLE 16 *(Continued)*

Models (Matches in %)	mod0.bmp (Hammer)			Mod1.bmp (Wrench)			Mod2.bmp (Clip)			Mod3.bmp (Cutter)		
	Translation		Rotation (Degree)	Translation		Rotation (Degree)	Translation		Rotation (Degree)	Translation		Rotation (Degree)
Scene (Area)	T_x	T_y	θ	T_x	T_y	θ	T_x	T_y	θ	T_x	T_y	θ
	Not present			28	–65	89.96	20	–1	0.00	Not present		
		47.37			66.67			45.83				
Scene 5 (5208)												
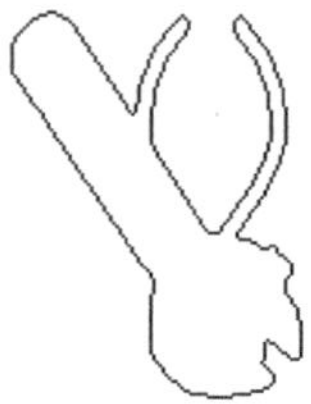	N o t present			88	–97	179.92	56	89	179.92	56	89	179.92
		63.16			33.33			41.67				
Scene 6 (4836)												

TABLE 16 *(Continued)*

Models (Matches in %)	mod0.bmp (Hammer)			Mod1.bmp (Wrench)			Mod2.bmp (Clip)			Mod3.bmp (Cutter)		
	Translation		Rotation (Degree)	Translation		Rotation (Degree)	Translation		Rotation (Degree)	Translation		Rotation (Degree)
Scene (Area)	T_x	T_y	θ	T_x	T_y	θ	T_x	T_y	θ	T_x	T_y	θ
	–3	–4	0.00	–11	1	0.00	–11	2	0.00	Not present		
Scene 7 (7517)	55.56			36.84			53.33			20.83		
	–39	–45	89.96	49	35	–89.96	34	–2	0.00	Not present		
Scene 8 (7497)	83.33			36.84			40.00			45.83		

TABLE 16 *(Continued)*

Models (Matches in %)	mod0.bmp (Hammer)			Mod1.bmp (Wrench)			Mod2.bmp (Clip)			Mod3.bmp (Cutter)		
	Translation		Rotation (Degree)	Translation		Rotation (Degree)	Translation		Rotation (Degree)	Translation		Rotation (Degree)
Scene (Area)	T_x	T_y	θ	T_x	T_y	θ	T_x	T_y	θ	T_x	T_y	θ
 Scene 9 (7725)	0	0	0.00	93	–106	179.92	Not present			–19	–46	–269.92
	44.44			78.95				29.17				
 Scene 10 (7953)	–114	–42	–179.92	–90	101	–179.92	52	49	86.96	Not recognized		
	50.00			21.05			26.67			29.17		

of development of this new concept. In this section we compare the performance of the present design study with two very prominent existing state of object recognition under occlusion, *viz.* Ayache and Faugeras [12] and Koch and Kashyap [74]. For comparison we consider the occluded scene of Figure 1. We see both the existing approaches fail to recognize the correct object (i.e. the cutter) under occlusion. Whereas, the present design study revises its belief that the scene consists of cutter and another object. Table 17 shows the result of comparison.

TABLE 17 Recognition result for the occluded scene of Figure 1.

Method	Recognition result before verification	Recognition result after verification	Belief revision
Koch and Kashyap [74]	The occluded scene of Figure 1 consists of clip and another object.	Same as before. The physical verification of the occluded scene says that the scene actually consists of cutter and another object.	No belief revision.
Ayache and Faugeras [12]	The occluded scene of Figure 1 consists of clip and another object.	Same as before. The physical verification of the occluded scene says that the scene actually consists of cutter and another object.	No belief revision.
Proposed method	The occluded scene of Figure 1 consists of clip and another object.	Doubt occurs at step 4 of the algorithm and vision process rolls back for belief revision.	Revised result says that the occluded scene consist of cutter and another object.

20.7 CONCLUSION

We present a model based vision scheme to locate and identify objects in a scene. The object may touch or overlap, giving rise to partial occlusion. The boundaries and the features of the objects play an important role in both initial and final hypothesis generation. Initial hypothesis generation algorithm depends on (ATM) system technique and final hypothesis algorithm depends upon the coordinate transformations of the model objects on to the scene to generate hypothesized scene. The present vision scheme is equipped with belief revision scheme based on ATM system.

We can extend the scheme of 2D world to 3D world. The validity of the present vision scheme is tested on a several scene examples and very promising results are obtained.

This vision scheme may be implemented in robotics system, military and so on. In automated military system or for efficient performance in robotics that is where the information is less or insufficient this vision scheme may give a satisfactory result.

KEYWORDS

- **Assumption-Based truth maintenance**
- **Environmental lattice**
- **Hypothesis generation**
- **Vision algorithm**
- **Vision system**

21 Neuro-fuzzy Reasoning for Occluded Object Recognition: A Learning Paradigm through Neuro-fuzzy Concept

CONTENTS

21.1 INTRODUCTION

In this chapter we design a scheme for recognition of 2D occluded objects in a scene using the concept of Neuro-Fuzzy learning. The basic recognition scheme is based on the design philosophy as stated earlier.

"We can recognize the objects which have seen earlier."

The terms "seen earlier" imply the notion of our knowledge base which contains information about the model objects. At this stage, we like to state that for our design study we adopt the following two assumptions:

1. We have limited number of model objects (say 2, 3 etc.)
2. Our recognition scheme fails to recognize an object which is occluded by other objects(s) in such a way that more than 50% of the dominant points of the said object are not visible in the scene. As all the points on the contour of a 2D object do not carry significant information to recognize it (the said object), we consider those points as dominant points which carry significant information for recognition. And in our knowledge base we store only the information of the dominant points of the model objects. Note that, if we represent the pattern of each dominant point by c number of features we can have c-dimensional pattern vector for each dominant point. Such c-dimensional pattern vector of each dominant point is our basic information to be stored.

But for sake of presentation (in the knowledge base) of the information about the dominant points of the model objects we do not directly consider the c-dimensional pattern vectors of each dominant point of the model objects. Instead, in the knowledge base, we try to store the gross information about the patterns of the dominant points which are scattered on the c-dimensional pattern space. Hence, we consider the fuzzy set and multi-dimensional fuzzy implication (MFI) [63, 178] as the primary tools of engineering. The basic motivation behind the selection of fuzzy tools of engineering is not to compete with the complex and efficient recognition mechanism of human being but to mimic the human cognitive process of knowledge representation and reasoning, for recognition of objects, in a simple computational form.

The present approach which realizes a new interpretation of MFI through back propagation-type neural network produces graded consequences which are most suitable for pattern recognition and object recognition problems.

The object recognition scheme (using MFI), proposed in this chapter is a model-based system in which recognition involves matching the input image with a set of predefined models of objects. In such a system, the known objects are precompiled creating a model database and this database is used to recognize objects in an image scene [133].

Existing object recognition methods can be categorized as either global or local in nature. Global methods are based on global features of the boundary or of an equivalent representation. Such techniques are the Fourier descriptors [178], the moments [64] and methods based on autoregressive models [69]. Local methods use local features such as critical points or holes and comers. They perform very well in the presence of noise, distortion or partial occlusion since such effects on an isolated region of the contour alter only the local features associated with that region, leaving all other local features unaffected. However, the choice of representative local features is not trivial and the recognition process based on local features is more computationally intensive and time consuming. On the other hand, global methods have the disadvantage that a small distortion in a portion of a boundary of an object will result in changes to all global features. In the present chapter, we use internal angles and curvatures of the significant points on the boundary of an object as local features for model-based recognition.

21.2 KNOWLEDGE REPRESENTATION

Human cognition, represented by production rule is widely used in the field of Artificial Intelligence (AI) as a modular knowledge representation scheme. Each production rule is in the form of a condition-action pair: "if this condition occurs perform this action". The condition consists of clauses connected by 'and' conjunctions. Rules are in the following form:

Rule 1: if a region is a long stripe region and it is classified as asphalt, then the region is a road.
Rule 2: If a region is a long stripe region and it is classified as water, then the region is a river and so on.

A set of production rules is a knowledge base. In the present work, we consider a new interpretation of MFI as declarative form of knowledge representation. The new interpretation is basically a collection of 1D fuzzy if then production rules. As human cognition very much depends on human perception which, for a given situation, is not unique and which, for a given situation, varies (from person to person) within certain limit we consider the new interpretation of MFI as the primary tool for the declarative form of knowledge representation of the present vision problem.

The basic motivation behind the use of the method of MFI is to take care of the uncertainties in the local features of the objects under different orientations and noisy environment. Thus, invariance property of the local features of the object holds good under the assumption that significant changes in internal angles and curvatures are invariant under rotation and translation. In our model-based object recognition scheme, object models are represented into a model database, using fuzzy If-Then rules. Generally, increasing the number of object models in the model database greatly increases the computational complexity and the time requirements of the system. However, implementation of a model-based object recognition scheme using back propagation type neural network seems to be very attractive. First of all, neural network provides its own way to represent the knowledge (in terms of data which are basically the local features) that it stores [173]. In addition, the complexity and the computational burden increase very slowly as the number of data models increases. In general, the performance of the neural network is very good but it takes lot of time for learning. However, in our case, learning time is comparatively less because at the time of learning, instead of feeding individual data, we feed a cluster of data occupied by the antecedent part of the fuzzy If-Then rules. Thus, the task of learning of the neural network about the total data set is completed much earlier.

21.3 MFI AND STATEMENT OF THE PROBLEM

In general, there are two methods on fuzzy reasoning: one is based on the compositional role of inference [164] and the other on fuzzy logic with such and such a base logic [164, 178]. Now, if we have a (MFI) such as:

"if (x is A, y is B) then z is C",

where, A, B, and C are fuzzy sets, we can have the following conventional interpretation taken in the multi-dimensional case. For example, in the compositional role of inference, the 2D implication is translated into:

(i) if x is A and y is B then z is C, or

(ii) if x is A then if y is B then z is C.

According to Tsukamoto [164, 178], the MFI can be represented as:

$$\text{if } x \text{ is } A \text{ then } z \text{ is } C, \tag{21.1}$$

$$\text{if } y \text{ is } B \text{ then } z \text{ is } C,$$

and the intersection C' ∩ C', where C' is the inferred value from the first implication and C'' is that from the second implication, is taken for the consequence of reasoning.

To tackle the object recognition problems using MFI, we provide the following new interpretation of the MFI:

$$\text{if } x \text{ is } A \text{ then } z \text{ is } C_1, \tag{21.2a}$$

$$\text{if } y \text{ is } B \text{ then } z \text{ is } C_2, \tag{21.2b}$$

where $C = C_1 \cap C_2$ and the intersection $C_1 \cap C'_2$, where C'_1 is the inferred value from the first implication and C'_2 is the inferred value from the second implication, is taken for the consequence of reasoning.

The essential difference between the Tsukamoto model and the newly proposed model occurs at the interpretation of the consequent part of each of the decomposed fuzzy implication (DFI) of the MFI. According to Tsukamoto, the consequent parts of the DFIs (see Equation (21.1)) of an MFI are same as the consequent part of the said MFI. Whereas, in the newly proposed model, the consequent part of MFI is the intersection of the consequent parts of the DFIs (see Equations (21.2a) and (21.2b)). Thus, the linguistic connective "and" of Equations (21.2a) and (21.2b) has a more meaningful logical interpretation "∩" than that of Equation (21.1).

Neural networks are generally robust static pattern classifiers. But, they are not effective in classifying patterns with inherent temporal variations. In order to compensate for temporal variations resulting from occlusion, a multilayer neural network system which realizes the new interpretation of MFI is proposed. The occluded object recognition approach is basically similar to the decision-theoretic approach to pattern recognition.

Under decision-theoretic approach, each pattern is represented by a vector of features. The feature space is divided into a number of regions, each of which represents a prototype pattern or a cluster of patterns. A decision function maps the given pattern to previously determined regions.

In the MFI approach to pattern recognition and occluded object recognition, each element of the (local) feature vector is represented by the fuzzy linguistic variable instead of a real number. For instance, suppose we have a (2 x 1) feature vector $F = (F_1, F_2)^T$ (T is transpose), where F_1 is the internal angle at a significant point on the boundary of an object and F_2 is the curvature at that significant point. In the decision theoretic approach to pattern recognition, F_1 and F_2 are represented by real numbers. Whereas, in the MFI approach to occluded object recognition F_1 and F_2 are local features represented by the fuzzy linguistic variables example F_1 is small and F_2 is medium. The elements of the feature vector which are represented by fuzzy linguistic variables are characterized by their membership functions. Thus, in the present approach, instead of a single pattern, a population of patterns is being represented by the fuzzy feature vector F (see Figure 1). These elements of the fuzzy feature vector actually constitute the antecedent part of the MFI. The consequent part of the fuzzy implication represents the possibility of occurrence of each class on the feature space.

The process of recognition is usually divided into two steps, learning and classification. The main stages in the learning process are feature extraction, feature selection, clustering and determination of the appropriate fuzzy If-Then rules which will constitute a decision function.

The main stages in classification are extraction of a selected set of features, application of If-Then rules and decision making based on the results of the application of the If-Then rules. In the classification process, first an unknown scene is presented to the system, then a set of predetermined features are extracted from the scene. Finally, a set of If-Then rules determines the possibility of occurrence of each object in the scene. Before we implement the new interpretation of MFI in back propagation type neural network we state some further aspects of fuzzy feature vector.

21.3.1 Aspects of Fuzzy Feature Vector

Since a (MFI) such as "if (x is A, y is B) then z is C" where A, B and C are fuzzy sets is not merely a collection of 1D implications, a conventional interpretation is usually taken in the multi-dimensional case. For example, according to the conventional interpretation, the 2D implication is translated into:

$$\text{if } x \text{ is } A \text{ and } y \text{ is } B \text{ then } z \text{ is } C \tag{21.3a}$$

or

$$\text{if } x \text{ is } A \text{ then if } y \text{ is } B \text{ then } z \text{ is } C \tag{21.3b}$$

Note that the interpretation of Equation 21.3(a) of a MFI means a fuzzy vector shown in Figure 1(a) and a fuzzy set C formed by the relative position of the fuzzy vector and the well-defined cover of the pattern space. The relation formed by the antecedent clauses of Equation 21.3(a) is represented at the tip of the fuzzy vector that is the area ABCD of Figure 1(a). In Figure 1(a) the relation contained in the area ABCD is obtained by considering 'min' operation between the corresponding elements of the two fuzzy sets namely M_1 and M_2 defined over two feature axes F_1 and F_2, respectively. That means, the relation formed by the antecedent clauses of Equation 21.3(a) is obtained by considering 'min' operator for the linguistic connective 'and' between two antecedent clauses. Instead of a 'min' operator we may consider other kind of operators like algebraic product, and so on, to replace the connective 'and' between the antecedent clauses of Equation 21.3(a). In that case, the tip of the fuzzy vector that is the area ABCD of Figure 1(a) represents a relation of different nature. Precisely speaking to deal with the relation formed by the antecedent clauses of Equation 21.3(a) means to deal with a fuzzy vector of the type shown in Figure 1(a). The relation which is contained in the area ABCD of Figure 1(a) implies the consequence C as indicated in Equation 21.3(a). The fuzzy set C of Equation 21.3(a) is obtained as discussed in Example 21.1.

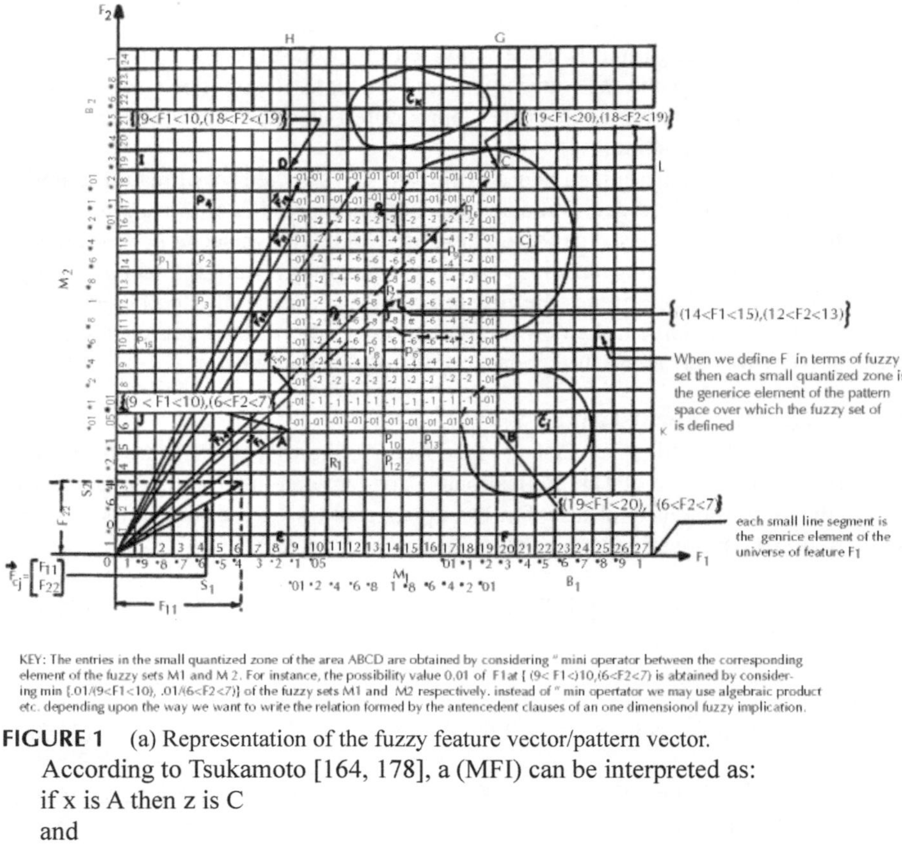

KEY: The entries in the small quantized zone of the area ABCD are obtained by considering " mini operator between the corresponding element of the fuzzy sets M1 and M 2. For instance, the possibility value 0.01 of F1at { (9< F1<)10,(6<F2<7) is abtained by considering min {.01/(9<F1<10), .01/(6<F2<7)} of the fuzzy sets M1 and M2 respectively. instead of " min opertator we may use algebraic product etc. depending upon the way we want to write the relation formed by the antencedent clauses of an one dimensionol fuzzy implication.

FIGURE 1 (a) Representation of the fuzzy feature vector/pattern vector.

According to Tsukamoto [164, 178], a (MFI) can be interpreted as:

if x is A then z is C

and

$$\text{if y is B then z is C} \tag{21.4}$$

and the intersection C' ∩ C" where C' is the inferred value from the first implication and C" that from the second implication, is taken for the consequence of reasoning.

To tackle the pattern classification problem using MFI we provide the following new interpretation of the (MFI),

if x is A then z is C_1

and

$$\text{if } y \text{ is } B \text{ then } z \text{ is } C_2 \qquad (21.5)$$

and the intersection $C_1' \cap C_2''$ where C_1' is the inferred value from the first implication and C_2'' is the inferred value from the second implication, is taken for the consequence of reasoning. Note that the fuzzy set $= C_1 \cap C_2$ and the fuzzy set C will carry the same defuzzy information which is ultimately needed for classification of patterns/objects (see Example 21.2).

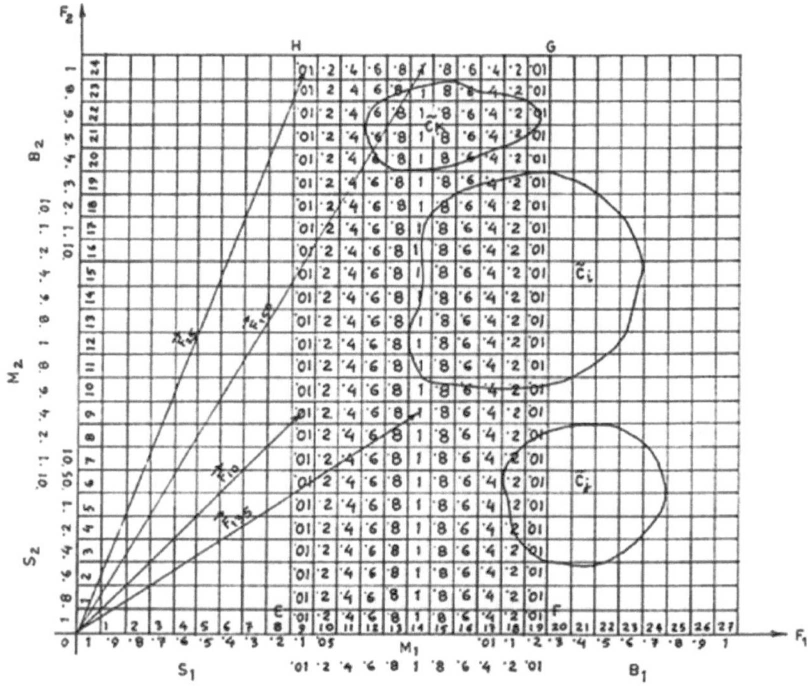

KEY : The entries in the small quantized zones of the area E F G H is obtained by considering the cylindrical extension of the fuzzy set M1 . The vectors starting from F 126 to F 150 are the elements of the fuzzy set which are obtained by defduzziliction of the fuzzy vector Ffg as mentioned in Example 2.2.

FIGURE 1 (b) Representation of fuzzy pattern vector indicated by the cylindrical extension of fuzzy set M_1.

To deal with the individual 1D fuzzy implication of Equation 21.5 means to deal with a single antecedent clause along with its cylindrical extension over the appropriate universe of the feature axis and the consequence C_i, where i varies from 1 to 2 (in case we consider pattern classification on R^2). The necessity of considering the cylindrical extension of the fuzzy set which represents the antecedent clause of one dimensional fuzzy implication is to induce a relation in the pattern space. The induced relation due to the cylindrical extension of the said fuzzy set is itself a fuzzy set (see Figure 1(b)) which is a fuzzy vector. Thus, we can locate patterns on the pattern space by a fuzzy vector F_{fj} induced by the cylindrical extension of the fuzzy set of the antecedent clause of each 1D fuzzy implication of Equation 21.5. The consequence C_i (i varies from 1 to 2) of each 1D fuzzy implication is formed by the relative position of the induced fuzzy vector and the well-defined cover of the pattern space (see Example 21.1). Therefore, when we consider the antecedent part of each 1D fuzzy implication of Equation 21.5 along with its consequence we implicitly consider an induced fuzzy vector as stated, which implies the consequence C_i, i varies from 1 to 2.

The essential difference between Tsukamoto model (Equation 21.4) and the newly proposed model (Equation 21.5) occurs at the interpretation of the consequent part of each of the 1D fuzzy implication of the (MFI). According to Tsukamoto, the consequent part of each 1D fuzzy implication (see Equation 21.4) of a MFI is same as the consequent part of the said MFI. Whereas, in the newly proposed model, the consequent part of each of the 1D fuzzy implication (see Equation 21.5) is different from the consequent part of the MFI. The linguistic connective "and" has a logical interpretation "$\cap$".

Based on the conventional interpretation of MFI, that is Equation 21.3(a), we introduce the notion of fuzzy pattern vector/feature vector (F_{fj}) (see Figure 1(a)) whose elements are the antecedent parts of the 1D fuzzy implications of Equation 21.5 which (i.e. the antecedent parts) are basically the linguistic features like F_1 is small and F_2 is medium and so on (see Figures 1(a)–(c)) where small, medium, and so on, are represented by fuzzy sets. The tip (ABCD) (see Figure 7(a)) of the fuzzy pattern vector/feature vector F_{fj} represents a population of patterns instead of a single pattern (see Example 21.1). In the context of pattern classification, the consequent part of a MFI which is a fuzzy set in the pattern space represents the possibility of occurrence of different classes of patterns which is determined by the relative position of the fuzzy feature vector/pattern vector F_{fj} with respect to the defined cover of the pattern space. According to a particular MFI, a particular class $_i$ may have higher possibility of occurrence in pattern space R^2 (for simplicity of discussion we consider R^2. Such concept holds good for R^n; $n \geq 2$) than that of a class $_j$, $i \neq j$, provided a larger area (we are talking in terms of "area" because for simplicity we restrict our discussion within R^2) of the class $_i$ is occupied by the tip (ABCD) of F_{fj} (see Figure 1(a)) than that of $_j$. Whereas, according to the same MFI, the class $_k$, $k \neq i \neq j$, whose area is not covered up by the tip (ABCD) of F_{fj} (see Figure 1(a)) is having zero possibility of occurrence. This kind of judgment of possibility for different classes of patterns in the pattern space is basically an outcome of subjective quantification of human perception. The fuzzy sets C_1 and C_2 of Equation 21.5 are not directly obtainable only from the fuzzy set C which is the consequent part of a MFI. The fuzzy set C is formed depending upon the relative position of the fuzzy feature vector/pattern vector F_{fj} with respect to the defined cover of the pattern space. On the other hand when we consider the elements of F_{fj} individually, which are also fuzzy sets on the respective feature axis, the cylindrical extension of them (the said fuzzy sets) and the well-defined cover of the pattern space provide the information about the fuzzy sets C_1 and C_2 (see Example 21.2).

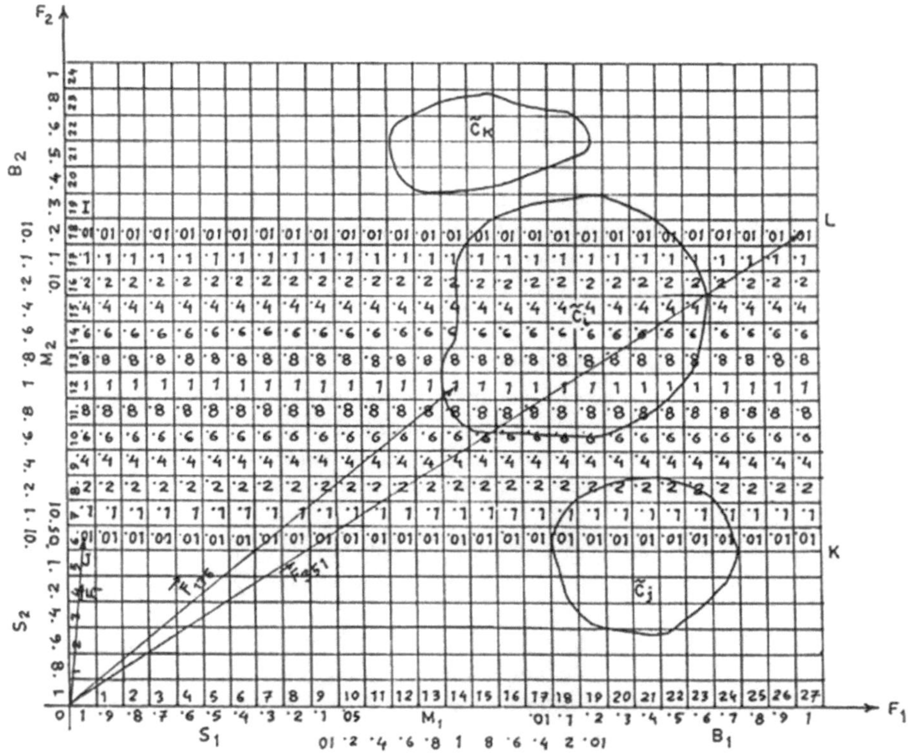

K E Y : The entries in the small quantized zones of the area 1 JKL is abtained by considering the cylindrical extension of the fuzzy set M 2.

FIGURE 1 (c) Representation of fuzzy pattern vector indicated by the cylindrical extension of fuzzy set M_2.

Example 21.1

Let us consider the fuzzy pattern vector F_{fj} of Figure 1(a). The position of F_{fj} on the pattern space means the area ABCD. The position ABCD of F_{fj} is obtained when

$$F_{fj} = \begin{bmatrix} F_1 \text{ is } M_1 \\ F_2 \text{ is } M_2 \end{bmatrix}$$

And the position of F_{fj} is changed when

$$F_{fj} = \begin{bmatrix} F_1 \text{ is } M_1 \\ F_2 \text{ is } S_2 \end{bmatrix} \text{ and so on.}$$

Now, if we try to compute the fuzzy set C, which is the consequent part of the following MFI

$$\text{if}\begin{pmatrix} F_1 \text{ is } M_1 \\ F_2 \text{ is } M_2 \end{pmatrix} \rightarrow C,$$

we have to consider the relative position of F_{fj} that is the area ABCD with respect to the defined cover $\hat{c}_i$, $\hat{c}_j$, and $\hat{c}_i$. For simplicity of demonstration we consider partial cover.

From Figure 7(a) it is obvious that the area of class $\hat{c}_i$ is substantially occupied by the tip ABCD. Looking at the possibility values of the small quantized zones of each $\hat{c}_i$ occupied by the tip ABCD we can have the following four types of estimate of class membership for the classes $\hat{c}_i$, $\hat{c}_j$, and $\hat{c}_k$.

For class $_i$:

1. Optimistic Estimate: The highest possibility value of the small quantized zones of the area of class $\hat{c}_i$ occupied by the tip ABCD. For instance, 1 indicated by F_{72} of Figure 1(a) (also see Example 21.1).
2. Pessimistic Estimate: The lowest possibility value of the small quantized zones of the area of class $\hat{c}_i$ occupied by the tip ABCD. For instance, 0.01 (see Figure 1(a)).
3. Expected Estimate: Average of the possibility values of all the small quantized zones of the area of class $\hat{c}_i$ occupied by the tip ABCD. For instance, 0.358 (see Figure 1(a)).
4. Most Likely Estimate: It comes from the subjective quantification of human perception. Here in this example the subjective quantification of belongingness of a population of patterns to a particular class (say $\hat{c}_i$) is achieved looking at the area of the said class occupied by the tip of the fuzzy vector. For instance, 0.6 (see Figure 1(a)) which is the subjective quantification of human perception as stated and which may vary like 0.5 or 0.7 as perception from one person to another varies within certain limit; but cannot be changed abruptly like 0.1 or 0.2, and so on, which is an indication of wrong perception. For further verification of the subjective quantification of our perception we may consider the following simple calculation.

Let the area of the class $\hat{c}_i$ of Figure 1(a) be approximated by the total number of small quantized zones covered up (partly or fully) by the contour of the class $\hat{c}_i$. From Figure 1(a) 90 such zones represent the area of the class $\hat{c}_i$. The area of the class $\hat{c}_i$ occupied by the tip ABCD of the fuzzy vector is 53 zones. Now, if we take the ratio (53 ÷ 90) = 0.588, which is the computed value of the belongingness of a population of patterns to class $\hat{c}_i$ with respect to the fuzzy pattern vector F_{fj} of Figure 1(a), then we see that the subjective quantification of belongingness of a population of patterns that is 0.6 or 0.5 or 0.7 lies close to that of the computed value.

For class $\hat{c}_j$:

1. Optimistic estimate = 0.2.
2. Pessimistic estimate = 0.01.
3. Expected estimate = (0.2 + 0.01 + 0.1 + 0.01 + 0.01 + 0.01)/6 = 0.34/6 = 0.0566.

4. Most likely estimate = 0.1. Note that here the computed value of belongingness is (6 ÷ 40) = 0.15.

For class $\hat{C}_k$:

All estimates are zero.

Thus, we get four fuzzy sets for the consequent part of a MFI (fuzzy set C)

$$C_{opt} = \{1.0/\hat{C}_i, 0.2/\hat{C}_j, 0.0/\hat{C}_k\},$$

$$C_{pess} = \{0.01/\hat{C}_i, 0.01/\hat{C}_j, 0.0/\hat{C}_k\},$$

$$C_{expt} = \{0.358/\hat{C}_i, 0.0566/\hat{C}_j, 0.0/\hat{C}_k\},$$

$$C_{most} = \{0.06/\hat{C}_i, 0.1/\hat{C}_j, 0.0/\hat{C}_k\}.$$

For all subsequent discussions for the design study of the classifier, we assume the optimistic estimate of the fuzzy set C (without mentioning anything like opt, pess, expt, most) which represents the consequent part of (a) of Equation 21.3. Instead of considering the optimistic estimate of the consequent part of (a) of Equation 21.3 in our design study we may go by other kind of estimates as stated. But whatever, may be the estimate of the fuzzy set C, the defuzzify decision remains same.

Next, with respect to Figure 1(a), let us consider the computation of the fuzzy set C_1 which is the consequent part of the first implication of Equation 21.5 is as follows:

$$C_{1_{opt}} = \{1.0/\hat{C}_i, 0.2/\hat{C}_j, 1.0/\hat{C}_k\},$$

$$C_{1_{press}} = \{0.01/\hat{C}_i, 0.01/\hat{C}_j, 0.01/\hat{C}_k\},$$

$$C_{1_{expt}} = \{0.469/\hat{C}_i, 0.105/\hat{C}_j, 0.543/\hat{C}_k\},$$

$$C_{1_{most}} = \{0.7/\hat{C}_i, 0.2/\hat{C}_j, 1.0/\hat{C}_k\}.$$

Similarly, with respect to Figure 7(c) the computation of the fuzzy set C_2 which is the consequent part of the second implication of Equation (4) is as follows:

$$C_{2_{opt}} = \{1.0/\hat{C}_i, 0.2/\hat{C}_j, 0.0/\hat{C}_k\},$$

$$C_{2_{press}} = \{0.01/\hat{C}_i, 0.01/\hat{C}_j, 0.0/\hat{C}_k\},$$

$$C_{2_{expt}} = \{0.508/\hat{C}_i, 0.086/\hat{C}_j, 0.0/\hat{C}_k\},$$

$$C_{2_{most}} = \{0.9/\hat{C}_i, 0.4/\hat{C}_j, 0.0/\hat{C}_k\}.$$

Note that, from now onwards whenever we will mention the fuzzy sets C_1 and C_2, we will assume the fuzzy sets C_1 and C_2 are obtained from the optimistic estimations of the membership value of the different class of patterns.

Now, we would like to mention that the relation represented in the area EFGH of Figure 7(b) which is obtained by the cylindrical extension of the fuzzy set M_1 over the universe of the feature axis F_2 is a fuzzy set which is fuzzy vector as per our earlier Definition 1. Note that in this case, if we defuzzify the fuzzy vector we get a fuzzy vector which is a fuzzy set and which is not fuzzy singleton. The fuzzy set obtained after the defuzzification of the stated fuzzy vector will be having elements of equal membership values as shown in Figure 1(b). The meaning of this defuzzified version, which is a fuzzy set, is that all the elements of the fuzzy set, as shown in Figure 1(b), are equally possible to occur under the cylindrical extension of the defuzzified information of the fuzzy set M_1.

Example 21.2

Let us consider Figure 1(a). For simplicity of demonstration, we do not consider the total cover of the pattern space.

Now, we get

$$\text{if}\begin{pmatrix} F_1 \text{ is } M_1 \\ F_2 \text{ is } M_2 \end{pmatrix} \rightarrow C = \{1.0/\hat{C}_i, 0.2/\hat{C}_j, 0.0/\hat{C}_k\},$$

where the fuzzy set C is obtained optimistically from the position of the fuzzy sets M_1 and M_2 on the universes of the feature axes F_1 and F_2, respectively, that means from the relative position of fuzzy pattern vector F_{fj} (i.e. the relative position of the area ABCD) with respect to the defined cover of the pattern space.

Whereas

$$\text{if } F_1 \text{ is } M_1 \rightarrow C_1 = \{1.0/\hat{C}_i, 0.2/\hat{C}_j, 1.0/\hat{C}_k\},$$

where the fuzzy set C_1 is obtained optimistically from the cylindrical extension of M_1 over the universe of the feature axis F_2 (i.e. from the position of the area EFGH on the pattern space of Figure 1(b)) and the well-defined cover of the pattern space,

$$\text{and if } F_2 \text{ is } M_2 \rightarrow C_2 = \{1.0/\hat{C}_i, 0.2/\hat{C}_j, 0.0/\hat{C}_k\},$$

where the fuzzy set C_2 is obtained optimistically from the cylindrical extension of M_2 over the universe of the feature axis F_1 (i.e. from the position of the area IJKL on the pattern space of Figure 1(c)) and the well-defined cover of the pattern space, then we get

$$\hat{C} = C_1 \cap C_2 = \{1.0/\hat{C}_i, 0.2/\hat{C}_j, 0.0/\hat{C}_k\},$$

which is accidentally same as C (but should be different in general) and the defuzzify information provided by and C are same, that is the class $\hat{C}_i$ (we go by selecting class having highest membership value: in case of the tie situation, we break the tie by taking arbitrary decision).

Now instead of considering the optimistic estimates of the fuzzy sets C, C_1, and C_2 if we go by most likely estimates of C, C_1, and C_2 then we get

$$\hat{C} = C_1 \cap C_2 = \{0.7/\hat{C}_i, 0.2/\hat{C}_j, 0.0/\hat{C}_k\},$$

which is in general different from $C = \{0.6/\hat{C}_i, 0.1/\hat{C}_j, 0.0/\hat{C}_k\}$: but the defuzzify information provided by and C are same as before, that is the class $\hat{C}_i$. Similarly, if we go by pessimistic estimates or expected estimates of C, C_1, and C_2 we will get in general two different fuzzy sets $\hat{C} = C_1 \cap C_2$ and C: but their defuzzify information remains the same.

Example 21.3

Part I: With respect to the data of Figure 1(a), let us consider a MFI of the following form:

$$\text{if}\begin{pmatrix}F_1 \text{ is } M_1\\ F_2 \text{ is } M_2\end{pmatrix} \rightarrow C.$$

If we go by interpretation 21.3(a) then we can write the MFI in the following 1D If-Then form:

$$\text{if } F_1 \text{ is } M_1 \text{ and } F_2 \text{ is } M_2 \rightarrow C.$$

Now, the antecedent clauses of the If-Then rule form the following relation which is a fuzzy set (see Definition 1):

$$\{0.01/F_1 + 1/F_{72} + 0.01/F_{143}\}.$$

Whenever, we consider interpretation 2(a) for reasoning (fuzzy reasoning/neuro fuzzy reasoning) for classification the mentioned fuzzy set acts as an input to the system at the learning stage. Thus, we inject information of a set of vectors $\{F_1\ F_{72},.\ F_{143}\}$ in terms of its possibilities to the system. At the learning stage the desired outputs of the system are equal to the possibility of each class of the mentioned fuzzy set C.

Part II: With respect to the data of Figure 1(b) and Figure 1(c), let us consider 1D fuzzy implications of the following form:

$$\text{if } F_1 \text{ is } M_1 \rightarrow C_1 \quad \text{and} \qquad F_2 \text{ is } M_2 \rightarrow C_2.$$

If we go by interpretation (4) and consider reasoning (fuzzy reasoning/neuro fuzzy reasoning) for classification then inputs to the system would be of the following form:

(a) The _first input$\{0.01/F_1 + 1/F_{135} + 0.01/F_{275}\}$ which implies the consequence C_1.

(b) The second input $\{0.01/F_1 + 1/F_{176} + 0.01/F_{351}\}$ which implies the consequence C_2.

The inputs of (a) and (b) are obtained by considering the cylindrical extension of M_1 and M_2 over the universes of F_2 and F_1, respectively. Similarly, the consequences C_1 and C_2 are obtained by considering the cylindrical extension of M_1 and the well-defined cover of the pattern space and the cylindrical extension of M_2 and the well-defined cover of the pattern space, respectively. As the impact of the cylindrical extensions of M_1 and M_2 are reflected on the fuzzy sets C_1 and C_2, respectively, we may consider the fuzzy set M_1 instead of the input (a) and the fuzzy set M_2 instead of the input (b) as the two independent inputs to the system at the learning stage. Note that, the fuzzy set M_1 is nothing but the projection of the relation of Figure 1(b) over the universe of the feature axis F_1 and the fuzzy set M_2 is the projection of the relation of Figure 1(c) over the universe of the feature axis F_2. Thus, we get a reduced form of inputs that is M_1 and M_2 which imply the fuzzy sets C_1 and C_2, respectively. The fuzzy sets C_1 and C_2 are two independent outputs of the system (at the learning stage). At the testing stage, the said outputs are combined by an intersection operator to obtain the fuzzy set = $C_1' \cap C_2'$ (see Figure 8) where C_1' is the inferred value at the first output and C_2' is the inferred value at the second output.

Thus, according to the newly proposed model (see Equation 21.5) a suitable representation of MFI by 1D fuzzy implications and the notion of induced fuzzy feature vector/pattern vector $\mathbf{F}_{fj}$ are possible. In the context of pattern classification such meaningful representation of MFI is not possible by the conventional model (see Equation 21.4) of Tsukamoto [164, 178]. In the later sections, we demonstrate that the notion of fuzzy pattern vector/feature vector and the newly modified MFI (i.e. Equation 21.4) are very effective for pattern classification.

For pattern classification, there are basically two approaches, namely the decision theoretic and the syntactic. The MFI approach to pattern classification is similar to the decision theoretic method of pattern classification.

21.4 IMPLEMENTATION OF THE NEW INTERPRETATION OF MFI IN BACK PROPAGATION-TYPE NEURAL NETWORK

Let us consider Equations (21.2a) and (21.2b). The first Equation (21.2a) is one law of implication and the second Equation (21.2b) is another law of implication. Both these laws of implication can be independently realized through two back propagation neural networks which are basically the conventional three-layered perceptrons (see Figure 2). The input of each neural network is the antecedent part of the If-Then clause. The antecedent part is represented by the fuzzy membership function. The reference output of the network is the consequent part of the clause. The consequent part is also represented by the fuzzy membership function which basically represents the possibility of occurrence of each class. Other features of the networks are same as the conventional back propagation type neural network. Each network is trained independently by a set of fuzzy If-Then statements using generalized delta rule [99]. For instance, the first multilayer perceptron (MLP) of Figure 2 is trained by (21.2a) and the second MLP of Figure 2 is trained by (21.2b). Thus, the antecedent part of (21.2a) is the input of the first MLP of Figure 2 and the consequent part of (21.2a) is the reference (target) output of the first MLP of Figure 2. Similarly, for the second MLP of Figure 2, the antecedent part of (21.2b) is the input of the second MLP of Figure 2 and the consequent part of (21.2b) is the reference (target) output of the second MLP of Figure 2. The antecedent parts of (21.2a) and (21.2b) are the first and second elements of the fuzzy feature vector shown in Figure 1.

Once the networks are trained, we can combine the output of each network by an intersection ($\cap$) operator (see Figure 2). Thus, if we have a 2D law of fuzzy implication as shown in Equation (21.2), we can realize MFI through a network configuration shown in Figure 2. If, we have an n-dimensional law of fuzzy implication, we can realize MFI through a network which consists of n number of independent three-layered perceptrons which are trained using the principle of back propagation neural networks. The outputs of n networks are combined through intersection ($\cap$) operator. We have considered object recognition on R^2, we always follow the network configuration of Figure 2.

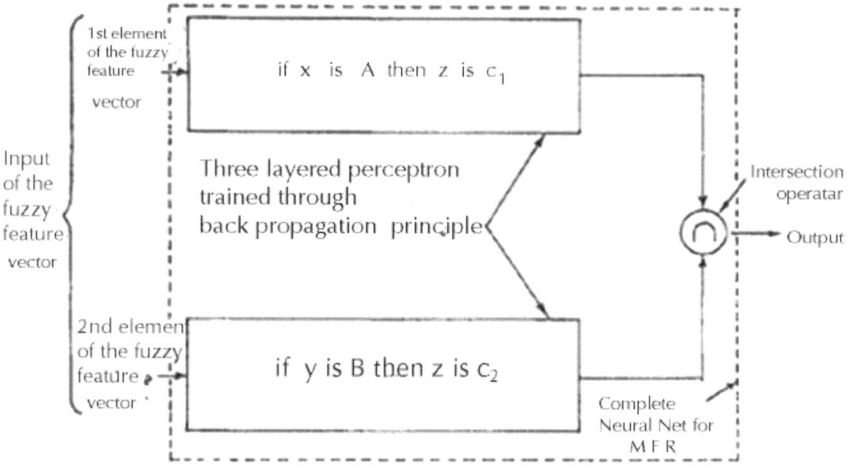

FIGURE 2 Realization of new interpretation of MFI through MLP-type neural network.

TABLE 1 QT matrix.

IF	Will *C*1 be		Will *C*2 be	
	Low	High	Low	High
***F*1 is Low**	~	~	~	~
***F*2 is High**	–	+	+	–
***F*1 is Medium and *F*2 is High**	–	+	–	+
***F*1 is high and *F*2 is Medium**	+	–	–	+

TABLE 2 Decision table.

Number of model features visible in the scene	Possibility of presence	Degree of presence
EQ 0	Nil	0
GT 0 but LT (50% of TNMF–σ)	Poor	0.3
EQ (50% of TNMF ± σ)	Fair	0.7
GT (50% of TNMF + σ) but		
LT (k*TNMF)	Good	1

Note: GT: greater than; LT: less than: EQ: equal to: TNMF: total number of features of a model object; σ represents the spread around the expected number of features under maximum allowable occlusion. The exact value of σ varies depending upon the dynamic ranges of the features. It (σ) is basically a design heuristic k represents the factor of safety of the entire design scheme. Its values are different under hard partitioning and fuzzy partitioning. The degree of presence (rightmost column of Table 2) is basically a scaling process [174].

21.5 DESIGN OF KNOWLEDGE-BASED OCCLUDED OBJECT RECOGNIZER

In the design of the object recognition scheme, we have basically two phases: namely the learning phase (training phase) where we learn the weights of the neural net from a set of training data and the testing phase (recognition phase) where we test the performance of the recognizer in terms of recognition scene.

Existing object recognition methods can be categorized as either global or local in nature. Global methods are based on the global features of the boundary or of an equivalent representation. Such techniques are the Fourier descriptors, the moments and methods based on auto regressive models. Local methods use local features such as critical points or holes and corners. They perform well in the presence of noise, distortion or partial occlusion since such effects on an isolated region of the contour alter only the local features associated with that region, leaving all other local features unaffected. However, the choice of representative local features is not trivial and the recognition process based on local features is more computationally intensive and time consuming. On the other hand, global methods have the disadvantage that a small distortion in a portion of a boundary of an object will result in changes to all global features. We use an internal angle and arc length of the dominant points on the boundary of an object as local features for knowledge-based recognition.

21.5.1 Local Feature Extraction

The image of the model object is defined by a image matrix as follows:

$$\begin{bmatrix} f(x_0,y_0) & \cdots & \cdots & f(x_0,y_{N-1}) \\ f(x_1,y_0) & \cdots & \cdots & f(x_1,y_{N-1}) \\ \cdots & \cdots & \cdots & \cdots \\ \cdots & \cdots & \cdots & \cdots \\ f(x_{M-1},y_0) & \cdots & \cdots & f(x_{M-1},y_{N-1}) \end{bmatrix}$$

and is called a digital image. Each element of the matrix is referred to as image element or pixel.

The image is transferred to a binary image by taking a threshold 126. Since $f(x_i, y_i)$ be the pixel value of the point with coordinate (x_i, y_i) then after transformation we get binary pixel value $p(x_i, y_i)$ that is:

$$p(x_i, y_i) = \begin{Bmatrix} 0 \; if \; 0 \leq f(x_i,y_i) \leq 126 \\ 1 \; if \; 126 < f(x_i,y_i) \leq 512 \end{Bmatrix}. \tag{21.3}$$

Now we find the boundary points of the object from the binary image. So we first chose a boundary point $P_0 = (x_i, y_i)$ with $p(P_0) = 1$.

Let N_{P_0} be the $8-nbd$ of the point P_0 defined by

$$N_{P_0} = \{Q_0, Q_1, Q_2, Q_3, Q_4, Q_5, Q_6, Q_7\} \tag{21.4}$$

where Q_k, $k = 0, 1, 2, 7$ are obtained from the Table 3. The next boundary point P_l can be find so that

$$P_1 = Q_S \text{ if } p(Q_k) = 0 \ \&\ p(Q_S) = 1 \text{ with } s = (k+1)(\text{mod } 8) \ \exists\ K \in \{0, 1, 2, 7\}. \quad (21.5)$$

TABLE 3 The 8 – nbd of the point P, Np.

$Q_0 = (x_{i-1}, y_{i-1})$	$Q_1 = (x_{i-1}, y_j)$	$Q_2 = (x_{i-1}, y_{j+1})$
$Q_7 = (x_i, y_{j-1})$	$P = (x_i, y_j)$	$Q_3 = (x_i, y_{j+1})$
$Q_6 = (x_{i+1}, y_{j-1})$	$Q_5 = (x_{i+1}, y_j)$	$Q_4 = (x_{i+1}, y_{j+1})$

Continuing this process we find b + 1 boundary points of the model object $\{P_0, P_1, P_2, P_b\}$.

Let $\vec{R}_i$ and $\vec{R}_{ij}$ be the two vectors defined at the point P_i that is:

$$\vec{R}_i = (x_{i+1} - x_i, y_{i+1} - y_i)$$

$$\vec{R}_{ij} = (x_j - x_i, y_j - y_i). \quad (21.6)$$

Let θ_k be the angle between the $_i$ and $_{ij}$ and is defined by:

$$\theta_k = \cos^{-1}\left(\frac{\vec{R}_i . \vec{R}_{ij}}{|\vec{R}_i||\vec{R}_{ij}|}\right) \quad (21.7)$$

where j = (i + k)(*modb*+1) with k = 2, 3, where θ_k is computed given increment to k until for some k, $\theta_{k-1} < \theta_k \geq \theta_{k+1}$.

When the last inequality holds for the consecutive values (j–1, j, j+1) the set of points $\{P_i, P_{i+1}, P_{i+2}, P_{i+r}\}$ is said to constitute the right arm of the point P_i and the length of the right arm is r.

Similarly we find the set points $\{P_i, P_{i-1}, P_{i-2}, P_{i-l}\}$ is said to constitute the left arm of the point P_i and the length of the left arm is l.

Thus $\{P_{i-l}, P_{i-2}, P_{i-1}, P_i, P_{i+1}, P_{i+2}, P_{i+r}\}$ is the region of the support at the point P_i and $L_i = l + r$ is called the arc length of the support of the region at the point P_i.

Let φ_i be the internal angle between P_{i-l} and P_{i+r} at P_i and is defined by:

$$\varphi_i = \cos^{-1}\left(\frac{\vec{L}A_l . \vec{R}\,A_r}{|\vec{L}A_l||\vec{R}\,A_r|}\right) \quad (21.8)$$

where

$$\vec{LA}_l = (x_{i-l} - x_i, y_{i-l} - y_i)$$

$$\vec{R}_{ij} = (x_{i+r} - x_i, y_{i+r} - y_i).$$

Processing in this way, we find $(\varphi_i, L_i)\ \forall_i \in \{1, 2, N\}$. Now we determine the dominant points of the model object by taking a threshold $\varphi_\in$ on the internal angle. Therefore, the dominant points are those P_i's which satisfy $\varphi_i \leq \varphi_\in$.

21.5.2 Training Phase

Here the pattern vectors of the training data, obtained from the dominant features of the model objects, are two dimensional. In the training phase with the training data set $(\varphi_i, L_i)\ \forall_i \in \{1, 2, N\}$, we discretize (quantize) the individual feature axis and the entire pattern space in the following way:

Determine the lower and upper bounds of the data of the ith feature value. Let is the ith data of the ith feature F_i and d_i is the length of segmentation along ith feature axis.

Minimum of the data f_j^i, j = 1, 2, of the ith feature is $f_{min}^i = min_j\,(f_j^i)$. Let r_{min}^i be the remainder when is divided by d_i. Therefore, the lower bound of the ith feature axis

$$LB_i = \begin{Bmatrix} f_{min}^i & \text{if } r_{min}^i = 0 \\ f_{min}^i - r_{min}^i & \text{otherwise} \end{Bmatrix} \tag{21.9}$$

This LB_i is taken as the ith coordinate of the origin.

Again, maximum of the data j = 1, 2, of the ith feature is = max(). Let be the remainder when is divided by d_i. Therefore, the upper bound of the ith feature axis:

$$LB_i = \begin{Bmatrix} f_{max}^i & \text{if } r_{max}^i = 0 \\ f_{min}^i + (d_i - r_{max}^i) & \text{otherwise} \end{Bmatrix} \tag{21.10}$$

The local features (internal angle (φ_i) and arc length (L_i) at dominant points of the model objects) are plotted on F_1–F_2 plane where F_1 axis represents the internal angle and F_2 axis represents the arc length. Depending upon the values of the local features of the dominant point of the model objects several clusters of the pattern are formed on F_1–F_2 plane.

Let U_i be the universe of discourse on the ith feature axis F_i then U_i has $m_i = (UB_i - LB_i)/d_i$ genetic elements and these are u_j^i j = 1, 2, m_i which we define as follows:

$$u_j^i = \begin{Bmatrix} [LB_i + (j-1).d_i, LB_i + j.d_i] \text{for } j = 1,2,\ldots\ldots,(m_i - 1) \\ [LB_i + (m_i - 1).d_i, UB_i] \quad \text{for } j = m_i \end{Bmatrix} \tag{21.11}$$

Therefore, the universe on the ith feature axis, $U_i = \{u_1^i, u_2^i, \dots\dots, u_{m_i}\}$. Now we define k_i fuzzy sets on U_i say D_l^i, l = 1, , l = 1, 2, k_i as shown in Table 4. So there are k_i fuzzy If-Then rules as follows:

$$R_l^i\text{: if } F_i \text{ is } D_l^i \text{ the } C \text{ is } C_l^i \in F(C_{class}),\text{ where } C_{class} = \{c_1, c_2, c_n\}\cdot$$

TABLE 4 Fuzzy sets in $F(U_i)$.

(*)	u_1^i	u_2^i	... ...	$u_{m_i}^i$
D_1^i	$\mu D_1^i\ (u_1^i)$	$\mu D_1^i\ (u_2^i)$	... ...	$\mu D_1^i (u_{m_i}^i)$
D_2^i	$\mu D_2^i\ (u_1^i)$	$\mu D_2^i\ (u_2^i)$	... ...	$\mu D_2^i (u_{m_i}^i)$
.	.	.	... ...	.
.	.	.	... ...	.
.	.	.	... ...	.
$D_{k_i}^i$	$\mu D_{k_i}^i (u_1^i)$	$\mu D_{k_i}^i (u_2^i)$	... ...	$\mu D_{k_i}^i (u_{m_i}^i)$

TABLE 5 Query table.

Antecedent clause of DFI	Consequent clause			
	C_1		C_2	
	high	low	high	low
F_1 is low	~	~	~	~
F_2 is high	–	+	+	–

Note that the cylindrical extension of the antecedent clause of each fuzzy If-Then rule locates a particular region on the pattern space. This particular region on the pattern space may be occupied by different classes of patterns. Depending upon the availability of the features of the dominant points each model object at that particular region, at the training phase, we develop the fuzzy set c_l^i to quantify our perception about the degree of occurrence of different model objects at a particular observation (see Figure 1).

In practice, during training phase, we generate the fuzzy set c_l^i based on a Query Table (QT). We explain this table with the help of an example. Let F_1 and F_2 be the names of two linguistic variables and high, medium, low are the three primary terms. Assume that F_1 and F_2 form the essential input features to the recognizer. Let there be two classes, C_1 and C_2. Suppose, the output linguistic variables are formed with the help of two primary terms low and high. Now, we construct the QT as shown in Table 5. If we consider Table 5 and read any of the rows first and then any of the columns of the QT, a question will be formed. The answer to this question has to be put in the corresponding elemental position of the matrix. If the answer is 'yes', we put '+', if 'no', a '–' and if not answerable a '~'. As for illustration, if we read the first row then the third column, we will have the question: if F_1 is low will {c_1 be {low/high}, c_2 be {low/high}}? If answer is yes for class c_1(low) put '+' and no for the class c_2(low) put '–' in (1,1) and (1, 4) position of Table 5. We quantify our perception about low/high of the consequent clause of If-Then rule by setting membership value within the range 0 to 1. The quantification of perception may not be unique for each person but it will lie, if it is correct based on the training set of data, within a threshold.

Thus, several If-Then rules are constructed to capture the information about the local features of the dominant points of the model objects. The fuzzy If-Then rules form the knowledge base of the object recognition scheme. As this knowledge is formed by the fuzzy If-Then rules, it has the inherent capacity for management of uncertainties in local features due to the presence of noise and other disturbances. Each rule has two components, one is the antecedent clause which is represented by a fuzzy set and other is the consequent clause which is represented by a fuzzy set. So based on the generated fuzzy rules, at the end of training phase we estimate the weights of the neural net.

At the end of the training phase we cross verify the quality of the trained network by checking the recognition score of the training data set (based on which the fuzzy If-Then rules are initially generated). If the recognition score of the training data set (which are fuzzified by fuzzy masking at the time of testing) does not reach to the satisfactory threshold (say 80% recognition score is set as threshold) we may have to modify the initial fuzzy If-Then rules to represent our knowledge about the model objects. After satisfactory training of the net we switch over to the testing phase (recognition phase) where we consider the recognition of data which do not belong to the model objects (see Figure 4).

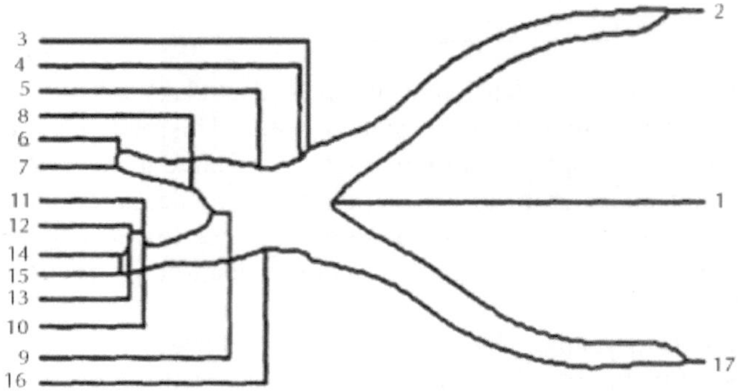

(a) Pliers

FIGURE 3 *(Continued)*

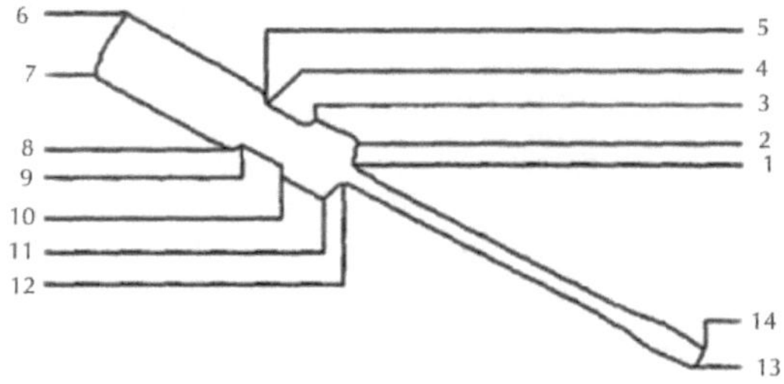

(b) Screw-driver

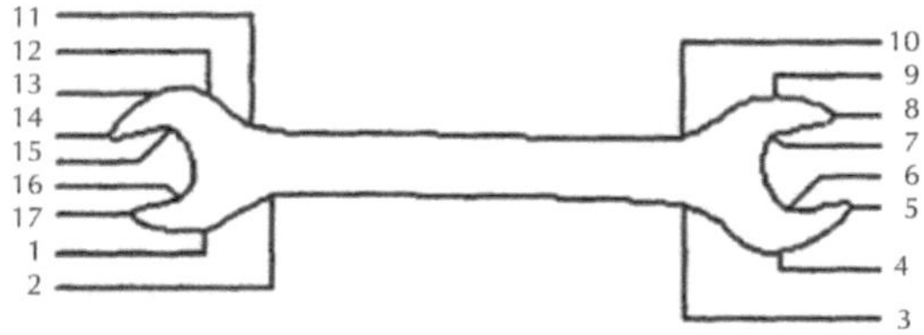

(c) Wrench

FIGURE 3 Two dimensional model objects.

21.5.3 Testing Phase

At the testing phase (recognition phase) the features of the selected patterns are fuzzified using the concept of the fuzzy masking. The recognition results obtained from the Equation (3) produces a fuzzy set $C = \sum_{i=1}^{n} \mu_C(c_i)/c_i$ which represents the degree of occurrence of each test pattern at different classes in the pattern space. Thus, we get a fuzzy recognition of a test pattern. Now to calculate the recognition score from the result we have to go through certain decision process. In the first stage of our decision process, we have to increase the level of confidence by prescribing a α-cut of the fuzzy set C that is:

$$C_\alpha = \{c_i \mid \mu_C(c_i) \geq \alpha;\ c_i \in C_{class}\}.$$

If $C_\alpha = \emptyset$ (empty set) then the given test pattern is not recognized by the present classifier. Otherwise:

$$hgt\,(C) = \bigvee_{c_i \in C_\alpha} \mu_C(c_i). \tag{21.12}$$

Now we get the set of recognized classes as,

$$Class_{recognize} = \{c_i \mid hgt(C) - \mu_C(c_i) \leq \theta,\ c_i \in C_\alpha\} \tag{21.13}$$

where θ is a small threshold prescribed by the designer to capture the relative change in membership values among the elements of the recognized classes $Class_{recognize}$

i. In case $Class_{recognize}$ is a singleton set then the given test pattern is recognized uniquely: otherwise
ii. Multiple recognitions of the given test pattern occurs: that means a test pattern may occur at different model objects.

The notion of multiple recognition is very natural in case of the test patterns occurring at the overlapped classes. Such choice of multiple recognitions is a kind of grace to take care of all uncertainties (e.g. uncertainties in the representation of knowledge about training patterns in the process of fuzzification, through fuzzy masking, of the test patterns etc.) in our recognition process.

21.5.4 Condition of Recognition

As the scene consist of the model objects which are placed in different orientations and which are partially occluded by each other, it is obvious that the number of local features visible in the scene is less than the number of local features visible in the corresponding model object. In our recognition scheme, we assume that under occlusion at least 50% of local features $\pm\sigma$ of the model object is tolerable to the recognition scheme. At maximum occlusion, occlusion of the model object is tolerable to the recognition scheme. At maximum occlusion, under different kind of uncertainties, σ is the spread around the "expected number" of features.

Note that the "expected number" is not in statistical sense: but in the following sense.

Let a model object has total 10 features. Under maximum occlusion (i.e. under 50% occlusion) 5 features are expected to be visible in a scene consisting of that model and other objects. During recognition process out of 5 features of the said model either:

i. Some of the features (say, one or two) may not be visible due to some error in computation or noise or
ii. Some additional features (say, one or two) which are newly generated features due to occlusion, and so on may be recognized as features of the said model.

Thus, the concept of spread σ around the expected number of features is generated.

After the recognition of the local features of the scene if the recognition scheme produces, for each model object, the number of model features visible in the scene then the decision of Table 6 linguistically states the possibility of presence as well as the degree of presence, which is logically a scaling process, for each model object in the scene.

Now for each model object 'M_i' in the scene, a rough estimate of the number of model features not visible in the scene is represented by the difference between the number of features (for occluded object 'M_i') correctly recognized minus k*T (T is the total number of model object). Here k represents the factor of safety of the entire design scheme. Depending upon the dynamic range of the feature values the value of k is chosen in an adhoc manner. In the occluded environment, under single choice of recognition, this said difference is non positive. And for asserting the degree of presence of an object in the scene we safely go by the decision Table 6. But under single choice of recognition if the said difference is positive then it becomes an alarming situation indicating that some of the features of the model object 'M_i' are recognized as the features of the model object 'M_i' where $i \neq j$. In that case the vision system should be intervened in an interactive mode. In occluded environment, the value of k is experimentally chosen as 1.5. Thus, the vision system is given a freedom to recognize the features of some newly generated points which are either formed due to overlap of different objects in a scene or created due to some environmental uncertainties.

TABLE 6 Decision table.

Number of recognized local features of the occluded model object (M_i)	Decision	
	linguistic value	numerical value
$Y = 0$	nil	0.0
$0 < Y \leq [N/2] - \sigma$	poor	0.3
$[N/2] - \sigma < Y \leq [N/2]$	medium	0.5
$[N/2] < Y \leq [N/2] + \sigma$	fair	0.7
$[N/2] + \sigma < Y \leq k * N$	good	1.0

Key: $Y \equiv$ number of recognized local features of the occluded model object M_i
$N \equiv$ total number of local features of the model object M_i
$\sigma \equiv$ Spread around N/2

21.6 CONTROL SCHEME OF THE VISION PROCESS

There are several existing control schemes of the vision process. For instance, in the following we provide representative control schemes for the vision process.

1. Bottom-up control
2. Top-down control
3. Feedback control
4. Hierarchical control

The recognition of two dimensional occluded objects, we follow the bottom-up control scheme. By adapting the scheme for the vision process we can have the following advantages:

i. Knowledge (model) of object is used only for matching scene descriptions.
ii. In Bottom-up control, raw data are gradually converted into more organized and useful information.
iii. In Bottom-up control objects may easily be changed simply by changing the models of the objects.

But the Bottom-up control scheme provides the following disadvantages also:

(a) Lower processes are not well adjusted to a given particular scene because they are domain independent. Errors may be inherited.
(b) Insufficient because it is always executed in the same manner regardless of the scene.

The present approach to the recognition to the recognition of occluded object is guided by a decision table (Table 6) which basically follows a voting scheme. As we do not consider the positioning problem and do not follow the hypothesis generation and verification paradigm, we simply stick to the bottom-up scheme. But depending upon the need of the problem we may follow any other control scheme.

21.7 EFFECTIVENESS OF THE PROPOSED CLASSIFIER

To test the effectiveness of our design study, we consider the recognition of six occluded objects, which are shown in Figure 4 and Figure 6. Here we take three models, namely plier(M_1), screw-driver(M_2), and wrench(M_3). According to the methods stated in section 21.4.1 we extract the two features (φ_i, L_i), i = 1, 2, N_j of the dominant points of the jth model M_j, j = 1, 2, 3, which are shown in the Table 7. Now we discuss two cases as follows:

TABLE 7 The dominant features of the models M1, M2, and M3.

Serial number	Pliers		Wrench		Screw-driver	
	φ	L	φ	L	φ	L
1	110.180519	8	153.373230	18	112.540939	58
2	60.397854	21	152.20459	51	110.180519	21
3	130.548752	6	150.194672	5	105.212776	31
4	134.945694	5	143.914703	8	138.758224	20
5	126.818855	4	47.400433	14	136.919830	31
6	50.772362	16	125.42858	19	99.872559	46
7	38.716938	47	150.194672	5	114.882240	37
8	66.568520	45	71.536263	9	136.680695	38
9	134.945694	5	154.377899	10	117.732239	30
10	134.945694	5	152.292740	55	124.330307	33
11	60.230877	12	154.496078	53	121.626518	17
12	33.676514	12	153.373230	10	134.521454	55
13	122.855797	8	147.035583	11	105.081718	21
14	59.834534	19	49.165131	12	71.536263	9
15	59.240669	14	139.342621	6		
16	134.945694	5	143.072525	4		
17	42.391071	30	49.378834	9		

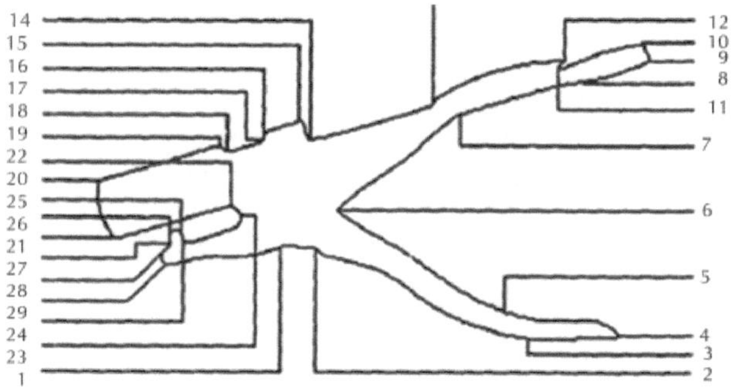

(a) Occlusion 1

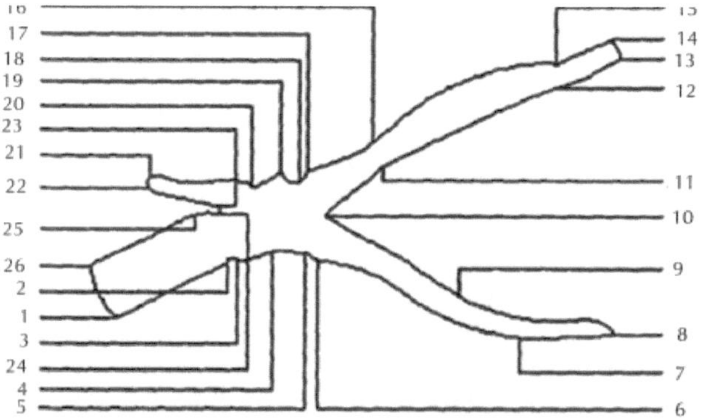

(b) Occlusion 2

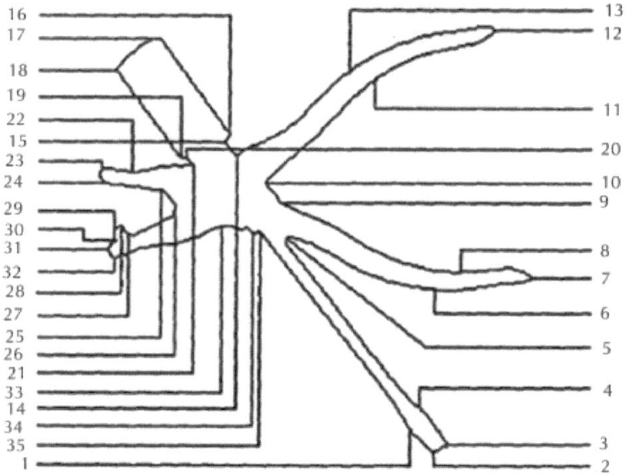

(c) Occlusion 3

FIGURE 4 Occluded objects of two models M_1 and M_2.

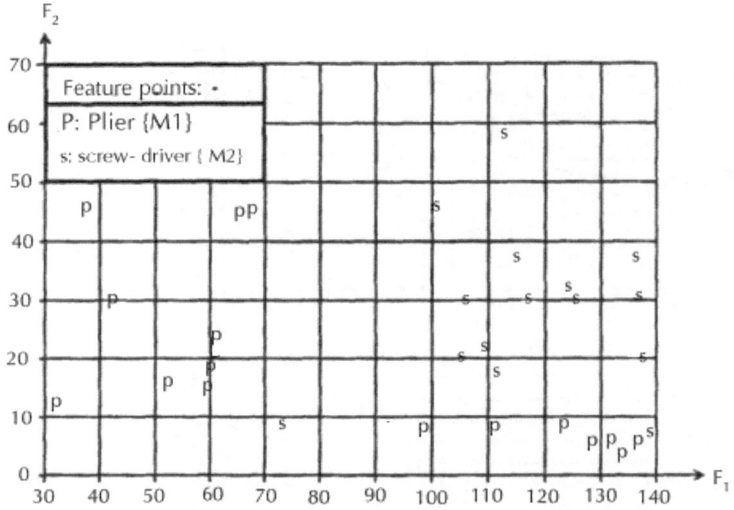

FIGURE 5 Significant features of model M_1 and M_2 on F_1–F_2 plane.

21.7.1 Recognition of the Scene Consist of Model Objects M_1 and M_2

Here we consider two model objects pliers (M_1) and screw-driver (M_2). The feature values are plotted on the F_1–F_2 plane in Figure 5. The number of local features of the model M_i is N_i and that of visible in occluded environment is V_i out of N_i. Let O_i be the total number of visible features of the model M_i in occluded situation. Note that in the number O_i there are some additional features which are generated under occlusion as stated earlier. Let Y_i be the number of correctly recognized features in occluded environment. The decision of the recognition of the model objects in the scenes shown in Figure 4 is based on the Table 6 shown in the Table 8.

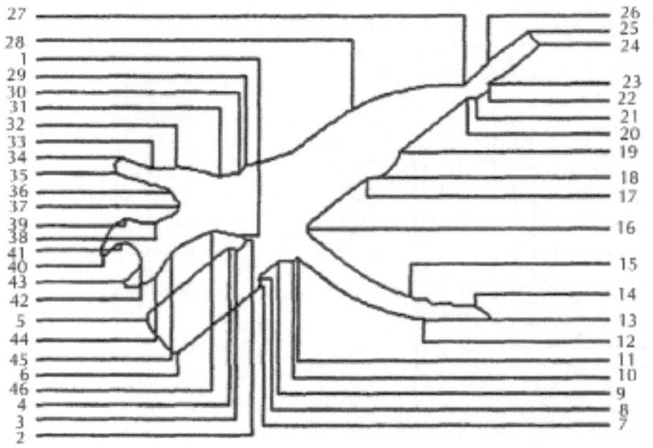

(a) Occlusion 1

FIGURE 6 *(Continued)*

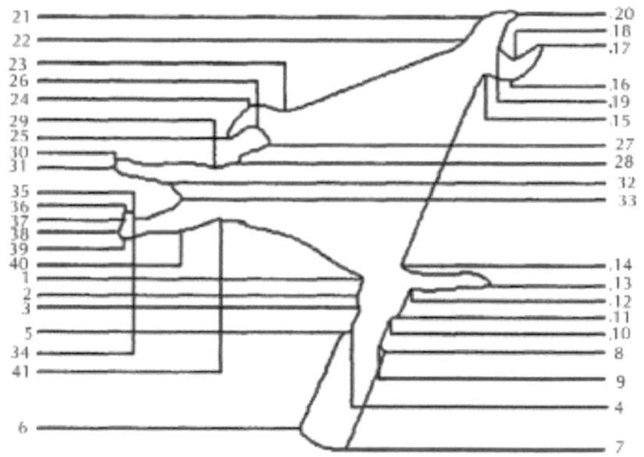

(b) Occlusion 2

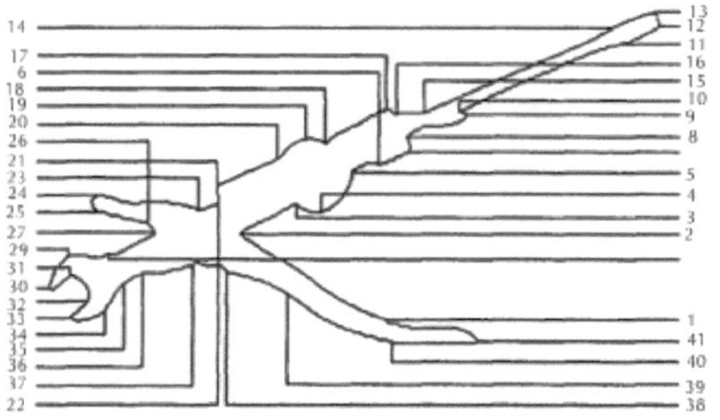

(c) Occlusion 3

FIGURE 6 Occluded objects of three models M_1, M_2, and M_3.

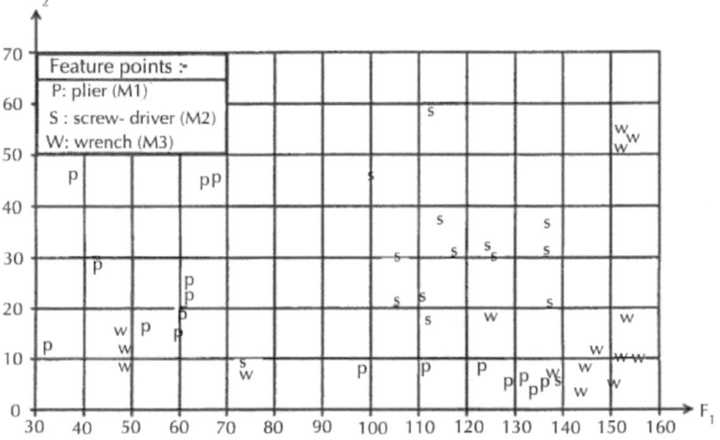

FIGURE 7 Significant features of model M1, M2, and M3 on F1–F2 plane.

TABLE 8 Decision of the occluded objects as shown in Figure 4.

Occluded objects	Model objects	No. of features				Decision	
		N_i	V_i	O_i	Y_i	linguistic	numerical
Figure 4(a)	Pliers (M_1)	17	11	14	10	good	1.0
	Screw-driver (M_2)	14	9	11	8	good	1.0
Figure 4(b)	Pliers (M_1)	17	7	12	9	medium	0.5
	Screw-driver (M_2)	14	7	8	6	good	1.0
Figure 4(c)	Pliers (M_1)	17	13	19	15	good	1.0
	Screw-driver (M_2)	14	7	10	8	fair	0.7

TABLE 9 Decision of the occluded objects as shown in Figure 6.

Occluded objects	Model objects	No. of features				Decision	
		N_i	V_i	O_i	Y_i	linguistic	numerical
Figure 6(a)	Pliers (M_1)	17	9	16	12	good	1.0
	Screw-driver (M_2)	14	9	11	8	good	1.0
	Wrench (M_3)	17	9	9	9	good	1.0
Figure 6 (b)	Pliers (M_1)	17	14	15	12	good	1.0
	Screw-driver (M_2)	14	8	10	7	good	1.0
	Wrench (M_3)	17	11	11	10	fair	0.7
Figure 6(c)	Pliers (M_1)	17	7	11	8	medium	0.5
	Screw-driver (M_2)	14	5	9	7	good	1.0
	Wrench (M_3)	17	10	10	9	good	1.0

21.7.2 Recognition of the Scene Consist of Model Objects M_1, M_2, and M_3

Here we consider two model objects pliers (M_1), screw-driver (M_2) and wrench (M_3). The feature values are plotted on the F_1–F_2 plane in Figure 7. The number of local features of the model M_i is N_i and that of visible in occluded environment is V_i out of N_i. Let O_i be the total number of visible features of the model M_i in occluded situation. Let Y_i be the number of correctly recognized features in occluded environment. The decision of the recognition of the model objects in the scenes shown in Figure 6 is based on the Table 6 shown in the Table 9.

21.8 CONCLUSION

We have successfully applied a Neuro-Fuzzy reasoning to recognize partially occluded objects. We have tested our proposed scheme with several complex scenes and we have obtained very promising results. Throughout the case studies, we have seen that our proposed vision system has not generated any alarming signal which says that features of model object j are detected as features of model object i, where $i \neq j$. Thus, we claim that our proposed vision system works as per the specifications of design study. At present, we have considered only the recognition of objects in a scene, but we have not considered the positioning of the objects in the scene. The present scheme will be very suitable for industrial robot vision.

KEYWORDS

- **Fuzzy feature vector**
- **Industrial robot vision**
- **Multi-dimensional fuzzy implication**
- **Neuro-fuzzy reasoning**
- **Proposed vision system**

22 Conclusion

CONTENTS

We have developed a series of algorithms for polygonal approximation of closed digital curves. The problem has been treated as the side detection as well as angle detection problem.

The split and merge technique introduced in Chapter 2 resolves the problem of initial segmentation. The initial segmentation technique is simpler than that of the Ansari-Delp algorithm [6] as it requires less number of arithmetic operations than that of Ansari-Delp. The procedure has been tested with a number of digital curves. The experimental results have been compared with the Ansari-Delp algorithm and have been found to produce more accurate results.

From Chapter 3 through 6 we have developed a series of sequential algorithms. The one developed in Chapter 3 is a one-pass algorithm which is simple and fast. It involves no arithmetic except subtractions. The approximation error is controlled by a counter. The higher the value of the counter is the higher is the approximation error. The procedure is purely a searching technique which is done in a sequential fashion. Though this algorithm is simple and fast but it holds for curves with uniformly spaced points only. It may be possible to make this procedure applicable to curves with non-uniformly spaced points also by replacing the finite differences by divided differences.

The algorithm developed in Chapter 4 is sequential one-pass and holds for curves with uniformly as well as non-uniformly spaced points. This algorithm is more efficient than the other sequential one-pass algorithm such as Williams [184] and Wall-Danielsson [180] algorithm in the sense that it involves less number of arithmetic operations than the others. It need a small and finite memory and detects the sides on the fly. The approximation error is controlled by the critical value of the area and perimeter of triangles. The approximation made in computing the side length of the triangles are taken from Wall-Danielsson and has also been used in many problems of image processing [53]. We have used this approximation because it produces good results reducing the computational load significantly. One can as well use Tchebycheff's approximation at the cost of higher computation load.

In each of these algorithms it is necessary to specify the maximum allowable error. In the algorithms developed in Chapters 3 and 4 the maximum allowable error is specified indirectly by the counter and the critical values of the area and perimeter of the triangles. In the other existing algorithms too, either the number of line segments or the maximum allowable error is specified (directly or indirectly). We have shown in Chapter 5 and 6 that how to make a polygonal approximation without specifying the maximum allowable error or the number of line segments. In these algorithms we keep both these parameters free and allow the procedure to determine it on the basis of local topography of the curve. So these procedures can run without operator's intervention. Integral square error has been used to measure the closeness of the polygon to the curve in Chapter 5. We have shown in Chapter 6 that though the most commonly used norms are integral square error and the maximum error but it is possible to use sum of absolute error as a measure of closeness. Both these algorithms are sequential one-pass and they hold for curves with uniformly spaced points as well as non-uniformly spaced points. Unlike the existing sequential algorithms neither do they round off sharp turnings nor do they miss corners.

Each of the sequential algorithms that we have developed has been compared with the Williams and the Wall-Danielsson algorithms because the later are sequential and one-pass in nature. In each case our algorithm produces more accurate results than either of the two. Our procedures neither miss corners nor do they round off sharp turnings. The Williams' algorithm misses corners and rounds off sharp turnings. The Wall-Danielsson algorithm does not round off sharp turnings but sometimes it detects false vertices. The peak-test succeeds to retain sharp turnings but it fails to locate a vertex at its actual position when the turning is not so sharp.

We have used distance to a point instead of perpendicular distance as a measure of collinearity in Chapter 7. Since the technique makes two-passes through data hence it can produce symmetrical approximation from symmetrical digital curve. The procedure retains the sharp turnings and does not dislocate the vertex points. This chapter also introduces a new metric to measure the quality of polygonal approximation.

We have shown in Chapter 8 that how apply reverse engineering on Breseham's algorithm for polygonal approximation. The curve points may be uniformly or non-uniformly spaced. There are three features of the proposed method, namely, it is sequential and runs in linear time, it produces symmetric approximation from symmetric digital curve and it is an automatic algorithm.

Polygonal approximation in Chapter 9 and 10 has been treated as an angle detection problem. In contrast to the other algorithms here all processing are locally and hence they are suitable for parallel processing. We have shown in Chapter 9 that is possible to use *k*-cosine to determine the region of support without using any input parameter. A new measure of significance called smoothed *k*-cosine has been introduced. This measure of significance is different from the smoothed *k*-cosine introduced by Rosenfeld and Wezska [142]. The procedure has been applied on a number of digital curves and promising results have been obtained. It is also found to perform well on curves which consist of features of multiple sizes.

We have introduced in Chapter 10 that the concept of asymmetric region of support and a new measure of significance called *k-l* cosine. The procedure has been tested with a number of digital curves and promising results have been obtained. The procedure does not need any input parameter and performs well on curves which consist of features of multiple sizes. Each of these algorithms has been compared with the Teh and Chin algorithm [170] and in each case our procedures are found to detect more significant/dominant points than the Teh and Chin algorithm.

We have made scale space analysis using one of our polygonal approximation schemes and have shown its application to corner detection in Chapter 12. Our procedure is shown to detect more corner points than the Rattarangsi-Chin algorithm [119]. This scale space analysis in contrast to the existing ones does not require convolution. It does not detect corners *via* curvature estimation. It holds for curves with uniformly spaced points only.

As the conventional scale space analysis technique involves convolution of a curve with a smoothing kernel, in Chapter 13 we have made scale space analysis by repeated convolution of a curve with Gaussian kernel with constant window size. This is in contrast to the existing Gaussian smoothing which makes use of the Gaussian kernel with varying window size. This technique reduces the space requirement to a small and finite quantity from $O(n^2)$ which is required in the existing Gaussian smoothing process. The computational load has also been shown to be reduced to $O(n)$ from $O(n^2)$. The smoothing process has been shown to enjoy scale space property. The scale space map has been used to detect corners on digital curves. The corner detector has been found to produce more corners than the Rattarangsi-Chin algorithm. The corner detector has also been found to be robust to noise.

We have used a discrete scale space kernel with a continuous scale parameter proposed by Lindeberg [79] to make scale space analysis and corner detection in Chapter 14. The numerical problems that arise in the implementation of the discrete kernel have been addressed and possible solutions are proposed. The corner detector has been found to produce the least number of corners that consists of circular arcs. The corner detector has also been found to be robust to noise.

From Chapters 12, 13, and 14 on scale space analyses we conclude that scale space analysis of digital curves without convolution with a smoothing kernel is feasible provided the smoothing technique preserves scale space property. Scale space analysis of discrete curves using continuous scale space kernel is also feasible provided the continuous kernel is discretized properly. Scale space analysis of discrete curves using discrete scale space kernel with a continuous scale parameter is also feasible with some suitable assumptions.

In contrast to the existing methods of smoothing an entire curve at various levels of detail in Chapter 15 we have suggested an adaptive method of corner detection. This procedure does not require construction of the complete scale space map and it is also not necessary to convert the map into a tree representation. The procedure has been applied on a number of digital curves and the experimental results have been compared with existing work.

From Chapter 17 to Chapter 21 we essentially consider the applications of polygonal approximation in structural pattern classifications and occluded 2D object recognitions. The content of Chapter 17 deals with dissimilarity measure between two polygons and classifications of structural patterns based on smoothed version of polygons are considered. Chapter 18 deals with matching of polygon fragments which can be utilized to recognize and locate occluded objects. Chapter 19 deals with hypothesis generation and verification paradigm to recognize and locate partially occluded 2D scene. Chapter 20 is a departure from the standard paradigm of recognition 2D occluded scenes and deals with a fusion technology which is based on hypothesis generation and belief revision scheme. Belief revision is basically a tool for Artificial Intelligence (AI) which has been fused with vision technology and a new computer vision paradigm has been developed. Chapter 21 is essentially based on the voting scheme for recognition of occluded scene. It is again basically a fusion technology,that is "neuro-fuzzy" to handle uncertainty in real application environment of recognition.

The part III of this book handles series of case studies with detail design parameters and several successful experimental results are reported.

22.1 APPENDIX

In Chapter 19, 20, and 21 we explore different approaches to recognize occluded objects. All the approaches essentially perform the task of recognition using local invariant features of the model objects and scene consists of model objects. The three different approaches of the three chapters use different types of algorithm for recognition but based on different types of invariant local features. In the following we provide two such local features, *viz.*, curvatures and internal angles of the dominant points of the model objects and scene consists of model objects as a set of local features which can be utilized by any interested user of the book for further verification of different existing algorithms for object recognition available in the literature including the algorithm of this book.

22.2 LOCAL FEATURE EXTRACTION

As the global object features, such as area and perimeter of boundary curve, centroid and shape moments are not suited for the recognition of partially occluded objects. We consider a set of invariant local features, such as curvature and internal angle of curved objects. The invariance property of such object features is used to generate the F_1–F_2 feature space. In our case, we have considered the internal angle of significant points on a curve as the feature F_1 and its corresponding curvature as the feature F_2. Now, curvature is a measure of the rate of change of orientation per unit arc length. The geometric interpretation for the curvature is depicted in Figure 1. Let P is a point on a curve, T be the tangent at that point and A be a neighboring point on the curve. Let α denote the angle between the line AP and T, and || the arc length between A and B. The curvature κ at P is the ratio $\alpha/\widehat{AB}$.

If we deal with digital curves, it is not immediately clear how to define a discrete curvature and internal angle. Suppose, a digital curve is defined as a sequence of integer coordinate points p_1, p_2 where p_{i+1} is a neighbor of p_i, $1 \leq i \leq n$. Now, on a digital curve, successive points can only differ in slope by a multiple of 45°. Hence, small changes in slope are impossible to define. This difficulty can be reduced using a smoothed slope measurement, example defining the slope at pi as $(y_{i+k} - y_i)/(x_{i+k} - x_i)$ for some $k > 1$. It is not obvious which value of k to choose in a given application. This expression becomes unstable as $(x_{i+k} - x_i)$ approaches zero: therefore direct differences in this discrete analog to slope cannot be used as estimators of point curvature.

An early model for points of inflection on a digital curve is as follows. The k-vectors at p_i are defined as:

$$a_{ik} = (x_i - x_{i+k}, y_i - y_{i+k}), \tag{22.1}$$

$$b_{ik} = (x_i - x_{i-k}, y_i - y_{i-k}), \tag{22.2}$$

and the k-cosine at p_i as:

$$C_{ik} = (a_{ik} \bullet b_{ik})/|a_{ik}|\ |b_{ik}|. \tag{22.3}$$

This c_{ik} is the cosine of the angle between a_{ik} and b_{ik}. Thus, $-1 \le C_{ik} \le 1$, where C_{ik} is close to 1 if a_{ik} and b_{ik} make an angle near 0° and C_{ik} is close to –1 if a_{ik} and b_{ik} make an angle near 180°: in other words, C_{ik} is larger when the curve is turning rapidly and smaller when the curve is relatively straight.

The smoothing factor k at each point Pi of the curve can be selected as follows. At each point P_i, compute C_{ik} for $1 \le k \le m$, for some fixed m. In our case, we have considered m to be 8. A size h and smoothed k-cosine C_{ih} is assigned to each point such that:

$$C_{im} < C_{im-1} < \ldots\ldots. < C_{ih} \not< C_{ih-1}. \tag{22.5}$$

C_{ih} is considered as the cosine at p_i and p_i is an "angle" point if C_{ih} is a local maxima in the sense that $|i - j| \le ½h$ implies $C_{ih} \ge C_{jh}$ with $C_{io} = -1$ to insure that there always is such an h between m and 1. Now, C_{ih} is the cosine at P_i is denoted by c_i which is a measure of the internal angle at p_i. Hence, the curvature at p_i is defined as $(\Pi - \cos^{-1}(c_i))/2hr$ (see Figure 1).

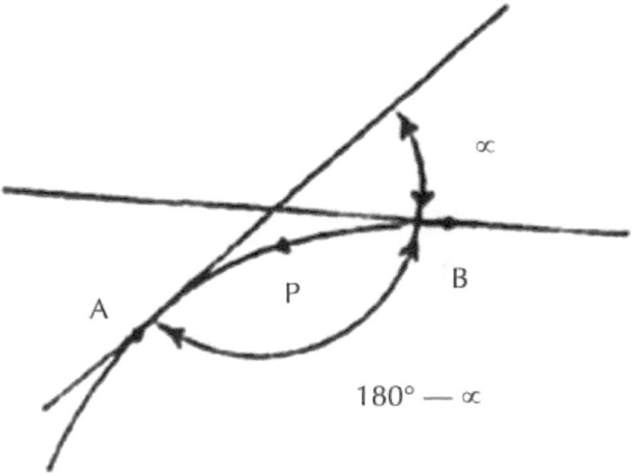

FIGURE 1 The internal angle at point P, with respect to its neighborhood, is 180° – α.

Based on the over concept now we consider a set of model objects and a set of occluded scenes consist of model objects and provide a set of values of curvatures and internal angles of the said models and scenes.

22.3 MODEL OBJECTS AND INVARIANT LOCAL FEATURES

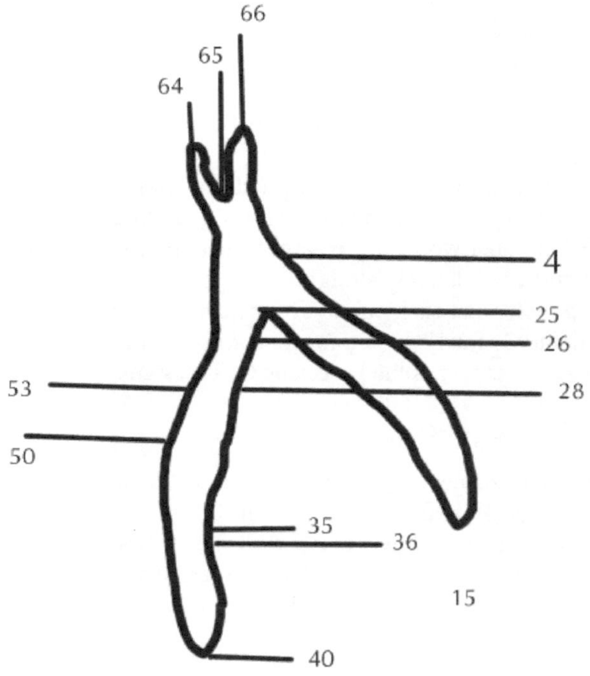

FIGURE 2 Model 1 (pliers).

TABLE 1 Internal angle *vs*. curvature of model object of Figure 2.

Model point	Internal angle	Curvature
15	56.07	7.74
25	54.44	8.96
40	63.98	7.25
64	39.07	10.06
65	46.89	9.50
66	32.64	9.20

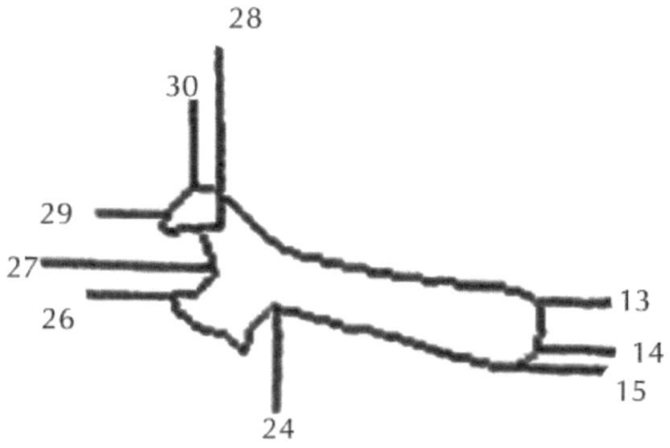

FIGURE 3 Model 2 (Wrench).

TABLE 2 Internal angle *vs*. curvature of model object of Figure 3.

Model point	Internal angle	Curvature
13	109.39	4.41
14	123.64	3.52
15	118.36	3.85
24	112.78	4.20
26	60.77	9.93
27	101.40	4.91
28	107.05	6.07
29	57.97	7.62
30	110.00	4.47

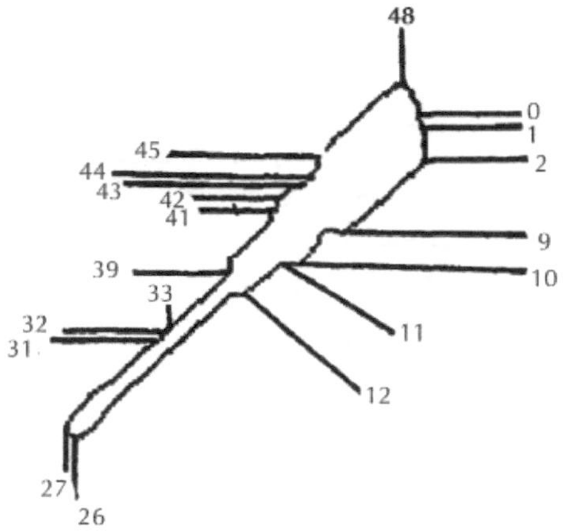

FIGURE 4 Model 3 (Screw-driver).

TABLE 3 Internal angle vs. curvature of model object of Figure 4.

Model point	Internal angle	Curvature
0	146.25	4.21
1	134.94	1.26
2	128.60	5.13
9	109.60	7.03
10	128.60	5.13
11	140.13	3.32
12	143.07	4.61
26	49.37	8.16
27	49.37	8.16
31	143.07	9.23
32	143.07	9.23
33	143.07	9.23
39	134.94	4.50
41	134.94	4.50
42	142.88	2.65
43	148.97	3.10
44	138.31	3.47
45	134.94	1.26
48	94.72	6.09

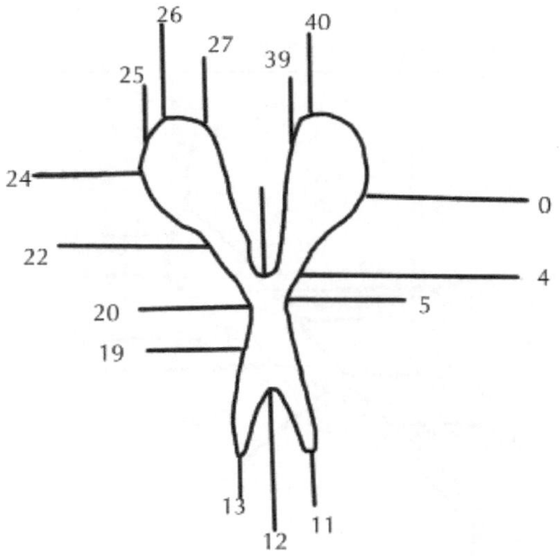

FIGURE 5 Model 4 (Scissors).

TABLE 4 Internal angle *vs*. curvature of model object of Figure 5.

Model point	Internal angle	Curvature
0	142.06	2.37
4	134.94	11.26
5	146.25	2.81
11	31.52	9.27
12	52.92	9.07
13	36.48	8.96
19	143.07	6.15
20	140.81	2.44
22	149.42	2.18
24	140.81	2.44
25	127.82	3.72
26	123.64	3.52
27	114.83	4.07
32	67.35	7.04
39	139.34	5.08
40	112.12	4.24

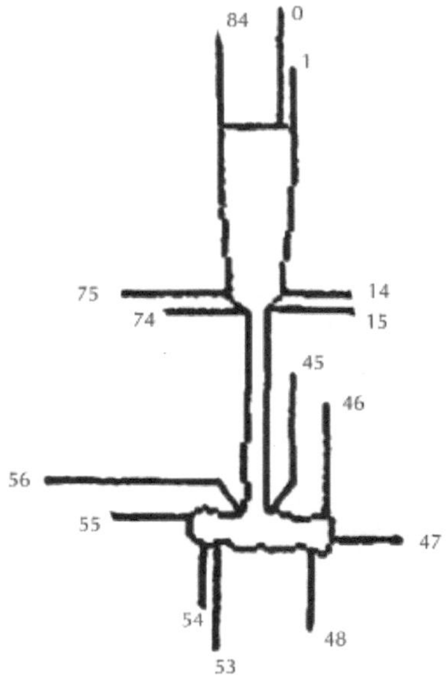

FIGURE 6 Model 5 (Hammer).

TABLE 5 Internal angle *vs*. curvature of model object of Figure 6.

Model point	Internal angle	Curvature
0	103.99	4.75
1	103.99	4.75
14	134.94	5.63
15	134.94	5.63
45	110.51	434
46	81.83	7.01
47	82.15	6.11
48	139.34	5.08
53	134.94	11.26
54	127.82	8.69
55	67.35	7.04
56	89.96	5.62
74	123.64	9.39
75	123.64	9.39
84	97.08	5.18

The distributions of different features of model objects are shown in Figure 7, Figure 8, and Figure 9. Different clusters of the model features give a rough impression about the overall complexity of the recognition of different occluded scenes consists of the model objects.

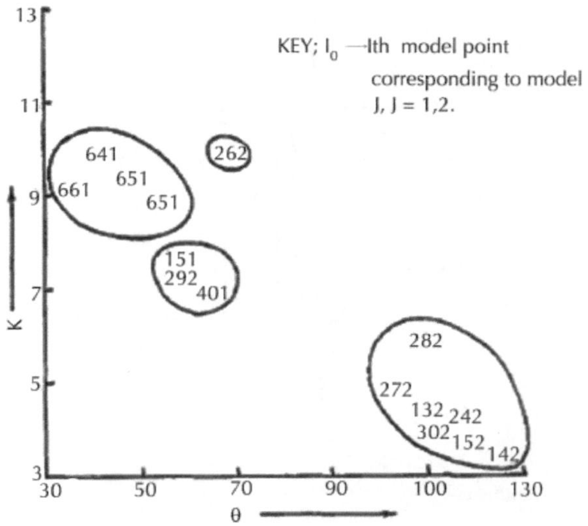

FIGURE 7 Significant features of pliers and wrench.

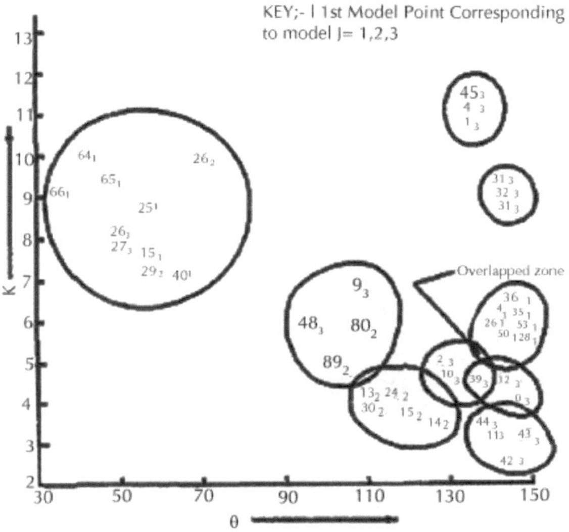

FIGURE 8 Significant features of pliers, wrench, and screwdriver.

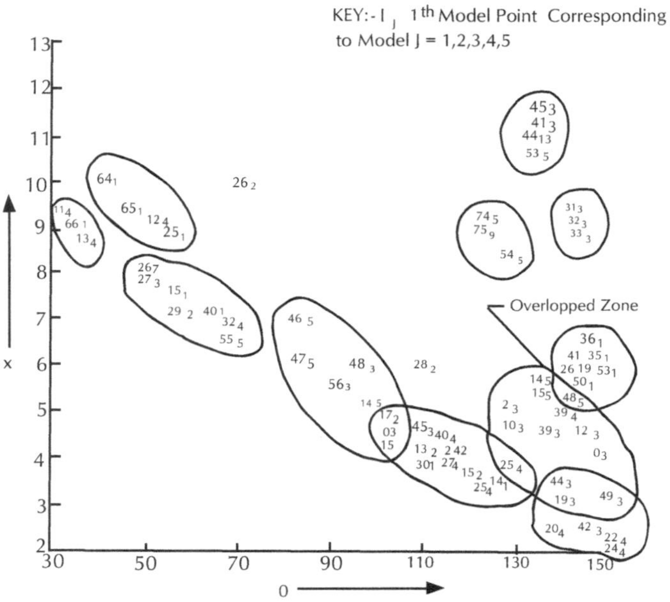

FIGURE 9 Significant features of pliers, wrench, screwdriver, scissors, and hammer.

In the following we provide a set of occluded scene consists of the model objects and the sets of local features of the different dominant points of the scenes which are to be recognized by process of recognition stated in Chapter 19, 20, and 21.

22.4 OCCLUDED SCENES AND INVARIANT LOCAL FEATURES

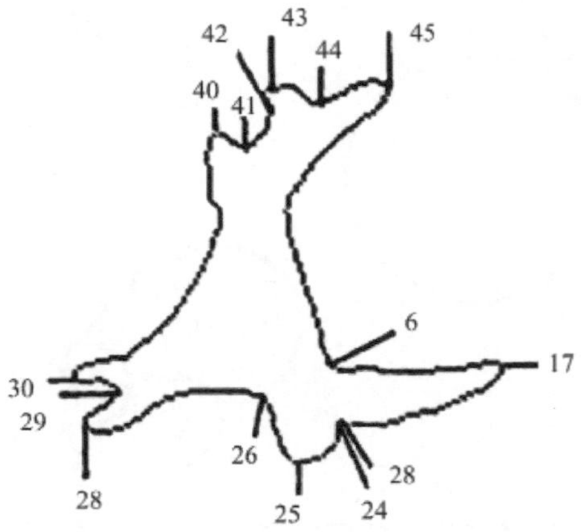

FIGURE 10 Occlusion 1.

TABLE 6 Internal angle *vs*. curvature of occlusion 1.

Scene point	Internal angle	Curvature
6	110.51	4.34
17	48.71	8.20
23	89.96	7.50
24	89.96	9.00
25	104.93	4.69
26	117.63	3.89
28	52.10	7.99
29	59.01	7.56
30	37.85	8.88
40	75.93	6.50
41	94.82	5.32
42	127.82	8.69
43	82.84	6.07
44	115.94	4.00
45	66.09	7.11

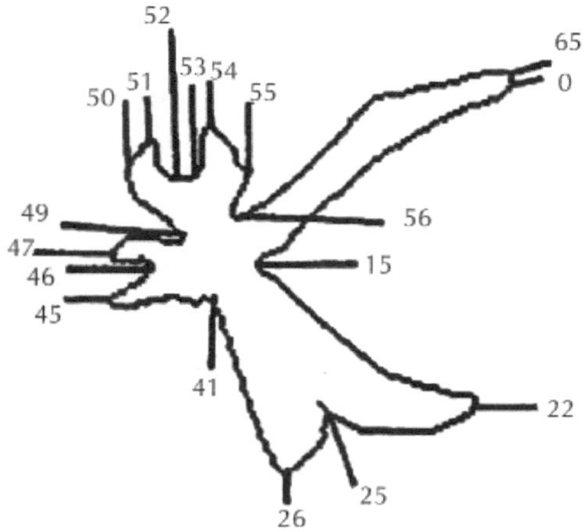

FIGURE 11 Occlusion 2.

TABLE 7 Internal angle *vs*. curvature of occlusion 2.

Scene point	Internal angle	Curvature
0	60.92	7.44
15	69.52	7.89
22	60.92	7.44
25	68.35	7.97
26	85.10	5.93
41	108.39	5.96
45	42.04	8.62
46	52.10	7.99
47	53.10	7.93
49	44.98	9.64
50	126.10	3.36
51	58.21	7.61
52	102.29	4.85
53	99.12	5.05
54	59.01	7.56
55	103.99	7.60
56	53.47	7.90
65	69.41	6.91

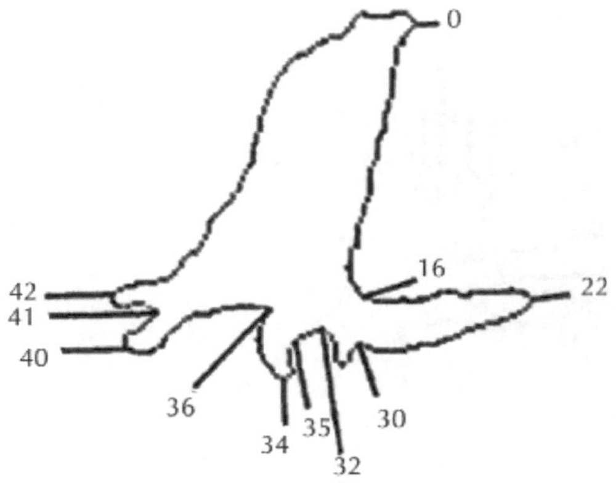

FIGURE 12 Occlusion 3.

TABLE 8 Internal angle *vs*. curvature of occlusion 3.

Scene point	Internal angle	Curvature
0	65.31	7.16
16	123.64	3.52
22	56.52	7.71
30	110.18	5.81
32	84.52	5.96
33	121.62	4.16
34	56.50	7.71
36	70.31	10.96
40	66.11	7.11
41	52.10	7.99
42	8.80	39.19

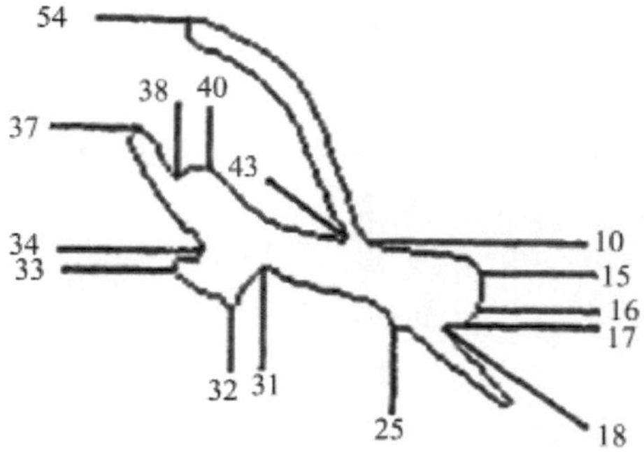

FIGURE 13 Occlusion 4.

TABLE 9 Internal angle *vs*. curvature of occlusion 4.

Scene point	Internal angle	Curvature
10	129.07	3.18
15	116.51	3.96
16	123.64	3.52
17	128.60	5.13
18	54.44	10.46
25	126.81	8.86
31	107.35	5.18
32	74.90	10.50
33	63.40	8.32
34	68.17	6.98
37	44.98	8.43
38	101.40	4.91
40	118.21	3.86
43	55.38	7.78
54	59.99	7.50

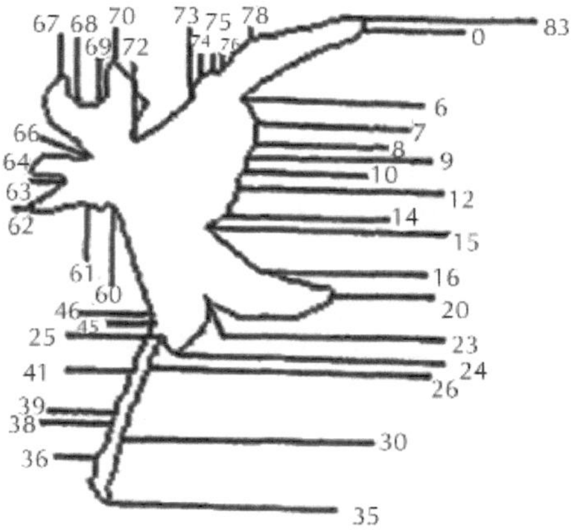

FIGURE 14 Occlusion 5.

TABLE 10 Internal angle *vs*. curvature of occlusion 5.

Scene point	**Internal angle**	**Curvature**
0	60.62	7.44
6	89.96	5.62
7	135.95	2.75
8	148.97	3.10
9	146.25	5.62

TABLE 10 *(Continued)*

Scene point	Internal angle	Curvature
10	143.07	4.61
12	143.07	4.61
14	130.54	6.18
15	74.26	7.53
16	149.59	2.53
20	60.92	7.44
23	68.35	7.97
24	88.81	7.59
25	52.10	10.65
26	146.25	5.62
30	146.25	5.62
35	42.49	8.59
36	146.25	2.81
38	146.25	5.62
39	146.25	5.61
41	134.94	11.26
45	138.83	2.57
46	138.83	2.57
60	108.39	5.96
61	146.82	3.31
62	42.64	8.58
63	52.10	7.99
64	51.46	8.03
66	44.98	9.64
67	59.01	8.64
68	102.48	4.84
69	96.48	5.21
70	02.82	7.32
72	48.86	8.19
73	146.25	3.37
74	134.94	3.75
75	134.94	3.75
76	139.91	4.00
78	138.25	2.60
83	69.41	6.91

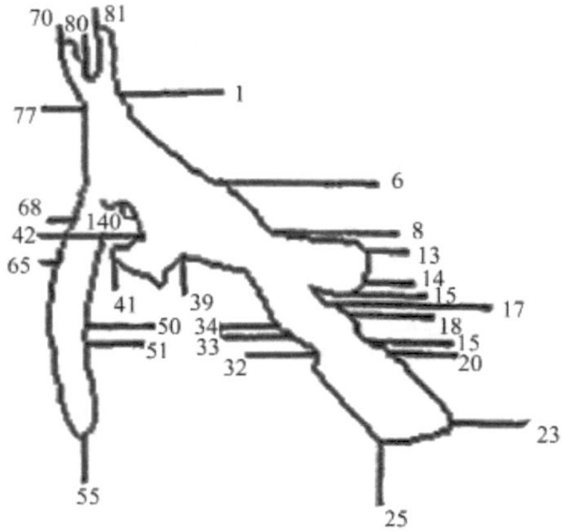

FIGURE 15 Occlusion 6.

TABLE 11 Internal angle *vs*. curvature of occlusion 6.

Scene point	Internal angle	Curvature
1	143.07	4.61
6	123.64	9.39
8	133.94	2.87
13	109.39	4.41
14	123.64	3.52
15	118.36	3.85
17	134.94	11.26
18	112.88	2.65
19	134.94	3.75
20	134.94	11.26
23	93.49	5.40
25	128.60	5.13
32	109.60	7.03
33	149.42	2.18
34	149.42	2.18
39	112.78	4.20
41	7063	6.83
42	101.40	4.91
43	79.57	6.27
50	143.07	6.15
51	143.07	6.15
55	63.98	7.25

TABLE 11 *(Continued)*

Scene point	Internal angle	Curvature
65	146.25	5.62
68	146.25	5.62
77	147.93	2.00
79	39.07	10.06
80	46.89	9.50
81	33.67	9.14

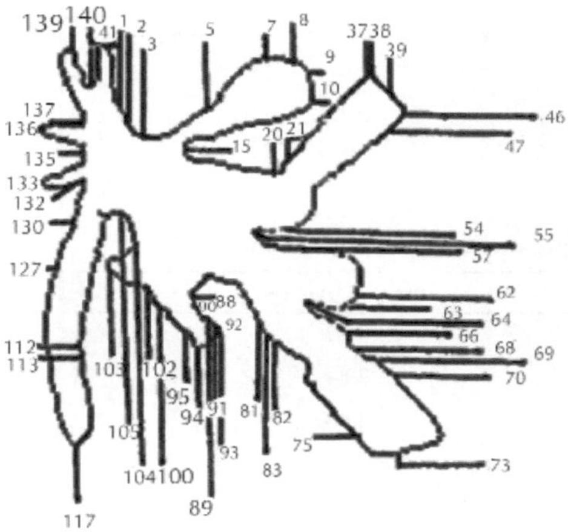

FIGURE 16 Occlusion 7.

TABLE 12 Internal angle *vs.* curvature of occlusion 7.

Scene point	Internal angle	Curvature
1	143.07	4.61
2	134.94	11.26
3	122.61	4.09
5	149.42	2.18
7	40.81	2.44
8	127.82	3.72
9	123.64	3.52
10	114.83	4.07
15	67.35	7.04
20	111.75	4.87
21	137.67	4.23
37	89.96	6.43

TABLE 12 *(Continued)*

Scene point	Internal angle	Curvature
38	89.96	5.62
39	89.96	6.43
46	103.99	4.75
47	103.99	4.75
54	119.69	4.30
55	129.07	3.18
57	34.68	9.08
62	109.39	4.41
63	123.64	3.52
64	118.36	3.85
66	134.94	11.26
68	134.94	3.75
69	134.94	3.75
70	134.94	11.26
73	93.49	5.40
75	128.60	5.13
82	109.60	7.03
83	149.42	2.18
84	149.42	2.18
88	110.87	4.93
89	103.99	4.75
90	103.39	11.93
91	89.96	5.62
92	89.96	5.62
93	89.96	6.43
94	84.99	5.93
95	130.54	4.94
100	138.54	3.47
102	130.54	3.09
103	70.63	6.83
104	101.40	4.91
105	79.57	6.27
112	143.07	6.15
113	145.07	6.15
117	63.98	7.25
127	146.25	5.62
130	146.25	5.62

TABLE 12 *(Continued)*

Scene point	Internal angle	Curvature
132	58.08	8.70
133	31.52	9.27
135	89.96	5.62
136	36.48	8.96
137	74.02	7.56
139	39.07	10.06
140	46.89	9.50
141	33.67	9.14

KEYWORDS

- **Approximation error**
- **Polygonal approximation**
- **Sequential algorithms**
- **Segmentation technique**
- **Wall-Danielsson algorithm**

Bibliography

1. Abramonowitz M. and I.A. Stegun.(1964). In: (2nd edn. ed.),Handbook of Mathematical Functions, Applied Mathematics Series **vol. 55**, Wat. Bureau Standards.
2. Aho A.V. and T. G. Peterson (1972). A minimum distance error correcting parser for context-free languages. SIAM J. Comput.1, 305–312.
3. Alt F.L. (1962). Digital pattern recognition by moments. J. Ass. Comput. Mach. 11, 240–258.
4. Anderberg, M.R., (1973). Cluster Analysis for Applications. Academic Press, New York, Chapter 7.
5. Anderson I.M. and J.C. Bezdek (1984). Curvature and tangential deflection of discrete arcs: a theory based on the commutator of scatter matrix pair and its application to vertex detection in plane shape data. IEEE Transactions on Pattern Analysis and Machine Intelligence, **PAMI-6,** pp. 27–40.
6. Ansari, N. and Delp, E.J.(1991). On detecting dominant points. Pattern Recognition vol. **24**, pp. 441–450.
7. Antoniou, G. "A tutorial on Default logic", Griffith University.
8. Aoyama H. and K. Kawagoe (1991). A piecewise linear approximation method preserving visual feature points of original figures. CVGIP: Graphical models and image processing, vol. 53, pp. 435–446.
9. Asada H. and M. Brady (1986). The curvature primal sketch. IEEE Trans. IEEE Transactions on Pattern Analysis and Machine Intelligence, **PAMI-8**, pp. 2–14.
10. Attneave F. (1954). Some informational aspects of visual perception. Psychological Review, vol. 61, pp. 183–193.
11. Ayache N. (1983). A model-based vision system to identify and locate partially visible industrial parts. Proc. Computer Vision and Pattern Recognition. June 19–23, pp. 492–494.
12. Ayache, N. and O.D. Faugerus (1986). HYPER: A new approach for the recognition and position of two-dimensional objects. IEEE Trans. Pattern Anal. Machine Intell. 8(1) 44–54.
13. Babaud J., A.P. Witkin, M. Baudin and R.O. Duda (1986). Uniqueness of Gaussian kernel for scale-space filters. IEEE Transactions on Pattern Analysis and Machine Intelligence, **PAMI-8**, pp. 26–33.
14. Baird, H.S. (1984). Model-based Image Matching Using location. MIT Press, Cambridge, MA.
15. Ballard D. H. (1981). Strip trees: A hierarchical representation of curves. Communication of the ACM, vol. 24, pp. 310–321.
16. Brady M. and H. Asada (1984). Smoothed local symmetries and their implimentation, Int. J. Robotics Res. 3 (3), pp. 36–61.
17. Ballard, D.H. (1983). Viewer independent shape recognition. IEEE Trans. Pattern Anal. Machine Intell.5(4), 653–660.
18. Bellman R. (1961). Dynamic programming, Princeton, N.J. Princeton University Press.
19. Bellman R. (1961). On the approximation of curves by line segments using dynamic programming. Communications of ACM, vol. 4, pp. 284.
20. Besl, P.J. and R.C. Jain (1984). An overview of three-dimensional object recognition. RSD-TR-19-84, Center for Robotics and Integrated Manufacturing, Univ. of Michigan, Ann Arbor, December.
21. Besl, P.J. and R.C. Jain (1984). Surface characterization for three-dimensional object recognition in depth map. RSD-TR-20-84, Center for Robotics and Integrated Manufacturing, Univ. of Michigan, Ann Arbor, December.
22. Bhanu, B. (1981). Shape matching and image segmentation using stochastic labelling. USCIPI Report 1030, Image Processing Institute, USC, Los Angeles, CA.
23. Bolles, R.C. and R.A. Cain (1982). Recognizing and locating partially visible objects: The focus feature method. Int. J. Robotics Res. 1(3), 57–81.

24. Bresenham J. E. (1965). Algorithm for Computer Control of a Digital Plotter, IBM System Journal, vol.. 4, pp.106 –111.
25. Bribiesca E. and A. Guzman (1980). How to describe pure form, and how to measure differences in shape using shape numbers. Pattern Recognition, 12, 101–112.
26. Burt P.J. (1981). Fast filter transforms for image processing. Computer Vision Graphics & Image Processing, vol. **16**, pp. 20–51.
27. Burt P.J. and E.H. Adelson(1983). The Laplacian pyramid as a compact image code. IEEE Transactions on Communications, **COM-31, no. 4.**
28. Cantoni A. (1971). Optimal curve fitting with piecewise polynomials. IEEE Transactions on Computers, vol. C-26, pp. 59–67.
29. Davis L.S. (1977). Understanding shape: angles and sides. IEEE Transactions Computers **vol. C-26**, pp. 236–242.
30. Davis L.S. (1979). Shape matching using relaxation techniques. IEEE Trans. Pattern Anal. Machine Intell., PAMI-1, 60–72.
31. Davis L.S. and T. C. Henderson (1981). Heirarchical constraint processes for shape analysis. IEEE Trans. Pattern Anal. Machine Intell., PAMI-3, 265–277.
32. De Kleer, J. (1986). An Assumption based TMS. Artificial Intelligence 28, pp. 127–162.
33. Debled-Rennesson I., F. Feschet, J. Rouer-Degli (2006). Optimal blurred segments decomposition of noisy shapes in linear time, Computers and.Graphics, vol. 30, pp. 30–36.
34. Douglas D. H. and T. K. Peucker (1973). Algorithm for the reduction of the number of points required to represent a line or its caricature. The Canadian Cartographer, vol. 10, pp. 112–122.
35. Doyle, J. (1979). A Truth Maintenance System. Artificial Intelligence 12, pp. 231–272.
36. Duchenne O., Bach, F. et. al. (2009). A Tensor based Algorithm for High Order Graph Matching. Proc. CVPR 2009.
37. Duda, R.D. and Hart, P.E.(1973). Pattern Classification and Scene Analysis. In: , Wiley, New York, pp. 328–339.
38. Dudani S.A., K.J. Breeding, and R.B. McGhee (1977). Aircraft identification by moment invariants. IEEE Trans. Comput. C-26, 39–45.
39. Dunham, J.G.(1986). Optimum uniform piecewise linear approximation of planar curves. IEEE Trans. IEEE Transactions on Pattern Analysis and Machine Intelligence. vol. PAMI-**8**, pp. 67–75.
40. Fahn, C.S. J. F. Wang and J.Y. Lee (1989). An adaptive reduction procedure for piecewise linear approximation of digitized curves. IEEE Transactions on Pattern Analysis and Machine Intelligence, vol PAMI-11, pp. 967–973.
41. Fan, N. (2010). Feature-Based Partially Object Recognition. 2010 International Conference on Pattern Recognition, pp. 3001–3004.
42. Feder J. and H. Freeman (1965). Segment fitting of curves in pattern analysis using chain correlation. March.
43. Feschet F and L. Tougne(1999), Optimal time computation of the tangent of a discrete curve: application to the curvature. DGCI, Lecture Notes in Computer Science, vol. 1568, pp.31–40.
44. Fischler, M. and R. Bolles (1981). Random sample consensus: A paradigm for model fitting with application to image analysis and automated cartography. Graphics and Image Processing 24(6), 381–395.
45. Freeman H. (1967). On the classification of line drawing data. In: W. Wather Dunn. Ed., Models for perception of speech and visual form; MIT press, Cambridge, MA, pp. 408–412.
46. Freeman H. (1974). Computer processing of line-drawing images. Comput. Surveys, 6, 57–97.
47. Freeman H. (1961). On the encoding of arbitrary geometric configurations. IRE Trans. Electron Comput., vol. EC-10, pp. 260–268.
48. Freeman H. and L.S. Davis (1977). A corner-finding algorithm for chain-coded curves. IEEE Transactions on Computers, vol. **C-26**, pp. 287–303.
49. Fu K.S. (1982). Syntactic Pattern Recognition and Applications. Englewood Cliffs, NJ: Prentice-Hall.

50. Gluss, B. (1962). Further remarks on line segment curve fitting using dynamic programming. Communications of ACM, vol. 5, pp. 441–443.
51. Gluss, B. (1962). A line segment curve fitting algorithm related to optimal encoding of information. Information and Control, vol. 5, pp. 261-267.
52. Gluss, B. (1964). An alternative method for continuous line segment curve fitting. Information and Control, vol. 7, pp. 200–206.
53. Gonzalez R. and P. Wintz (1987). Digital Image Porcessing. Addision-Wesley, Reading, MA.
54. Grandlund G.H. (1972). Fourier preprocessing for hand print character recognition. IEEE Trans. Comput. C-21, 195–201.
55. Grimson, E. and T. Lozano-Pirez (1985). Recognition and localization of overlapping parts from sparse data in two and three dimensions. Proc. IEEE Int. Conf. Robotics, St. Louis, MO, March, 61–66.
56. Grimson, E. and T. Lozano-Pirez (1984). Model-based recognition and localization from sparse range or tactile data. Int. J. Robotics Res. 3(3), 3–35.
57. Gupta L. and A.M. Upadhye (1991). Nonlinear alignment of neural net outputs for partial shape classification. Pattern Recognition 24, pp. 943–948.
58. Gupta L. and K. Malakapalli (1990). Robust partial shape classification using invariant break points and dynamic alignment. Pattern Recognition, vol. 23, pp. 1103–1111.
59. Iijima T (1962). Basic theory on normalization of pattern (in case of typical one dimensional pattern). Bull. Electrotech. Lab., vol. 26, pp.368–388.
60. Hakalahti, H., D. Harwood and L. Davis (1984). Two-dimensional object recognition by matching local properties of contour points. Pattern Recognition Letters 2(4), 227–234.
61. Hayashi I., H. Nomura, H. Yamasaki and N. Wakami (1992). Construction of fuzzy inference rules by NDF and NDFL. Internat. J. Approx. Reason. 6, pp. 241–266.
62. Ho S.Y. and Y.C. Chen (2001). An efficient evolutionary algorithm for accurate polygonal approximation, Pattern Recognition, vol. 34, pp. 2305–2317.
63. Hsiung, C.C. (1981). A First Course in Differential Geometry. Wiley-Interscience, New York.
64. Hu, M. (1962). Visual pattern recognition by moment invariants. IRE Trans. Information Theory 8, pp.179–187.
65. Imai H. and M. Iri (1986). Computational geometric methods for polygonal approximation of a curve. Computer Vision, Graphics and Image Processing, vol. 36, pp. 31–41.
66. Jana P. and K. S. Ray (2011). Object recognition with belief revision. An internal report of Electronics and Communication Sciences Unit, Indian Statistical Institute, Kolkata-700108.
67. Kashyap R. L. and B. J. Oommen (1982). A geometrical approach to polygonal dissimilarity and the classification of closed boundaries. IEEE Trans. Pattern Anal. Machine Intell., vol. PAMI-4, pp. 649–654.
68. Kashyap, R.L. and B.J. Oommen (1983). Scale preserving-smoothing of polygons. IEEE Trans. Pattern Anal. Machine Intell. 5 (6), 667–671.
69. Kashyap R. and R. Chellapa (1981). Stochastic models for closed boundary analysis: representation and recognition, IEEE Trans. Inform. Theory 27, pp. 109–119.
70. Klir G.J. and T. Folger (1989). Fuzzy Sets, Uncertainty and Information (Addison-Wesley, Reading, MA).
71. Knoll, T.F. and R.C. Jain (1986). Recognizing partially vision objects using feature indexed hypotheses. IEEE J. Robotics and Automation 2(1), pp. 3–13.
72. .Knuth, D. (1983). Fundamental Algorithms. Addison-Wesley, Reading, MA.
73. Koch, M.W. and R.L. Kashyap (1989). Matching polygon fragments. Pattern Recognition Letters 10 (November), pp. 297–308.
74. Koch, M.W. and R.L. Kashyap (1987). Using polygons to recognize and locate partially occluded objects. IEEE Trans. Pattern Anal. Machine Intell. 9 (4), 483–494.
75. Koenderink J.J. and A.J. van Doorn (1984). The structure of images. Biol. Cybernetics, vol. **50**, pp. 363–370.
76. Kurozumi Y. and W. A. Davis (1982). Polygonal approximation by minimax method. Computer Graphics and Image Processing, vol. 19, pp. 248–264.

77. Lee, C.C. (1990), Fuzzy logic in Control Systems: Fuzzy logic controller - Part I & II, IEEE Trans. Systems Man Cybernet. SMC 20(2), pp. 404–435.
78. Leordeanu, M. and Hebert M. (2005). A Spectral Technique for Correspondence Problem using Pairwise Constraints. Proc. ICCV.
79. Lindeberg T. (1990) Scale-space for discrete signals. IEEE Transactions on Pattern Analysis and Machine Intelligence,. vol. **PAMI-12**, pp. 234–254.
80. Lootsma, F.A. (1991). Scale sensitivity and rank preservation in a multiplicative variant of the analytic hierarchy process, Rep. 91–20, Faculty of Technical Mathematics and Informatics, University of Delft, The Netherlands.
81. Maps of the Great Lakes Region, Nat. Geographic, Dec. 1953.
82. Marji Majed and Pepe Sipy (2004). Polygonal representation of digital planar curves through dominant point detection—a nonparametric algorithm. Pattern Recognition, vol. 37, pp. 2113 – 2130.
83. Masood A (2008). Dominant point deletion by reverse polygonization of digital curves. Image and Vision Computing, vol. 26 pp. 702–715.
84. Masood A. (2008) Optimized polygonal approximation by dominant point deletion. Pattern Recognition, vol. 41, pp. 227–239.
85. McKee J. W. and J. K. Aggarwal. (1977). Computer recognition of partial views of curved objects. IEEE Trans. Computers, vol. C-26 (August), pp. 790–800.
86. Medioni, G. and A. Huertas (1985). Edge detection with subpixel precision. Techn. Rep. USCISG 106, University of Southern California.
87. Medioni G. and Y. Yasumoto (1987). Corner detection and curve representation using cubic B-splines. Computer Vision, Graphics and Image Processing, vol. 31, pp. 267–278.
88. Meer P., E.S. Bauger and A. Rosenfeld (1987). Frequency domain analysis and synthesis of image pyramid generating kernels. IEEE Trans. IEEE Transactions on Pattern Analysis and Machine Intelligence, **PAMI-9**, pp. 512–522.
89. Meer P., E.S. Bauger and A. Rosenfeld (1988). Extraction of trend lines and extrema from multiscale curves. Pattern Recognition, vol. **21**, pp. 217–226.
90. Mizumoto, M. (1985). Extended fuzzy reasoning, in: M.M. Gupta, A. Kandel, W. Bandler and J.B. Kiszka, Eds., Approximate Reasoning in Expert Systems (North-Holland, Amsterdam) 71–85.
91. Mokhtarian, F. and A. Mackworth (1986). Scale-based description and recognition of planar curves and two-dimensional shapes. IEEE Trans. IEEE Transactions on Pattern Analysis and Machine Intelligence, **PAMI-8**, pp. 34–43.
92. Mokhtarian, F., M. Bober (2003). Curvature scale space representation: Theory, Application and MPEG-7 standardization. Norwell, MA: Kluwer.
93. Mokhtarian, F., Suomela, R. (1998). Robust image corner detection through curvature scale space. IEEE Trans. IEEE Transactions on Pattern Analysis and Machine Intelligence, vol. PAMI-20, 1376–1381.
94. Montanari, U. (1970). A note on minimal length polygonal approximation to a digitized contour. Commun. ACM 13(1), pp. 41–47.
95. Nafarieh, A. (1988). A new approach to inference in approximate reasoning and its applications to computer vision, Ph.D. Dissertation, University of Missouri-Columbia.
96. Nguyen T. P. and I Debled-Rennesson (2011) A discrete geometry approach for dominant point detection. Pattern Recognition, vol. 44 pp. 32–44.
97. Nguyen T. P. and I. Debled-Rennesson (2007). Curvature estimation in noisy curves. CAIP, Lecture Notes in Computer Science, vol. 4673, pp.474–481.
98. Oommen, J.B. (1982). Pattern recognition with strings, substrings, and boundaries. PhD Thesis, Purdue University.
99. Pao, Y.H. (1989). Adaptive Pattern Recognition and Neural Networks (Addison-Wesley, Reading, MA).
100. Parui, S. and D. Dutta Majumder (1983). Shape similarity measures for open curves. Pattern Recognition Letters 1 (3), pp. 129–134.

101. Pavlidis T. (1973). Waveform segmentation through functional approximation. IEEE Transactions on Computers, vol. C-22, pp. 689–697.
102. Pavlidis T. (1977). Polygonal approximation by Newton's method. IEEE Transactions on Computers, vol. C-26, pp. 800–807.
103. Pavlidis T. (1977). Structural Pattern Recognition. New York: Springer-Verlag.
104. Pavlidis, T. (1979). A hierarchical syntactic shape analyzer. IEEE Trans. Pattern Anal. Machine Intell. 1 (1).
105. Pavlidis T. (1977). Syntactic pattern recognition of shape. Proc. IEEE Comput. Soc. Conf. Pattern Recognition Image Processing. June 98–107.
106. Pavlidis, T. (1982). Algorithms for Graphics and Image Processing. Computer Science Press, New York, pp. 281–297.
107. Pavlidis T. and F. Ali (1979) A hierarchial syntactic shape analyser. IEEE Trans. Pattern Anal. Machine Intell., PAMI-1, pp. 2–9.
108. Pavlidis, T. and Horowitz, S.L. (1974). Segmentation of plane curves. IEEE Transactions on Computers, vol. C-**23**, pp. 860–870
109. Perez J. C. and E. Vidal (1992). An algorithm for the optimum piecewise linear approximation of digitized curves, In Proc. 11th IAPR International Conference on Pattern Recognition, The Hague, The Netherlands, Aug. 30 – Sep. 3, vol. 3, pp. 167–170.
110. Perkins W. A. (1978). A model-based vision system for industrial parts. IEEE Trans. Computers, vol. C-27 (February), pp. 126–143.
111. Perkins W. A. (1980). *Simplified model-based part locator*. Proceedings of the 5th International Conference on Pattern Recognition, pp. 260–263.
112. Persoon E. and K.S. Fu (1977). Shape discrimination using Fourier descriptors. IEEE Trans. Syst. Man. Cybern. SMC-7, 170–179.
113. Pikaz Arie and Its'hak Dinstein (1995). An algorithm for polygonal approximation based on iterative point elimination. Pattern Recognition Letters, vol. 16, pp. 557–563
114. Pikaz Arie and Its'hak Dinstein (1995). Optimal Polygonal Approximation of Digital Curves. Pattern Recognition, vol. 28, pp. 373–379.
115. Phillips, T.Y., Rosenfeld (1988) A. An ISODATA algorithm for straight line fitting. Pattern Recognition Letters, vol. 7, pp. 291–297.
116. Philippe Cornic (1997), Another look at the dominant point detection of digital curves. Pattern Recognition Letters, vol.8, pp. 13–25.
117. Press W.H., B.P. Flannery, S.A. Teukolsky and W.T. Vetterling (1986). Numerical Recipes: The Art of Computer Programming. , Cambridge University Press, London.
118. Ramer, U. (1972). An iterative procedure for polygonal approximation of plane curves. Computer Graphics and Image Processing, **vol. 1**, pp. 244–256.
119. Rattarangsi A. and R.T. Chin (1992). Scale-based detection of corners of planar curves. IEEE Trans. IEEE Transactions on Pattern Analysis and Machine Intelligence, vol. **PAMI-14**, pp. 430–449.
120. Ray, B.K. and K.S Ray (1991). A new approach to polygonal approximation. Pattern Recognition Letters. vol. **12**, pp. 229–234.
121. Ray, B.K. and K.S Ray (1992). An algorithm for polygonal approximation of digitized curves. Pattern Recognition Letters. vol. **13**, pp. 489–496.
122. Ray, B.K. and K.S Ray (1994). A non-parametric sequential method for polygonal approximation of digital curves. Pattern Recognition Letters, vol. 15, pp. 161–167.
123. Ray, B.K. and K.S Ray (1993). An optimal algorithm for polygonal approximation of digital curves using L_1 norm. Pattern Recognition, vol. **26**, pp. 505–509.
124. Ray, B.K. and K.S Ray (1992). Detection of significant points and polygonal approximation of digitized curves. Pattern Recognition Letters, vol 13, pp. 443–452.
125. Ray, B.K. and K.S Ray (1992). An algorithm for detection of dominant points and polygonal approximation of digitized curves. Pattern Recognition Letters, vol 13, pp. 849–856.
126. Ray, B.K. and K.S Ray (1994).A new approach to scale space image. Pattern Recognition Letters, vol 15, pp. 365–372.

127. Ray B.K. and K.S. Ray (1995). A new split-and-merge technique for polygonal approximation of chain coded curves Pattern Recognition Letters, vol. 16 pp. 161–169.
128. Ray B. K. and K. S. Ray (1995). Corner detection using iterative Gaussian smoothing with constant window size. Pattern Recognition, vol. 28, pp.1765–1781.
129. Ray B. K. and K. S. Ray (1997) Scale-space analysis and corner detection on digital curves using a discrete scale-space kernel. Pattern Recognition, vol. 30 pp. 1463–1474.
130. Ray B. K. and R Pandyan (2003). ACORD--an adaptive corner detector for planar curves. Pattern Recognition, vol. 36, pp. 703–708.
131. Ray K.S. and D. Dutta Mazumder (1989). Application of differential geometry to recognize and locate partially occluded objects, Pattern Recognition Lett. 9, pp. 351–360.
132. Ray K.S. and D. Dutta Mazumder (1991). Recognition and positioning of partially occluded 3-D objects, Pattern Recognition Lett. 12, pp. 93–108.
133. Ray K.S. and J. Ghoshal (1998). Neuro-fuzzy reasononing for occluded object recognition. Fuzzy Sets and Systems 94, pp.1–28.
134. Reiter, R. (1980). A Logic for Default Reasoning. Artificial Intelligence 13, pp. 81–132.
135. Richard C.W. and H. Hemami (1974). Identification of three-dimensional objects using Fourier descriptors of the boundary curve. IEEE Trans. Syst. Man. Cybern. SMC-4, pp. 371–378.
136. Roberge J. (1985). A data reduction algorithm for planar curves. Computer Vision, Graphics and Image Processing, vol. 29, pp. 168–195.
137. Rosenfeld A. (1979). Survey, picture processing: 1978. Comput. Graphics Image Processing, 9, pp. 354–393.
138. Rosenfeld, A. and A. Kak (1982). Digital Picture Processing (2nd edition). Academic Press, New York.
139. Rosenfeld, A. and Johnston, E. (1973). Angle detection on digital curves. IEEE Transactions on Computers, vol. **22**, pp. 875–878.
140. Rosenfield, A. and Kak, A.(1982). Digital picture processing. (2nd edition). Academic press, NewYork.
141. Rosin P.L. (1997). Techniques for assessing polygonal approximations of curves, IEEE Transactions on Pattern Analysis and Machine Intelligence, vol. 19, pp. 659–666.
142. Rosenfeld A. and J.S. Weszka (1975). An improved method of angle detection on digital curves. IEEE Transactions Computers, **C-26**, pp. 940–941.
143. Rosenfeld A. and M. Thruston (1971). Edge and curve detection for digital scene analysis. IEEE Transactions on Computers, vol. C-24, pp. 940–941.
144. Rosenfeld, A., R.A. Hummel, and S.W. Zucker (1976). Scene labeling by relaxation operations. IEEE Trans. Syst. Man. Cybern. SMC-6, pp. 420–433.
145. Rumelhart D.E., G.E. Hinton and R.J. Williams (1986). Learning internal representations by error propagation, in: D.E. Rumelhart and J.L. McClelland, Eds., Parallel Distributed Processing: Explorations in the Microstructures of Cognition, Vol. I: Foundations (MIT Press, Cambridge, MA) 318–362.
146. Saaty T. L. (1977), A scaling method for priorities in hierarchical structures, J. Math. Psychol. 15 (3), 57–58.
147. Saint-Marc P., Jer-Sen Chen and G. Medioni (1991) Adaptive-Smoothing: a general tool for early vision. IEEE Transactions Pattern Analysis and Machine Intelligence, vol. **PAMI-13**, pp. 514–529.
148. Sankar P.V. and C.V. Sharma (1978). A parallel procedure for the detection of dominant points on a digital curve. Computer Graphics Image Processing vol. **7**, pp. 403–412.
149. Salotti M.(2001). An efficient algorithm for the optimal polygonal approximation of digitized curves. Pattern Recognition Letters, vol 22, pp. 215–221.
150. Sarkar D.(1993). A simple algorithm for detection of significant points for polygonal approximation of chain-coded curve, Pattern Recognition Letters, vol. 14, pp. 959 – 964.
151. Scarborough, Mames B. (1966). Numerical mathematical Analysis, Oxford and IBH Publishing Company.
152. Segen J. (1983). Locating randomly oriented objects from partial view. SP/E Intelligent Robots: 3rd /nt. Conf. on Robot Vision Sensory Controls, vol. 449 (November), pp. 676–684.

153. Selby, S. and R. Weast (1975). CRC Standard Mathematical Tables. CRC Press Boca Raton El.
154. Serra, J. (1982). Image Analysis and Mathematical Morphology. Academic Press, London.
155. Shamos, M.I. and D. Hoey (1975). Closest point problems. 16th Annual IEEE Symp. Foundations of Computer Science, October, 157–162.
156. Sklansky J., R. L. Chazin, and B. J. Hansen (1972). Minimum-perimeter polygons of digitized silhouettes. IEEE Trans. Comput., vol. C-21, pp. 260–268.
157. Sklansky J. and D. F. Kibler (1976). A theory of non uniformly digitized binary pictures. IEEE Trans. Syst., Man, Cybern., vol. SMC-6, pp. 637–647.
158. Sklansky J. and V. Gonzalez (1980). Fast polygonal approximation of digitized curves. Pattern Recognition, vol. 12, pp. 327–331.
159. Smith S.P. and A.K. Jain (1981). Chord distributions for shape matching. Proc. IEEE Conf. Pattern Recognition Image Processing, 168–170.
160. Späth, H. (1974). Spline Algorithms fro Curves and Surfaces. Utilitas Math., Winnipeg.
161. Stockman, G., S. Kopstein and S. Benett (1982). Matching images to models for registration and object detection via clustering. IEEE Trans. Pattern Anal. Machine Intell. 4 (3), 229 241.
162. Stone H. (1961). Approximation of curves by line segments. Mathematics of Computation, vol. 15, pp. 40–47.
163. Sugeno M. and G.T. Kang (1988). Structure identification of fuzzy model, Fuzzy Sets and Systems 28 (1) pp. 15–33.
164. Sugeno M. and T. Takagi (1983). Multi-dimensional fuzzy reasoning, Fuzzy Sets and Systems 9, pp. 313–325.
165. Sun Y.N. and S.C. Huang (2000). Genetic algorithms for error-bounded polygonal approximation," International Journal of Pattern Recognition and Artificial Intelligence, vol.14, pp. 297–314.
166. Takagi H. and I. Hayashi (1991). NN-driven fuzzy reasoning, Internat. Z Approx. Reason. 5 pp. 191–212.
167. Takagi H., N. Suzuki and Y. Kojima (1992). Neural networks designed on approximate reasoning architecture and their applications, IEEE Trans. Neural Networks 3 (5), pp. 752–760.
168. Tamir D.E., D.G. Schwartz and A. Kandel (1991). A pattern recognition interpretation of implications, Inform. Sci. 57-58, pp. 197–215.
169. Tang, J., Liang D. et. al. (2007). Spectral Correspondense using Local Similarity Analysis. Proc. ICCIS 2007.
170. Teh C.H. and R.T. Chin (1989). On the detection of dominant points on digital curves. IEEE Transactions Pattern Analysis Machine Intelligence, vol. **PAMI-11**, pp. 859–872.
171. Tejwani Y. J. and R. A. Jones (1985). Machine recognition of partial shapes using feature vectors. IEEE Trans. Syst. Man, Cybern., vol. SMC- 15 (July), pp. 504–516.
172. Tomek I. (1974) Two algorithms for piecewise linear approximation for functions of one variable. IEEE Transactions on Computers, vol. C-23, pp. 445–448.
173. Touretzky D. and D. Pomerleau (1989). What's hidden in the hidden layers?, Byte Mag. pp. 227–233.
174. Triantaphyllou E. (1993). A quadratic programming approach in estimating similarity relations, IEEE Trans. Fuzzy Systems 1 (2), pp. 138–145.
175. Tsai Yao-Hong and Hsuan Chuang(2006) Proceedings of the 5th IEEE/ACIS International Conference on Computer and Information Science and 1st IEEE/ACIS International Workshop on Component-Based Software Engineering, Software Architecture and Reuse (ICIS-COMSAR'06), pp. 322–326.
176. Tsai W.H. and K.S. Fu (1980). A syntactic-statistical approach to recognition of industrial objects. Proc. Int. Conf. Pattern Recognition, pp. 251–259.
177. Tsang P.W.M., P.C. Yuen and F.K. Lam (1992). Recognition of occluded objects, Pattern Recognition 25, pp. 1107–1117.
178. Tsukamoto Y. (1979). An approach to fuzzy reasoning method, in: M.M. Gupta, R.K. Ragade and R.R. Yager, Eds., Advances in Fuzzy Set Theory and Applications (North-Holland, Amsterdam) pp. 137–149.

179. Turney J. L., T. N. Mudge, and R. A. Volz (1985). Recognizing partially occluded parts. IEEE Trans. Patt. Anal. Mach. Intell., vol. PAMI-7 (July), pp. 410–421.
180. Wall, K. and Danielsson, P.E.(1984). A fast sequential method for polygonal approximation of digitized curves. Computer Vision, Graphics, and Image Processing **28**, pp. 220–227.
181. Wallace T. P. and P. A. Wintz (1980). An efficient three-dimensional aircraft recognition algorithm using normalized Fourier descriptors," Computer Graphics Image Processing. vol. 13, pp. 99–126.
182. Wallace T. P., O. R. Mitchell, and K. Fukunaga (1981). Three-dimensional shape analysis using local shape descriptors. IEEE Trans. Patt. Anal. Mach. Intell., vol. PAMI-3 (May), pp. 310–323.
183. Watson L.T. and L. G. Shapiro (1982). Identification of space curves from two-dimensional perspective views. Dep. Comput. Sci. Virginia Polytechnic Inst. and State Univ., Blacksburg, Tech. Rep. CS 80004-R.
184. Williams, C.M. (1978). An efficient algorithm for piecewise linear approximation of planar curves. Computer Graphics and Images Processing, vol. **8**, pp. 286–293.
185. Williams C. M. (1981). Bounded straight line approximation of digitized planar curves and lines. Computer Graphics and Image Processing, vol. 16, pp. 370–381.
186. Wang Yu-Zhu1, Yang Dan and Zhang Xiao-Hong (2007) Robust Corner Detection Based on Multi-scale Curvature Product in B-spline Scale Space. Acta Automata Sincia, vol. 33.
187. Witkin A. P. (1983). Scale space filtering. Proceedings of the 8th International Joint Conference on Artificial Intelligence, Karlsruhe, Germany, pp. 1019–1021.
188. Wu J. S. and J. J. Leou (1993). New Polygonal approximation schemes for object shape recognition. Pattern Recognition, vol. 26, pp. 471–484.
189. Wu L. D. (1984). A piecewise linear approximation based on statistical model. IEEE Transactions on Pattern Analysis and Machine Intelligence, vol. PAMI-6, pp. 41–45.
190. Yachida M. and S. Tsuji (1977). A versatile machine vision system for complex industrial parts. IEEE Trans. Computers, vol. C-26 (September), pp. 882–894.
191. Yang Y.H. and T.W. Sze (1981). A simple contour matching algorithm. Proc. IEEE Conf. Pattern Recognition Image Processing, pp. 562–564.
192. Yin Peng-Yeng (1998) Algorithms for straight line fitting using *k*-means. Pattern Recognition Letters, vol. 19, pp. 31–41.
193. Yin P.Y. (1998). A new method of polygonal approximation using genetic algorithm, Pattern Recognition Letters, vol. 19, pp. 1017–1026.
194. Yin P.Y. (2000). A tabu search approach to polygonal approximation of digital curves, International Journal of Pattern Recognition and Artificial Intelligence, vol.14, pp. 243–255.
195. You K.C. and K.S. Fu (1980). Distorted shape recognition using attributed grammars and error-correcting techniques. Comput. Graphics Image Processing, 13, pp. 1–16.
196. Yuilli A.L. and T.A. Poggio (1986). Scaling theorems for zerocrossings. IEEE Transactions Pattern Analysis Machine Intelligence **PAMI-8**, pp. 15–25.
197. Yuan Jianxing and Ching Y. Suen (1995). An Optimal *O(N)* Algorithm For Identifying Line Segments From A Sequence Of Chain Codes, Pattern Recognition, vol. 28, No. 5, pp. 635–646.
198. Zadeh L.A. (1970), Theory of approximate reasoning, in: J.E. Hayes, D. Michie and L.I. Mikulich, Eds., Machine Intelligence (Ellis Harwood, Chichester, UK) pp. 149–194.
199. Zahn C. and R. Roskies (1972). Fourier descriptors for plane closed curves, IEEE Trans. Comput. 21, pp. 269–281.
200. Zhong B. and Kai-Kuang Ma (2010). On the convergence of planar curves under smoothing. IEEE Transactions on Image Processing, vol. 19, pp. 2171–2189.
201. Zhong B. and W. Liao(2007). Direct curvature scale space: Theory and corner detection. IEEE Transactions on Pattern Analysis Machine Intelligence, vol. PAMI-29, pp. 508–512.
202. Zhang X., M. Lei, D.Yang, Y.Wang, and L. Ma (2007). Multi-scale curvature product for robust image corner detection in curvature scale space, Pattern Recognition Letters., vol. 28, no. 5, pp. 545–554.

Index

D

E

F

G